Springer
Berlin
Heidelberg
New York
Barcelona
Budapest
Hong Kong
London
Milan
Paris
Santa Clara
Singapur
Tokyo

H. Spiecker K. Mielikäinen M. Köhl J. Skovsgaard (Eds.)

Growth Trends in European Forests

Studies from 12 Countries

with 127 Figures

European Forest Institute Research Report No. 5

Springer

Heinrich Spiecker
Institut für Waldwachstum
Universität Freiburg
Bertoldstr. 17
D-79085 Freiburg
Germany

Michael Köhl
WSL Birmensdorf
Zürcherstr. 111
CH-8903 Birmensdorf
Switzerland

Kari Mielikäinen
Forest Research Institute
PL 18
SF-01301 Vantaa
Finland

Jens Peter Skovsgaard
Danish Forest and Landscape
Research Institute
Hørsholm Kongevej 11
DK-2970 Hørsholm
Denmark

ISBN 3-540-61460-5 Springer-Verlag Berlin Heidelberg New York

Cataloging-in-Publication applied for

Die Deutsche Bibliothek - CIP-Einheitsaufnahme
Growth trends in European forests: studies from 12 countries / H. Spiecker... (ed.). - Berlin ; Heidelberg ; New York ; Barcelona ; Budapest ; Hong Kong ; London ; Mailand ; Paris ; Santa Clara ; Singapur ; Tokio : Springer 1996
ISBN 3-540-61460-5
NE: Spiecker, Heinrich [Hrsg.]

Cover Design: Erich Kirchner, Heidelberg
Typesetting: PihkaPojat Oy
SPIN 10522046 31/3137 5 4 3 2 1 0 - Printed on acid free paper -

Acknowledgement

This book is the result of a project initiated by the European Forest Institute (EFI). The success of the project depended on the input and expertise of numerous scientists who contributed to the project. We would like to thank all project participants for their efforts. In total, 61 scientists independently reviewed the individual contributions presented in this book. Their valuable comments and suggestions, which we thankfully acknowledge, helped to ensure the scientific standard.

We would also like to thank Ms. Leena Roihuvuo, EFI, for her assistance in administrating the project and her outstanding help to finalise this volume. We thank our publisher, Springer-Verlag Heidelberg, for the friendly and helpful co-operation in publishing this book.

List of reviewers:

P. Abetz,	Germany
P. Bachmann,	Switzerland
M. Becker,	France
O.U. Bräker,	Switzerland
P. Brassel,	Switzerland
R.L. Czaplewski,	USA
E.R. Cook,	USA
I. Danilin,	Russia
P.C. Van Deusen,	USA
P.H. Dong,	Germany
B. Elfving,	Sweden
A. Franc,	France
K. von Gadow,	Germany
J. Gal,	Hungary
O. García,	Chile
F. Gemmel,	Sweden
G.Z. Gertner,	USA
T.G. Gregoire,	USA
P. Hari,	Finland
H. Hasenauer,	Austria
H. Henttonen,	Finland
E. Holmsgaard,	Denmark
R.F. Hüttl,	Germany
J.L. Innes,	Switzerland

H.-P. Kahle,	Germany
G. Kenk,	Germany
K. Kuusela,	Finland
J. Lappi,	Finland
J.B. Larsen,	Denmark
R.A. Leary,	USA
J.W. Leech,	Australia
M. Lindholm,	Finland
H.G. Lund,	USA
D. Mandallaz,	Switzerland
J. Molenaar,	The Netherlands
R.A. Monserud,	USA
H.T. Mowrer,	USA
M. Neumann,	Austria
P. Nöjd,	Finland
R. Ojansuu,	Finland
A.F.M. Olsthoorn,	The Netherlands
R. Päivinen,	Finland
H. Pretzsch,	Germany
T. Preuhsler,	Germany
G.A. Reams,	USA
F.A. Roesch,	USA
H. Röhle,	Germany
K. Schadauer,	Austria
K. Schieler,	Austria
H.T. Schreuder,	USA
C.T. Scott,	USA
H. Spellmann,	Germany
A.R. Stage,	USA
H. Sterba,	Austria
M. Tomé,	Portugal
E. Tomppo,	Finland
B. Tveite,	Norway
H. Untheim,	Germany
H. Utschig,	Germany
J.K. Vanclay,	Indonesia
P.W. West,	Australia

Foreword

The European Forest Institute (EFI) has five Research and Development priority areas: forest sustainability, forestry and possible climate change, structural changes in markets for forest products and services, policy analysis, and forest sector information services and research methodology. In the area of forest sustainability our most important activity has been the project "Growth trends of European forests", the results of which are presented in this book.

The project was started in August 1993 under the leadership of Prof. Dr. Heinrich Spiecker from the University of Freiburg, Germany, and it is one of the first EFI's research projects after its establishment in 1993. The main purpose of the project was to analyse whether site productivity has changed in European forests during the last decades. While several forest growth studies have been published at local, regional and national levels, this project has aimed at stimulating a joint effort in identifying and quantifying possible growth trends and their spatial and temporal extent at the European level.

Debate on forest decline and possible climate change, as well as considerations related to the long term supply of wood underline the importance of this project, both from environmental and industrial points of view. Knowledge on possible changes in growth trends is vital for the sustainable management of forest ecosystems.

From a methodological point of view, the leading idea in the project has been to utilise existing research plots and data from many countries to increase the empirical base and validity of the analysis. 44 scientists from 12 countries have contributed to this report. As such, the project is an excellent example of the basic idea behind EFI and its mode of operation - to create added value to existing data by pan-European research networking.

On behalf of EFI I would like to express my warmest thanks to all those who have contributed to the project and to this report. In particular, I would like to thank Prof. Dr. Heinrich Spiecker for his outstanding leadership of the project, and the Editorial Board of this report: Prof. Kari Mielikäinen from Finnish Forest Research Institute, Dr. J.P. Skovsgaard from the Danish Forest and Landscape Research Institute, Dr. Michael Köhl from the Swiss Federal Institute for Forest, Snow and Landscape Research, and Prof. Dr. Spiecker as the Chairman of the Editorial Board. I would also like to thank Ms. Ursula Gramm at Springer Verlag for her help in publishing this volume.

On Europe Day, 9 May 1996
Birger Solberg
Director, European Forest Institute

Contents

Introduction

Heinrich Spiecker (project coordinator)
Institut für Waldwachstum
Bertoldstr. 17
D-79085 Freiburg, Germany

Debates on forest decline and possible climatic changes emphasize the need for information regarding environmental changes and their effects on forests. In recent years, the condition of European forests has been systematically surveyed. For evaluation and interpretation of actual forest conditions long-term observations are needed. Only a few forest characteristics have been monitored over long periods of time. One of these characteristics is forest growth. For many decades, forest research institutions have installed permanent research plots and conducted tree ring and stem analyses to investigate long-term growth of forests. The aim of such research is to ensure sustainable forest production.

Recent investigations in European forests indicate both positive and negative changes in forest growth. In contrast to discussions about forest decline in the 1980s, several studies show that forest growth may have actually increased on specific sites (e.g. Abetz 1988; Kenk and Fischer 1988; Pretzsch 1992; Eriksson and Johansson 1993; Spelsberg 1994; Becker et al. 1994; Röhle 1995). A remarkable increase in growing stock of European forests has been reported (UN-ECE/FAO 1992; Kauppi et al. 1992; Kuusela 1994) which may have partly been caused by an increase in site productivity. Other possible causes for this observed increase are changes in tree species composition, age structure, silvicultural practices, spatial distribution of forest land and assessment methods.

Through these findings foresters were reminded that one could not simply rely on past growth observations to predict future growth. Even without human influences forests have always been exposed to environmental changes. As a result, growth rates have always been changing over time (Spiecker 1995). In addition, human activities are increasingly influencing the environment. Changes in litter raking, grazing, historical management (incl. regeneration methods, weed control, tending, thinnings and their effects on genetic resources, species composition, age structure and forest structure), fertilization and liming, climate (including air chemistry), atmospheric deposition, occurrence of pests and diseases, disturbances caused by fire, storm and snow, as well as browsing have influenced site conditions considerably. The global increase of atmospheric CO_2 and atmospheric deposition of nitrogen have an impact on forests. The level of CO_2 is predicted to double within the next century (Houghton et al. 1990). In consequence, more changes in climatic conditions are expected.

European Forest Institute Research Report No. 5
Growth Trends in European Forests
Edited by H. Spiecker, K. Mielikäinen, M. Köhl and J. P. Skovsgaard

The above mentioned environmental changes vary in space and time. Growth responses to these changes are modified by site and stand conditions. An impact of environmental changes on forest growth has to be assumed and therefore past growth observations may not reflect actual growth potential.

Growth reactions of trees are useful in identifying and quantifying possible environmental changes in their spatial and temporal extent while information about the effects of environmental changes may be used for the prediction of future forest development. In this context forest ecosystems could serve as bio-indicators. Furthermore, knowledge about forest growth trends is necessary for sustainable forest ecosystem management and for prognoses of wood supply. For these reasons the European Forest Institute initiated the project entitled "Growth Trends in European Forests - Has Site Productivity Changed?". In a planning meeting in 1993 the project's concept was developed. In 1994 a workshop was held in Joensuu/Finland, during which the scientific approach was discussed. The research was given a frame of objectives, data and methods to facilitate the comparison of results on a European scale. This frame was published as the EFI Working Paper No. 4 "Growth Trends of European Forests - Has Site Productivity Changed?" (Spiecker et al. 1994). Preliminary results of the project were presented at the IUFRO World Congress in Tampere, Finland in 1995 (Korpilahti et al. 1995, Spiecker et al. 1996). Finally, a research network was developed to evaluate growth data on a European level. The results of the joint effort of the participants are presented in this book.

While individual reports on growth and yields of forests have been published on local, regional and national levels, this project aimed at stimulating a joint effort in aggregating existing growth information and in identifying and quantifying possible growth trends and their spatial and temporal extent on a European level. The main objective of the project was to provide a retrospective view of the growth of European forests over the last decades. In most cases, data originally sampled for other purposes were used. This was possible because a large amount of information on forest growth has been accumulated in forest research institutions.

Growth of forests is modified by various natural and human influences. The main interest in growth trends in this project refers to changes in site productivity, which may have been caused by land use history, climate changes or atmospheric deposition.

The terms "site","forest site productivity" and "growth trend" have been defined in the context of this project as follows (EFI Working Paper No. 4):

- **Site**
  The term site is used to describe the sum of environmental conditions (biotic, edaphic, topographic and climatic conditions, including atmospheric composition) in existence at a particular location.

- **Forest Site Productivity**
  Forest site productivity is defined as the woody biomass production potential of a site. In this project the term site productivity is limited to the wood production potential of a site for a particular tree species, provenance or forest type. For exam-

ple, growth of volume, basal area or height may serve as an indicator of site productivity.

- **Growth Trend**
  A trend is a long-term change of a mean level. Growth trends within this project are indicated by long-term site-related deviations from expected growth. This project particularly focuses on the detection of trends during recent decades. For interpretation of growth trends 50-year-periods, preferably even longer periods, should be investigated.

It is difficult to separate growth trends which reflect long-term changes in site productivity from episodic changes caused by extreme events such as frost, drought, snow and storm damage, fire, insect or fungal diseases or by a combination of several events, which are often followed by a reverse change. Changing weather conditions typically cause short- and medium-term growth variation. However, effects of extreme climatic events on growth may have an impact on growth for up to several decades. Such an event may have an effect on the foliage, the root system and the water transportation system within a tree. In addition, it may change tree competition as trees with greater vitality gain growing space and resources from their weakened neighbours. These potential changes to stand dynamics initiate long lasting changes of individual tree and stand growth.

Quality data and evaluation methods are of central importance for reliable growth trend analysis (Sterba, in this volume). Three data sources were included in the project: research plot data, inventory data and tree ring or stem analysis data. Each of these three data sources has its own advantages and limitations. Stand history and site conditions on long-term permanent plots are generally well-known. Tree competition over time can be determined when the locations of individual trees are mapped. Volume growth per ha can also be derived from permanent plot data. However, permanent plots are not usually representative of larger areas because they have been established for other purposes and they may lack replication (Köhl et al. 1995). Trees on permanent plots are remeasured at intervals of several years. The time of the consecutive measurements has an effect on identification and description of short- and medium-term growth changes. Therefore, references to changes in relation to the findings of periodic measurement should be interpreted cautiously.

Forest inventory data, on the other hand, are usually sampled on a consistent statistical basis and are, therefore, representative for the entire inventory area (Köhl, in this volume). Even so, they lack detailed information on stand history, management and site conditions. Stem analysis, as the third source of data, allows the reconstruction of diameter and height growth over long time periods with annual resolution. However, it is not possible to find individual trees which represent site productivity over a long time period (Spiecker 1992) because stand density and aging have an effect on individual tree growth. The density effect is less pronounced for dominant trees although their growth is affected by neighboring trees as long as the trees have not grown free of any competition. Tree-ring data are appropriate for detecting agents with short- and medium-term influence on tree growth. Analysis of height increment is preferable in long-term observations since height growth is less influenced by the effects of competition. In such cases reliability of height growth as an indicator of site productivity has to be evaluated by long-term permanent plot data which provide information about volume increment per ha. A combination of different data types and variables may finally provide the best results. Combining single tree data with plot data helps to overcome some limitations of single tree growth data.

In some countries no long-term growth data were available in a form appropriate for analyzing possible growth trends. Furthermore, the time span of existing data varied from a few decades to several centuries. Some data reported in this book referred to case studies on rather small areas. Others, for example some inventory data, to larger regions or nations. Besides original studies, some summarizing papers are included in the book.

Since the data analysed were usually not collected for analysis of long-term growth trends, a variation in the applied methods was unavoidable. Some data sets could not be used for the comparative analysis: for example permanent plot data from stands that were considerably damaged by storm, snow, insects, fungi or fire. Likewise, fertilized plots could not be included in the trend analysis either. Information on stand structure and stand history was not always available in full detail. In some cases only data for rather short periods were at hand. In Southern Europe, for instance, trees are generally managed in short rotation periods and data from previous generations were at hand only in rare cases. Other data, such as inventory data, can be obtained for many decades in, for example, Finland and Sweden. In contrast, the first sample based inventory on forest district level in Switzerland was conducted in the early 1970s. In Germany the first nation wide inventory was conducted in the mid 1980s.

Data were checked, evaluated and interpreted by each author individually according to the recommendations of the EFI Working Paper No. 4. Due to the different data sources available different methods had to be used. Because of these shortcomings this book does not present growth development of forests in Europe in a uniform and statistically representative way. However, the results facilitate conclusions at the European level.

To determine possible growth trends, a growth reference describing expected growth is needed. Growth references have to be based on past experience. Preferably growth data from earlier growth periods are used. In some cases yield tables and other growth models are used as references. A reference is valid only when it represents the site and stand conditions as well as the management regime of the investigated forest in the reference period.

By using yield tables or other growth models as growth references the following problems have to be considered:

a) How do management practices (regeneration methods, tending of young stands, thinning regime etc.) of the reference stands influence reference values?
b) How do fluctuations in the growth conditions influence reference values? For example, unfavourable weather conditions lasting several years during the observation period may have an effect on the growth model used as a reference.

c) References may be biased due to an uneven distribution of site types in different age classes. If, for example, the oldest stands used in yield tables are grown on relatively poor sites, a comparison using these tables as a reference may lead to bias in estimation of actual productivity.

Twenty-two papers written by 43 scientists from 12 countries are presented in this book (Fig. 1). Most studies were conducted in Northern and Central Europe, only two studies refer to Southern Europe.

Fig. 1. Map showing the locations of the studies.

By analysing the spatial and temporal patterns of growth trends some hints regarding possible causes were expected. The results of the individual studies are discussed and summarized at the end of this book. Consequences for decision-makers and researchers in the field of forest growth and other fields are described.

References

Abetz P (1988) Untersuchungen zum Wachstum von Buchen auf der Schwäbischen Alb. Allgemeine Forst- und Jagdzeitung 159: 215-223

Becker M, Bert GD, Bouchon J, Picard J-F, Ulrich E (1994) Tendances à long terme observées dans la croissance de divers feuillus et résineux dur nord-est de la France depuis le milieu du XIXe siècle. Revue forestière francaise 46: 335-341

Eriksson H, Johansson U (1993) Yield of Norway spruce [Picea abies (L.) Karst.] in two consecutive rotations in southwestern Sweden. Plant and Soil 154: 239-247

Houghton JT; Jenkins, GJ and Ephraums JJ (1990) Climate change: The IPCC scientific assessment. Cambridge: University Press: 364 p.

Kauppi PE, Mielikäinen K, Kuusela K (1992) Biomass and carbon budget of European forests, 1971 to 1990. Science 256: 70-74

Kenk G, Fischer H (1988) Evidence from nitrogen fertilisation in the forests of Germany. Environmental Pollution 54: 199-218

Köhl M, Scott CT, Zingg A (1995) Evaluation of permanent sample surveys for growth and yield studies: a Swiss example. Forest Ecology and Management 71: 187-194

Korpilahti E, Salonen T, Oja S eds. (1995) Abstracts of invited papers, IUFRO XX Congress, 6-12 August 1995, Tampere, Finland: 270-277

Kuusela K (1994) Forest resources in Europe 1950-1990. European Forest Institute Research Report 1. Cambridge University Press: 154 p.

Pretzsch H (1992) Zunehmende Unstimmigkeit zwischen erwartetem und wirklichem Wachstum unserer Waldbestände - Konsequenzen für zukünftige ertragskundliche Informationssysteme. Forstwissenschaftliches Centralblatt 111: 366-382

Röhle H (1995) Zum Wachstum der Fichte auf Hochleistungsstandorten in Südbayern. Ertragskundliche Auswertung langfristig beobachteter Versuchsreihen unter besonderer Berücksichtigung von Trendänderungen im Wuchsverhalten. Mitteilungen aus der Staatsforstverwaltung Bayerns. 48: 272 p.

Spelsberg G (1994) Zum Höhenwachstum der Fichte in Nordrhein-Westfalen. Allgemeine Forst- und Jagdzeitung 165: 77-80

Spiecker H (1992) Which trees represent stand growth? In: Tree Rings and Environment, Proceedings of the International Dendrochronological Symposium, Ystad, South Sweden, 3-9 Sept. 1990, Lundqua Report 34: 308-312

Spiecker H (1995) Growth dynamics in a changing environment - long-term observations. Plant and Soil 168-169: 555-561

Spiecker H, Mielikäinen K, Köhl M, Untheim H (1994) Growth trends of European forests - Has site productivity changed? European Forest Institute, Working Paper 4: 68 p.

Spiecker H, Mielikäinen K, Köhl M, Skovsgaard JP (1996) Growth trends and site productivity: Has site productivity changed? Proceedings of two Sessions of S4.01-00 at the IUFRO XX World Congress, 6 - 12 August 1995 Tampere, Finland

UN-ECE/FAO (1992) The forest resources of the temperate zones, UN-ECE/FAO 1990 forest resource assessment , United Nations, ECE/TIM/62: 348 p

Construction of a 7500-Year Tree-Ring Record for Scots Pine (*Pinus sylvestris*, L.) in Northern Fennoscandia and its Application to Growth Variation and Palaeoclimatic Studies

Pentti Zetterberg[1], Matti Eronen[2] and Markus Lindholm[3]
[1]Karelian Institute, Univ. Joensuu, P.O. Box 111, FIN-80101 Joensuu, Finland
[2]Dept. of Geology, Univ. Helsinki, P.O. Box 11, FIN-00014 University of Helsinki, Finland
[3]Univ. Joensuu, FIN-57130 Savonlinna, Finland

Abstract

Precisely dated subfossil trees and long chronologies of tree growth provide a unique calendrical record for studies on growth variations in the past and climate history. A total of 1465 subfossil Scots pines (*Pinus sylvestris* L.) has been sampled from 42 sites in northern Fennoscandia. The sampling sites are mostly small lakes in the timberline zone located between 68°30' N and 70°00' N and at elevations ranging from 75 m a.s.l. to 515 m a.s.l. A total of 1023 subfossil trees have been dated by dendrochronology. From the dated trees we have constructed a chronology for each growing site and also a common chronology for the entire area. In this study, subfossil pines are used as "growth reference" of forest site productivity in the past. The abundance of subfossil pines makes it possible to explore the long-term stand history of pine at each site over thousands of years. The unbroken tree-ring record extends from the present back until 165 B.C. and after a 150-200-year gap until about 7500 years before the present time. The continuous part of the master curve is 664 years longer than the earlier published pine chronology for northern Sweden. This 7500-year pine master chronology can be used for dating subfossil pines from a wide area in northern Fennoscandia.

The subfossil pine material has been used for studies on the Holocene climatic history, growth variations, year-to-year variability in ring widths, periods of germination and mortality, population size and age structure at the sampling sites and tree-line changes. These combined data are valid indicators of diverse environmental changes in the past. A detailed register of the growth variations is available for two periods in mid- to late-Holocene times. The data for the period 4500-3000 B.C. suggest increased variability in tree growth after 3800 B.C. The mid-Holocene climatic change at this time was thus largely the result of a shift towards less stable

European Forest Institute Research Report No. 5
Growth Trends in European Forests
Edited by H. Spiecker, K. Mielikäinen, M. Köhl and J. P. Skovsgaard

growing conditions. For the late Holocene period from 165 B.C. to the present, the dating accuracy of variations in tree growth and climate is 1 year. The mean annual growth has temporarily varied significantly, being highly dependent on growing conditions at each site, but when the entire material is considered, in the long term the mean ring width has remained rather constant, being approximately 0.6 mm. The available data from the Holocene climatic optimum and late Holocene do not differ from each other in this respect.

1. Introduction

Precisely dated subfossil trees and long chronologies of tree growth can significantly increase the knowledge of growth variations in the past and climate history. The abundance of subfossil pines makes it possible to explore the long-term stand history of pine over thousands of years. The comparison of data from several sites will identify common changes in tree growth, which represents important evidence of large, regionally representative past climate-based growth changes. For the interpretation of growth trends in the past few decades, also long-term changes should be investigated as background information.

Pine (*Pinus sylvestris* L.) spread over northern Fennoscandia by 8000-7500 B.P. (conventional radiocarbon years). Sites well beyond the limits of present distribution of pine were populated under favourable climatic conditions in 7000-5000 B.P. The maximum spread of pine occurred between 6000 to 4000 B.P. and was followed by a retreat during the successive millennia (Hyvärinen 1975, 1976, Eronen 1979, Eronen and Huttunen 1987, 1993). Pine is a relatively long-living tree species and the ring-width variability is clear, making reliable dendrochronological dating possible. The structure of subfossil wood several thousands of years old is sufficiently well preserved to allow the detection of cell walls and boundaries of tree rings for measuring.

In areas where trees grow in marginal environments, the control of annual growth by environmental factors, particularly climate, is strong and clearly discernible (Fritts 1976). In the polar/alpine pine forest limit in northern Fennoscandia, the variations in tree-ring width are very similar within wide areas. This makes the correlation of tree rings over large areas possible and allows covering of short gaps in the site record with data from another growing site. Regional differences between sites indicate local differences in growing conditions.

The subfossil pine material has been used for studies on the Holocene climatic history, growth variations, year-to-year variability in ring widths, periods of germination and mortality, population size and age structure at the sampling sites (Zetterberg et al. 1994) and tree-line changes (Eronen and Zetterberg, in press). These combined data are valid indicators of diverse environmental changes in the past.

2. Material and Methods

A total of 1465 subfossil Scots pines (*Pinus sylvestris* L.) has been sampled from 42 sites in northern Fennoscandia. The sampling sites are mostly small lakes in the timberline zone located between 68°30' N and 70°00' N and at elevations ranging from 75 m a.s.l. to 515 m a.s.l. (Fig. 1). Individual lakes normally yielded several tens of subfossil logs. Over 100 pine trunks were collected from five small lakes, the record being a lake where a total of 219 subfossil pine trees were sampled. Best sites for the preservation of subfossil trees are small ponds with a thick layer of mud on the bottom. The majority of the trees are buried in the bottom sediments. The trunks, totally embedded in the soft sediment, can be located by the help of divers. The longest trunks reach a length of 15 m and the maximum diameter is more than 50 cm. Most of the sampled trees are well preserved, so that nearly all tree rings of the cut discs could be measured. Badly decayed trunks, stumps with only the base of the trunk preserved, and young trees containing below 50 annual rings were rejected as unsuitable for dendrochronological studies. Sample discs were cut by chainsaw

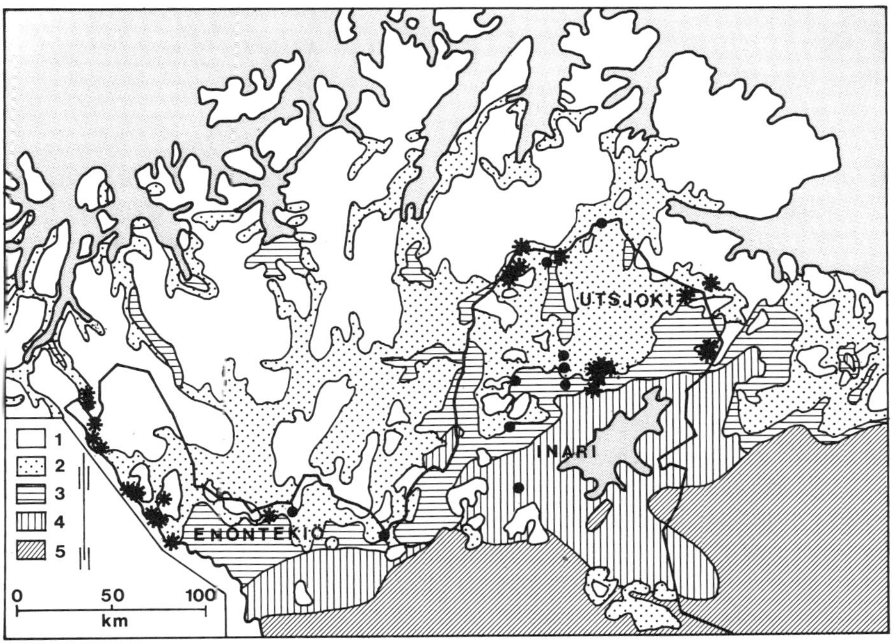

Fig. 1. Findings of subfossil pines in Finnish Lapland and adjacent areas. The asterisks indicate sites, mainly small lakes, where dendrochronological samples have been collected and black dots show the sites of other pine subfossil dates. The forest vegetation zones of northern Fennoscandia are also given: 1. Tundra, 2. Birch woodland, 3. Birch and pine forest, 4. Pine forest, 5. Pine and spruce forest.

about 1 - 1.5 m above the original root plate and higher up from the trunk when the lower part of tree was not preserved. The sampling methods are described in detail in Zetterberg et al. (1994).

The discs were dried at room temperature in the laboratory and the upper surface polished for ring-width measurement. Measurements were made with microscope along two to four radii and the mean value of these measurements was used to produce the tree-ring curves for individual trees. The replication facilitates the identification of partially absent rings and averaging the data across radii reduces the "within-tree" noise. The establishment of the chronologies requires the linking of tree-ring records together by means of the cross-dating. The cross-dating was first done by statistical correlation between series and it was always confirmed by visual inspection of the ring-width graphs. The most common reason for failure in dating was the occurrence of compression wood in trees which had grown at the shore and tilted towards the lake during their lifetime. Tree rings grow wider and more dense on the tilting side of the trunk in order to compensate for the lean and in an attempt to right the trunk. Difficulties also arose when the number of annual rings in a tree was less than 60.

3. Dendrochronological Dating and Development of the Long Tree Ring Record

A total of 1023 subfossil trees have been dated by dendrochronology in this study. The age distribution of the dated samples is given in Fig. 2. In this figure the material has been divided in three parts based on the origin of the samples: Enontekiö, Utsjoki and Inari areas. Enontekiö lies in the westernmost part of our study area, while Utsjoki is in the northernmost and Inari in the easternmost part.

The continuous tree-ring curve based on the dendrochronological dating extends back more than 2000 years, to the year 165 B.C. (Zetterberg et al. 1994). Over 200 subfossil trees from several sites in northernmost Finland are bound to the absolute chronology, and sample replication is more than 20 trees in most parts of the time period. Thus, the northern Finland absolute chronology can be regarded as a master curve and as such it is the longest for northern Europe so far.

Most of the dendrochronologically dated trees (approx. 800 samples) belong to the older part of the long master chronology, which is fixed to the time scale by many radiocarbon dates made from samples of subfossil wood. This continuous, more than 5000-year-long part of the master chronology now extends approximately to the year 7500 B.P. The 150-200-year-long gap, separating the older part from the younger absolute chronology, can probably be bridged in the near future. The age of the older chronology is known within the error limits of the radiocarbon method (several decades), but inside the master chronology the accuracy of the tree-

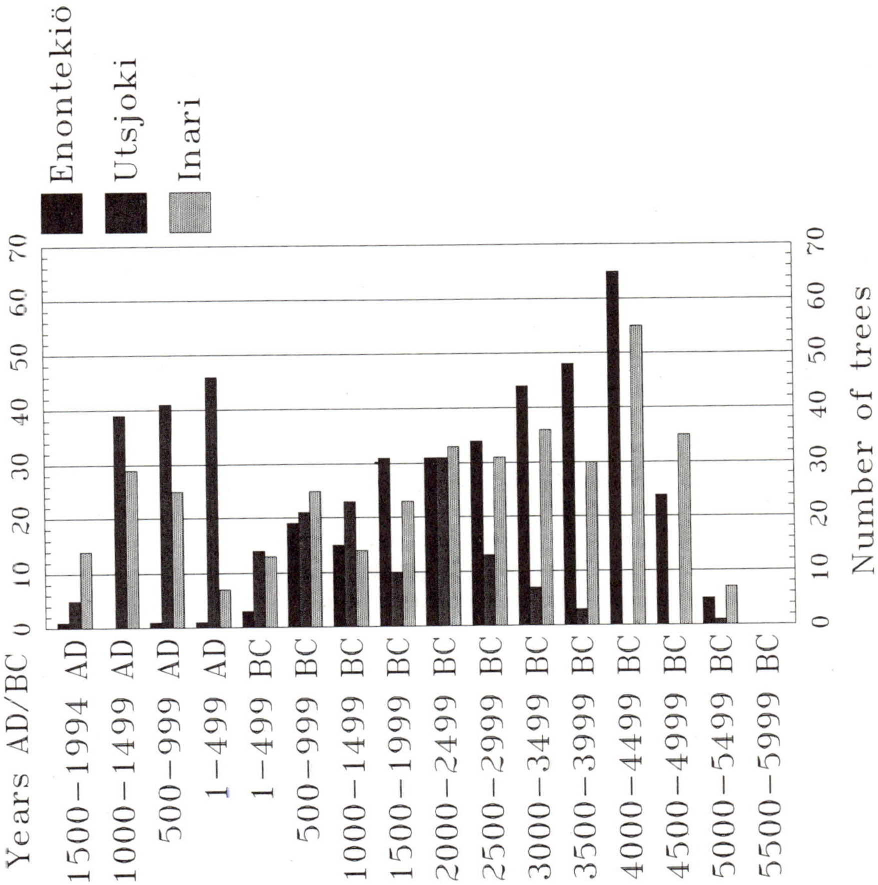

Fig. 2. Age distribution of dendrochronologically dated subfossil pines from the forest limit in subarctic Fennoscandia. For each sample the age is the mean value of the beginning and the end of dated tree-ring series.

ring series is 1 year. Thus, it can be used in studies of long- and short-term growth variations in northern Fennoscandia.

Together with the subfossil trees, we also collected tree-ring samples from a large number of living forest-limit pines, which have been used for construction of the youngest part of the Finnish Lapland master chronology. Tree-ring measurements from beams of old houses in northern Lapland (Zetterberg 1990) were also of great importance in cross-matching the curves derived from living and subfossil trees.

4. Growth Trends and Cycles, Examples from Utsjoki, Northern Finland

We used the long tree-ring records to study the growth trends (i.e. long-term change of a mean level) at several growing sites in northern Fennoscandia. One of the sites is Lake Ailigas in the Teno (Tana) River valley, near the main village of the Utsjoki commune (69°54'N, 27°04'E, 75 m a.s.l.). When developing the chronology for the Lake Ailigas, we did not employ any "standardization" (i.e. correction to remove the age-related trends in individual series) in the chronology construction in order to maintain all possible long-timescale variation (e.g. see Cook and Briffa 1990; Cook et al. 1995). Thus, the data contain some sample-age-related bias in different periods. The site chronology constructed from raw ring-width data of 78 subfossil pines is shown in Fig. 3. The general curve from 100 B.C. to A.D. 1250 is mostly based on several tree-ring series. Sample depth is above ten trees for half of the time and from five to nine trees during the remainder. Between the late 800s and the 930s the curve is based on only two to four trees. The data are also scarce from 1250 to the early 1400s. In the following, we will summarize the interpretations on growth trends and climate changes made from Lake Ailigas subfossil pines given in detail in Zetterberg et al. (1994).

All pines living between 130 and 120 B.C. indicate very poor growth followed by warming and pronounced radial growth after 120 B.C. In the beginning of the first millennium B.C., tree rings were relatively wide, while in the latter part of the century the growth decreased. More favourable conditions prevailed again around the beginning of the Christian era and in the first century the growth reached peak values around 90 A.D. Reduced growth occurred between A.D. 8 and 22 and again in A.D. 40. Fairly constant tree growth in relatively favourable conditions prevailed during the second century. This clearly favourable growth period was interrupted by years of poor growth, first in 111 - 114 and again 150, 168 and 187. During the third century, radial growth decreased, and the fourth century was even less favourable for the growth of pine. The fifth and sixth centuries are characterized by significant growth variations, among which the decline in growth in the year 536 A.D. was extremely abrupt. The strong variations in tree growth then subsided during the seventh century. The eighth century was characterized by a reduced trend in growth, in some respects resembling that of the fourth century. The periods of minimal tree growth for the time-span from 100 B.C. to 1250 A.D. occurred in the early and middle part of the century, while growth improved again in the late ninth century. There were no strong growth variations in the 10th century and in the early 11th century. From 1044-1060 A.D. growth was exceptionally poor, but after this cool phase, growth rose for three decades. The oscillations in tree growth had started in the latter part of the previous century and continued during the first half of the 12th century with growth peaking in the 1160s. Early 13th century summers were mostly cool, but later tree growth increases to a maximum in 1227. The tree-ring curve showing decreased growth, albeit based on few data, indicates that the period from

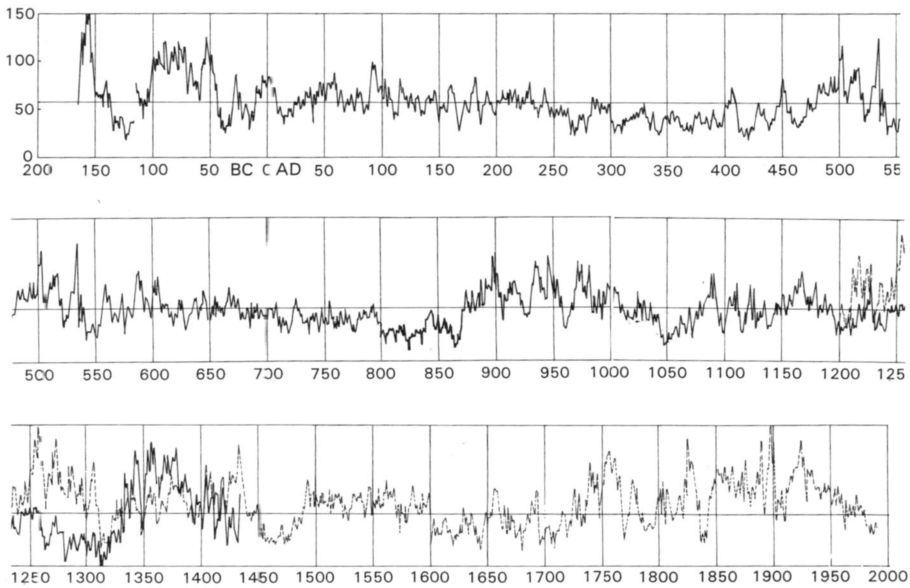

Fig. 3. Mean ring width chronology from the Lake Ailigas subfossil pines (wide line) with the mean ring width chronology of the Utsjoki-area for the period 1200-1991 AD. (broken line). The horizontal line shows the average of annual radial growth of pines in the forest limit zone in the mid Holocene time 4500-2900 B.C. (Zetterberg et al. in press).

1260 to 1320 was characterized by unfavourable conditions. After 1320, the curve demonstrates a large increase in tree growth. For this period the curve is based only on two trees, but the same rise can be seen in other data from Utsjoki (Fig 3). This additional tree-ring data was needed to connect the Lake Ailigas chronology to the present time. The other curve is based on tree-ring samples taken from living pines and old wooden buildings in Utsjoki (Zetterberg 1990) and subfossils from other small lakes in the Tana river valley. The previously published tree-ring curve from northern Sweden (Torneträsk) also shows a short but distinct peak in the beginning of the 14th century (Briffa et al. 1992). In these data, the peak was followed by a decrease and then average growth during the rest of the century. Our new data from the Inari area (Zetterberg, unpublished), 70 km south of Tana River valley, also shows a clear rise in growth after 1320. The growth peaked during 1350 and was followed by a decrease. The distinct growth reduce in the Utsjoki regional growth curve after the 1450s is also obvious in the data from Inari (Zetterberg, unpublished), but cannot be detected in the data from northern Sweden. In Utsjoki, the 16th century was a period of increased growth and was followed by a drastic decrease after 1600. The 17th century was mainly a period of very low growth. After the 1730s, the mean growth curve from Utsjoki shows strong fluctuations with low mean growth during the end of the 18th century, 1840s and in the beginning of the

20th century. Periods of increased growth occurred during the 1750s, the 1830s, in the latter part of the 19th century and in the 1920s.

From the tree-ring data collected from different parts of Utsjoki commune, we produced a cubic smoothing spline standardized chronology (Fig. 4, A). The chronology is based on 100 subfossil trunks (including 78 trees from Lake Ailigas), 10 historical beams from buildings and 20 living pines. The growth trends of the Lake

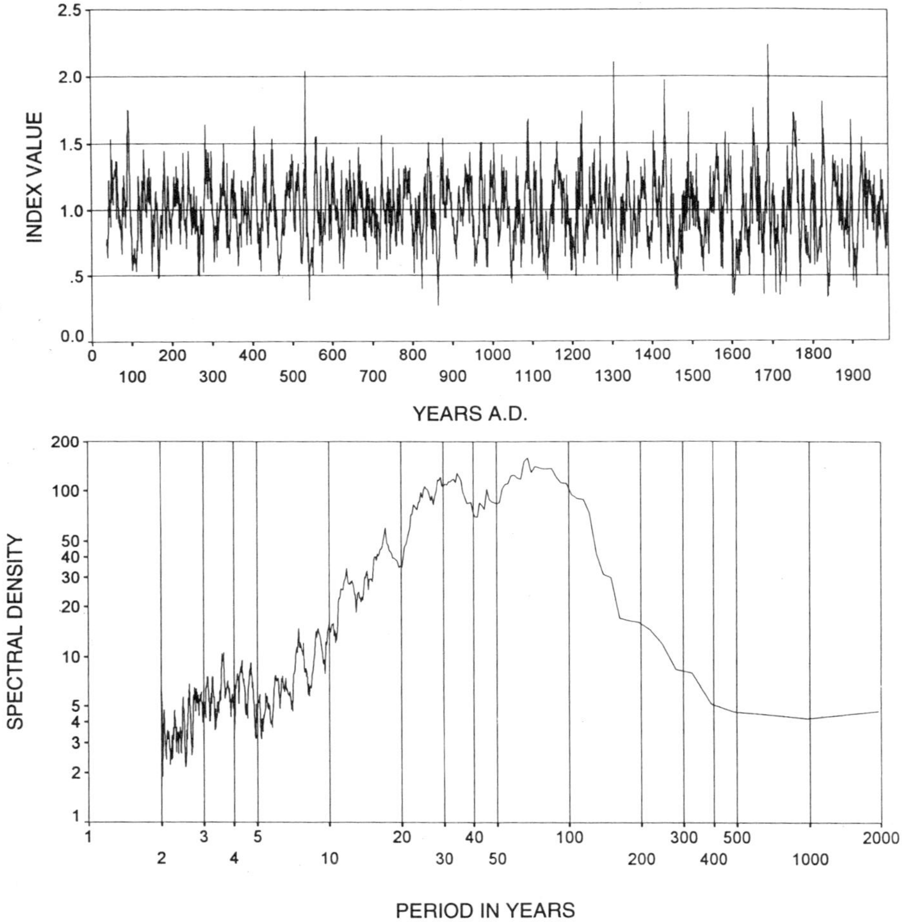

Fig. 4. *a*. The standardized mean ring-width chronology for Utsjoki, northern Finland. The chronology is based on subfossil trunks, historical beams and living pines. *b*. In the periodogram the horizontal axis shows the periods into which spectral analysis has decomposed the series, the vertical axis shows the relative importance of each period. The presentation is in logarithmic scale. This periodogram is the result of a preliminary test; further studies are needed to confirm the validity of this curve.

Ailigas one-site original chronology listed above, are much less obvious in the standardized curve, but this curve is more suitable for dendroclimatic reconstructions. A spectral analysis was carried out to check preliminarily if there are any strong periodicities in the Utsjoki tree-ring material. The relationships of index values of the chronology are shown at different points in time (Fig. 4, B). The periodogram presents the chronology in terms of cycles of different length that generate the time series. The variations are described in terms of cycles of sinuses and cosinuses. The highest frequency has half as many cycles as the number of tree rings. A smoothing transformation for the periodogram was obtained by a Tukey-Hamming smoothing procedure with a 15-year span.

4. Discussion and Conclusions

The subfossil pine material from northern Lapland has been used for studies on the Holocene climatic history, growth variations, year-to-year variability in ring-widths, periods of germination and mortality, population size and age structure at the sampling sites and tree-line changes. These combined data have given diverse indications of environmental changes and changes in the site productivity in the past.

Detailed tree-ring register of the growth variations in northern Fennoscandia is already available for two periods in mid- to late-Holocene times. The data for the period 4500-3000 B.C. suggest increased variability in tree growth after 3800 B.C. The mid-Holocene climatic change at this time was thus largely the result of a shift towards less stable growing conditions (Zetterberg et al. in press). For the late Holocene period from 165 B.C. to the present, growth and climate variations have been studied to the accuracy of 1 year (Zetterberg et al. 1994). The mean annual growth temporarily varied significantly, being highly dependent on growing conditions at each site, but when the entire material is considered, the mean ring width remained rather constant in the long term, being approx. 0.6 mm. The mean ring width for the Holocene climatic optimum (0.55 mm, Zetterberg et al. in press) and the late Holocene do not differ from each other in this respect.

In our procedure we used tree-ring chronologies based on raw unstandardized data in order to maintain all possible long-timescale variations. The traditional approach to the palaeoclimatological interpretation of old trees is to develop long series of averaged ring width or standardized ring indices and to reconstruct palaeotemperatures using regression equations calibrated against modern instrumental records (e.g. Fritts 1976). In northern Fennoscandia the growth of tree-limit pines is strongly correlated with summer temperatures (Hustich 1948, Mikola 1950). This situation has permitted the reconstruction of past summer temperature variations (Aniol and Eckstein 1984; Briffa et al. 1988, 1990, 1992). These interpretations are based on the principle of uniformitarianism; i.e. on the assumption that the environmental/climatic signal represented in the tree-ring time series has not changed over time. Apart from the climate, other factors such as site conditions, plant diseases, flowering and seed maturation, fire etc. influence tree growth. Detailed descriptions of representative sites are necessary to allow us to distinguish between local tree growth variations and large scale variations which are mainly caused by climatic changes in the forest-limit zone.

The periodicity of pine growth in northern Fennoscandia has been studied by Sirén and Hari (1971) and Briffa and Schweingruber (1992) for example, who both found periodicities of different length in the tree-ring data. Periodicity is obvious also in our data (Fig. 4, B); especially between 3 and 50 years, there are several peaks. On the other hand, periods of 2 - 5 years show only weak powers. However, there remain a number of uncertainties in the data. Several frequencies tend to occur together; in addition, there is a considerable amount of random noise evident in the time series. An areal chronology may also reveal purely local cycles. This random periodicity may be detected and possibly eliminated when several areal chronologies are compared. Spectral analysis may be applied to pairs of areal chronologies to examine their covariation at each period or reciprocally at each frequency.

Climatic stress, such as extremely cold conditions, influences the number of dying trees. Above-average warmth can increase the number starting their growth. Therefore, during periods of high inter-annual temperature variability (or some other climatic factor), a greater numbers of trees are likely to die and others germinate than would be the case under more stable conditions (cf. Karlén 1993, Kullman 1993). The variations in germination and dying-off of trees can be used as a complementary source of information on past growing conditions (Zetterberg et al. in press). The size of the pine population at a sampling site is also indicative of past variations in growing conditions and site productivity. In marginal environments, such as the polar/alpine tree line, the tree population usually grows larger during favourable, relatively warm periods than during unstable or cold times.

It was shown that in the Lake Ailigas data there were only four main regeneration periods between 300 A.D. and 1350 A.D. (Zetterberg et al. 1994). The sparsity of pine generations over almost 1000 years indicates that conditions were seldom favourable for flowering and maturation of pine seeds. It is known from modern forestry studies that the regeneration of pine in the forest-limit zone requires at least two consecutive summers warmer than the "average" and this occurs, at present, probably only a few times in a century (Sirén 1961). The Lake Ailigas subfossil pines show that conditions were similarly harsh for the northern pines in the past.

Subfossil logs have been found even tens of kilometres beyond the present limit of pine and at altitudes above the highest present occurrences of pine trees, indicating more favourable conditions thousands of years ago. Eronen (1979) surveyed earlier studies on the distribution of subfossil pines in Finnish Lapland. These, together with pollen studies, show that the maximum extension of pine in Finnish Lapland occurred during the mid-Holocene time (Eronen and Huttunen 1987, 1993). The dendrochronological dating and development of the 7500-year tree-ring record resulted in a large number of new dates of pine subfossils. The 1023 subfossil trunks dated in this study have multiplied the number of dated subfossil trees in the area.

The new dated trees confirm the earlier conclusion that the pine forests reached their maximum extent in northern Fennoscandia 4900 - 2500 B.C. (= 6000 - 4000 B.P.). The subsequent retreat shows somewhat different patterns in different regions, but the general trend is a gradual dying-off and receding from the outmost growing sites (Eronen and Zetterberg in press).

The future aims of this research are: 1) bridging the short gap in the tree-ring data before the Christian Era and lengthening the existing chronology to cover 8000 years, 2) developing several continuous site chronologies as long as possible in order to study changes in the forest site productivity, 3) detailed study of the growth variations and their climatic interpretations for the entire regional chronology. The 7500-year pine master chronology is suitable for dating of subfossil pines and archaeological pine wood from a wide area in northern Fennoscandia even though it is not yet fully complete.

Acknowledgements

This research is funded by the Academy of Finland (SILMU, Finnish Research Programme on Climate Change), the European Science Foundation (Project European Palaeoclimate and Man), and EC Environment Research Programme in Climatology and Natural Hazards contract EV5V-CT94-0500: Tree-ring evidence of climate change in Northern Eurasia during the last 2000 years.

References

Aniol RW, Eckstein D (1984) Dendroclimatological studies at the northern timberline. In: Mörner N-A, Karlén W (eds) Climatic changes on a yearly to millennial basis. D. Reidel Publishing Company, pp 273-279

Briffa KR, Bartholin TS, Eckstein D, Jones PD, Karlén W, Schweingruber FH, Zetterberg P (1990) A 1400-year tree-ring record of summer temperatures in Fennoscandia. Nature 346: 434-439

Briffa KR, Bartholin TS, Eckstein D, Schweingruber FH, Karlén W, Zetterberg P, Eronen M (1992) Fennoscandian summers from AD 500: temperature changes on short and long timescales. Climate Dynamics 7: 111-119

Briffa KR, Jones PD, Schweingruber FH (1988) Summer temperature patterns over Europe: a reconstruction from 1750 A.D. based on maximum latewood density indices of conifers. Quaternary Research 30: 36-52

Briffa KR, Schweingruber FH (1992) Recent dendroclimatic evidence of northern and central European summer temperatures. In: Bradley RS, Jones PD (eds) Climate since A.D. 1500. Routledge, London, pp 366-392

Cook ER, Briffa KR (1990) A comparison of some tree-ring standardization methods. In: Cook ER, Kairiukstis L (eds) Methods of dendrochronology. Applications in the environmental sciences. Kluwer/IIASA, Dordrecht, pp 104-123

Cook ER, Briffa KR, Meko DM, Graybill DA, Funkhouser G (1995) The 'segment length curse' in long tree-ring chronology development for palaeoclimatic studies. The Holocene 5: 229-237

Eronen M (1979) The retreat of pine forest in Finnish Lapland since the Holocene climatic optimum: a general discussion with radiocarbon evidence. Fennia 157(2): 93-114

Eronen M, Huttunen P (1987) Radiocarbon-dated subfossil pines from Finnish Lapland. Geografiska Annaler 69A(2): 297-304

Eronen M, Huttunen P (1993) Pine megafossils as indicators of Holocene climatic changes in Fennoscandia. Paläoklimaforschung - Palaeoclimate research 9: 29-40

Eronen M, Zetterberg P (1996) Expanding megafossil data on Holocene changes at the polar/alpine pine limit in northern Fennoscandia. Paläoklimaforschung - Palaeoclimate Research 20 (in press)

Fritts HC (1976) Tree-rings and climate. Academic Press, London, p 567

Hustich I (1948) The Scotch pine in northernmost Finland and its dependence on the climate in the last decades. Acta Botanica Fennica 42: 1-75

Hyvärinen H (1975) Absolute and relative pollen diagrams from northernmost Fennoscandia. Fennia 142(1): 1-23

Hyvärinen H (1976) Flandrian pollen deposition rates and tree-line history in northern Fennoscandia. Boreas 5: 163-175

Karlén W (1993) Glaciological, sedimentological and palaeobotanical data indicating Holocene climatic change in northern Fennoscandia. Paläoklimaforschung - Palaeoclimate Research 9: 69-83

Kullman L (1993) Dynamism of the altitudinal margin of the boreal forest in Sweden. Paläoklimaforschung - Palaeoclimate Research 9: 41-55

Mikola P (1950) Puiden kasvun vaihteluista ja niiden merkityksestä kasvututkimuksissa. Summary: On variations in tree growth and their significance to growth studies. Communicationes Instituti Forestalis Fenniae 38(5): 1-131

Sirén G (1961) Skogsgränstallen som indikator för klimatfluktuationerna i norra Fennoskandien under historisk tid. Communicationes Instituti Forestalis Fenniae 54(2): 3-65

Sirén G, Hari P (1971) Coinciding periodicity in recent tree rings and glacial clay sediments. Reports from the Kevo Subarctic Research Station 8: 155-157

Zetterberg P (1990) Lapin mäntyjen dendrokronologiaa. In: Holopainen IJ, Simola H (eds) Ruijan retki 1989. University of Joensuu, Joensuu, pp 93-99

Zetterberg P, Eronen M, Briffa KR (1994) Evidence on climatic variability and prehistoric human activites between 165 B.C. and A.D. 1400 derived from subfossil Scots pines (Pinus sylvestris L.) found in a lake in Utsjoki northernmost Finland. Bulletin of the Geological Society of Finland 66: 107-124

Zetterberg P, Eronen M, Lindholm M (1996) The mid-Holocene climatic change around 3800 B.C: tree-ring evidence from northern Fennoscandia. Paläoklimaforschung - Palaeoclimate Research 20 (in press)

Growth Trends of Forests in Finland and North-Western Russia

Kari Mielikäinen[1] and Svetozar N. Sennov[2]
[1]Finnish Forest Research Institute. FIN-1301 Vantaa, Finland
[2]St. Petersburg Forest Academy. SU-194018 St. Petersburg, Russia

Abstract

Several studies concerning growth trends in Finland and north-western Russia are cited in this paper. Many of these studies are from the northern part of the area, close to the Arctic timber line. The forests in that region are especially suitable for trend studies due to strong reactions of tree growth to the environmental changes and lack of competition due to low stand density.

In the North, both negative and positive trends have been found. Local forest growth decline of unusual extent was observed around non-ferrous smelters in the Kola Peninsula, north-western Russia. In addition to the total dieback of forests near the smelters, reduced growth of Scots pine was detected up to 30-35 kilometres away from the sources of pollution.

The area of local forest decline covers only a small fraction of the total forest area around pollution sources. In the major part of forests in Lapland, the radial increment of Scots pine has been at a clearly higher level than in the 19th century. This is mainly due to more favourable climatic conditions in this century, especially in the 1930s. This conclusion is supported by studies conducted in other parts of Finland. The nitrogen deposition, many times higher than in the North, has namely caused no measurable trends in the radial increment of Scots pine in southern Finland.

In the region of Sankt Petersburg in Russia, ageing study stands show no decrease in current annual volume increment during the last few decades. This abnormal behaviour of mature stands can be a sign of increased nitrogen deposition (25 kg ha^{-1} annually) or of normal long-term climatic variation.

1. Introduction

A key question in the discussion on forest decline in northern latitudes of Finland and Russia has been the doubt about the sensitivity of trees to decline due to stress

European Forest Institute Research Report No. 5
Growth Trends in European Forests
Edited by H. Spiecker, K. Mielikäinen, M. Köhl and J. P. Skovsgaard

caused by climate, gaseous pollutants and soils with low buffering capacity against acid deposition. The effects of pollutants, coming from the North (Kola peninsula) or from the South (Central Europe) have been widely discussed during the last 10 years. The existence and direction of these effects has, however, remained unclear.

The national forest inventories (eight NFIs since 1923) conducted in Finland show an increase of more than 40% in forest volume growth from 1950 to 1990 (Tomppo and Siitonen 1991). The main reasons for this increase are supposed to be changes in silvicultural practices and stand structure. Among these practices, the drainage of peatlands, regeneration of low-productive stands, forest fertilisation and the change from selection fellings to low thinnings are the main causes behind the growth increase. Deposition of man-made pollutants, originating mainly from the burning of fossil fuels and agriculture cannot, however, be excluded from the list of possible causes for increasing productivity of forests.

Fire has always played a decisive role in boreal forests. On the one hand, it has improved soil productivity by releasing nutrients from the thick humus layer, typical of old spruce-dominated stands. On the other hand, fire has in the past caused disturbances in individual tree growth, which is difficult to detect and analyse.

In the northern latitudes of Fennoscandia, the variation in tree growth between years and decades can be very high, thanks to the harsh climate and long-lasting lag effects (Siren 1961). This variation makes the detection of possible growth trends difficult.

2. Trends Showing Decreasing Growth

There are few publications showing clear decreasing trends in tree growth. Nöjd and Reams (1995) have published results on the basis of data measured around the big nickel smelters in the Russian Kola Peninsula, north of the Polar Circle, and along a gradient line transversing into Finnish Lapland. The annual deposition in 1990 of sulphur dioxide (SO_2) has been 230 kT from Monchegorsk and 190 kT from Nikel. The depositions of nickel and copper originating from Monchegorsk are 2-3 kT.

There is a circle of 5 km in radius of dead forest around the Monchegorsk smelter. Growth decline can be detected up to 30-40 km from the smelter. The variation in pine growth (annual rings) is presented in Fig. 1 as a function of time and distance from the pollution source. The growth decline began at the end of the 1940s and the area is constantly expanding. The distance from Monchegorsk to the border of Finland is more than 100 km.

The total emission of sulphur in Finland is less than half of that emitted by the two above-mentioned smelters. The emissions have diminished by 80% since the beginning of the 1970s. Thus the growth decline caused by sulphur cannot be detected in Finland.

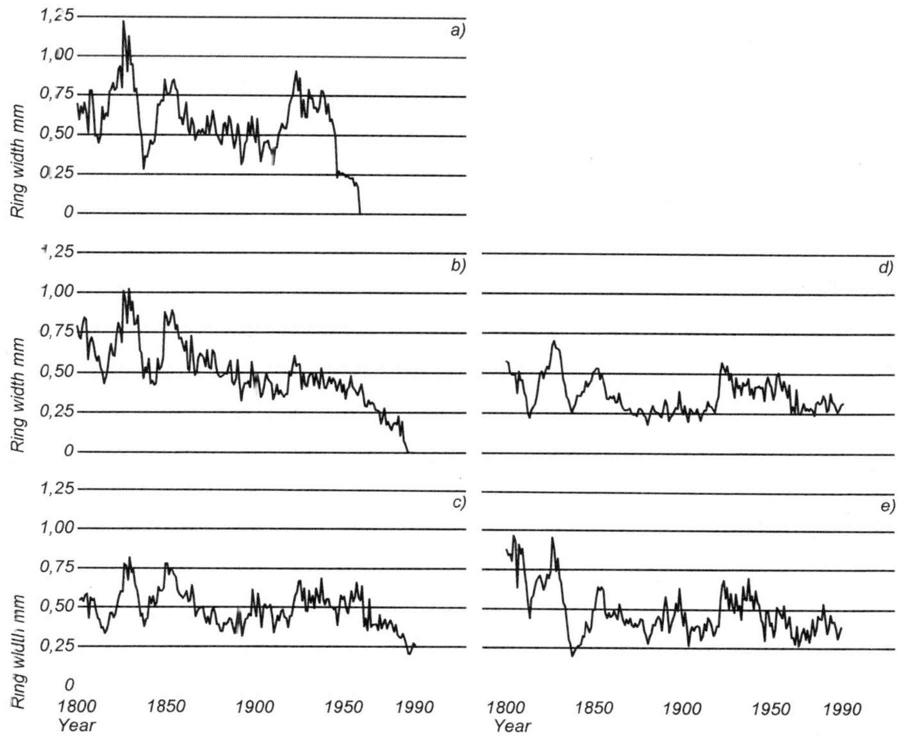

Fig. 1. Radial increment of Scots pine at different distances from the Monchegorsk smelter. Dead trees at distances of 4.5 km (*a*) and 12 km (*b*). Sample plots in Kola Peninsula 28-40 km (*c*) and 60-90 km (*d*) and in Finnish Lapland 130-150 km (*e*) from the pollution source according to Nöjd (1995). The results are based on increment cores taken from 300 standing trees.

3. Trends Showing Increasing Growth

3.1. Inventory Material and Permanent Experiments

Arovaara et al. (1984) have analysed data from three national forest inventories (1950-1980) by modelling stand growth as a function of stand age and volume. Comparing the model estimates with actual field measurements in successive inventories showed that pine stands in the 1970s grew better than in earlier times. This increase could not be directly explained by climatic variables. The trend disappeared partly or totally when the growth values were adjusted by existing annual ring indices.

Henttonen (1990) found a positive trend in the width of annual rings of Norway spruce in southern Finland between 1967 and 1987. This short-term trend could be explained by increasing temperature and precipitation during that period. Since 1987, successive dry summers, followed by abundant seed years, have caused the positive trend to end.

Sennov (1995) has studied the culmination of annual volume growth in stands dominated by pine, spruce or birch. Long-term experiments, established in the 1930s, did not show a clear culmination point at the age of 40 to 50 years as was supposed to happen (Fig. 2). On the contrary, some of these 9 study plots showed a constant or increasing growth in the 1970s and 1980s at a relatively high age. The author suggests that this increase in stand increment may be a sign of the beginning of climate change. The annual nitrogen deposition, about 25 kg ha^{-1} in that region, can have caused the trend. The short duration of the trend and the lack of climatic data in the analysis make the conclusions still uncertain. The trend can at least partially be caused by natural climatic variation or stand dynamics of mixed stands.

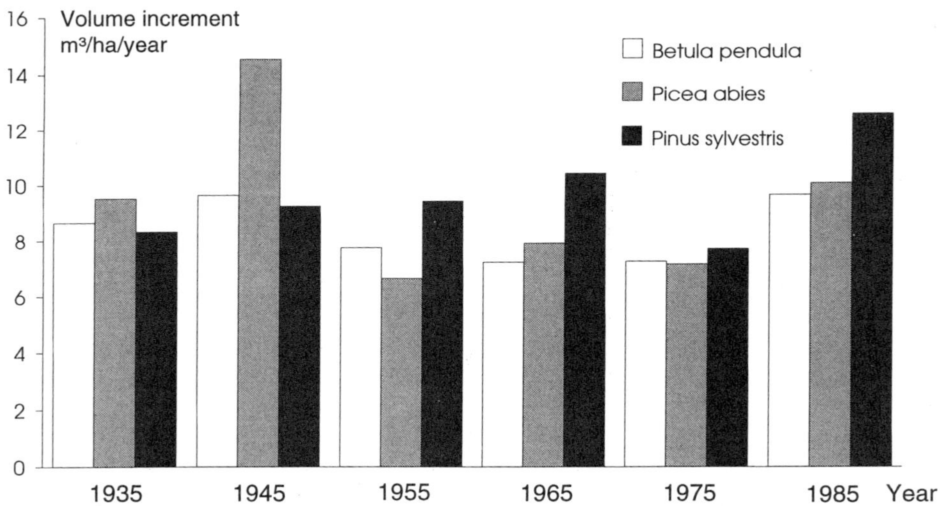

Fig. 2. Current annual volume increment of *Betula pendula, Pinus sylvestris* and *Picea abies* on the basis of long-term experiments in the region of Sankt Petersburg. The age of the stands is about 90 years (Sennov 1995).

3.2. Annual Ring Analyses

Hari et al. (1984) searched for a trend using annual ring data, measured in strict nature reserves in southern Finland. The material consisted of old stands (established between 1680 and 1800) for reference and younger stands (established between 1800 and 1900) for studying the trend. The existence of the trend was tested by comparing the growth of the young trees with the age trend curve estimated using the old trees. The total number of sample trees was about 400.

The results suggested that since the 1940s the annual rings of pine have become increasingly wider than the reference curve suggests. The authors concluded that this was probably caused by an increase in the concentration of atmospheric CO_2.

Based on annual rings of 98 old Scots pines (ages 300 to 400 years) close to the northern timber line in Finnish Lapland, Hari and Arovaara (1988) studied the trend in conditions where the depositions of sulphur and nitrogen were at minimum levels. The method used was to develop a climatic growth model for Scots pine by using the measured climate data from 1906 to 1940 as predictor variables. This model was then used to estimate the width of the annual rings between 1940 and 1983 (Fig. 3).

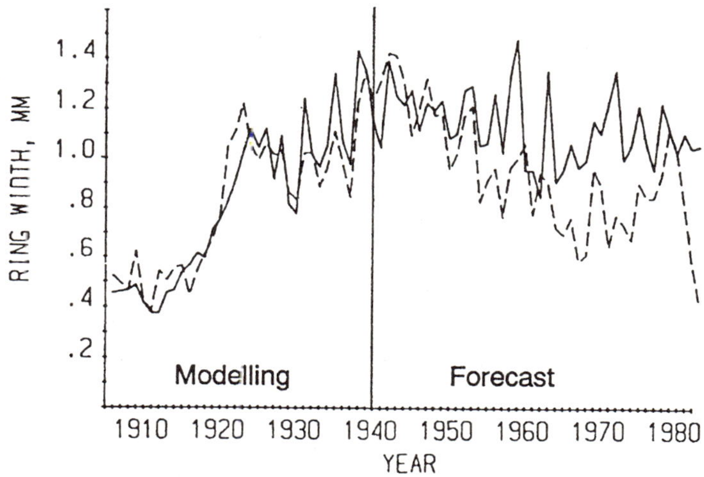

Fig. 3. The radial increment of old Scots pine in Finnish Lapland close to the northern timber line. Measured (solid line) and calculated (dashed line) ring widths from 1906 to 1983 according to Hari and Arovaara (1988).

The results showed that the actual width of annual rings has been higher than the model estimate during the last few decades. This trend was also found by Alekseev (1995) in the Russian Kola Peninsula. The uncertainty lies in the fact that the period 1906-1940 was climatically highly unusual in the whole of Lapland. After the severe frost damage of 1902, large numbers of pines died. Those that survived took as much as 20 years to recover.

The authors caution that the model is sensitive to slight changes in selecting a parameter describing the autocorrelation in the ring width series. Recovery from exceptional damage may also have affected the model parameters and estimates.

4. No Trend

On the basis of annual rings from old, unmanaged stands, Sinkevich (1995) studied the growth variation of Scots pine in Russian Karelia. He modelled the long-term cyclic increment fluctuation using the polyharmonic model. Data on solar activity, average temperature of the growing season and precipitation during July were also used in the model. The model was also used for making prognosis about future growth variation.

The results suggested that there is a high cyclic fluctuation in the radial increment of Scots pine. This fluctuation makes it impossible to find trendlike changes caused by the greenhouse effect in this data.

Eronen and Zetterberg (1992) have constructed an annual ring calendar of almost 8000 years using subfossil trunks, found from lakes close to the northern timber line in Lapland. Their results show that 4000-6000 years ago the climate was more favourable for tree growth than today. The variation in radial growth has been greater during the last 3000 years than before. Variation or trends, different from those in former times, cannot be detected this century (Fig. 4).

Mielikäinen and Timonen (1995) have studied the variation in the radial increment of Scots pine in southern Finland. Their first data set consists of increment cores, gathered from old, untouched forests in strict nature reserves. This set of data suggests no decline or increase in the radial increment of different age classes during the 20th century. This result is supported by increment cores taken from permanent experiments. The residual variation around the individual tree growth model developed was independent of the calendar year.

The height development of dominant trees differed from the site index curves published by Gustavsen (1980). This difference can mainly be attributed to differences in stand history between the two sets of data. Old stands in the Gustavsen´s data may have been affected by selective felling some 50 years ago.

5. Discussion and Conclusions

Different, and partly contradicting, growth trends have been reported for Finland and north-western Russia. This may be caused by differences in the data used. The often unknown stand history (wind, fire, management) makes conclusions uncertain. The recovery of trees after human or natural disturbances may have been interpreted as a trend caused by climatic change. Thus all growth variation should be compared to the variation in climate. Another problem in interpreting the results is the different time scales of the studies. A scale of a couple of decades may include trends which in a longer time span would turn out to be the result of natural growth variation.

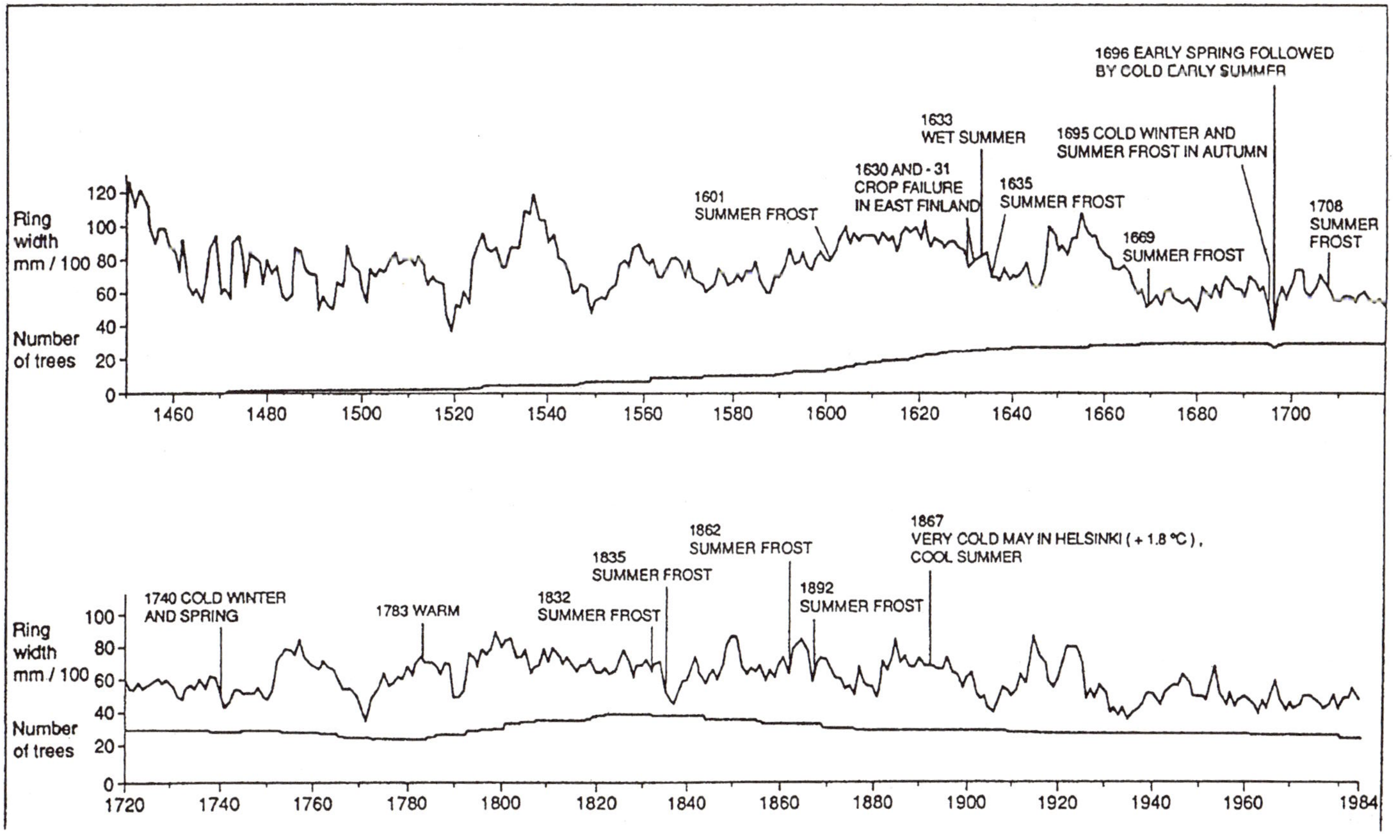

Fig. 4. Tree-ring curves of Scots pine for northern Karelia, eastern Finland. Some historical climatic events have been added to the curve according to Eronen et. al. (1994).

Our conclusion is that there is no clear growth trend visible in most parts of Finland and north-western Russia attributable to air pollution or changing climate. A possibly positive trend was found only in the region of Sankt Petersburg. The causes for our conclusion are:

1. The annual deposition of nitrogen, about 5 kg ha^{-1} in Central Finland (and much less in northern parts of Russia and Finland) during the past 20 years, is unlikely to cause abnormally high growth to be distinguished from long-term growth variation. The corresponding deposition, 25 kg ha^{-1} in the region of Saint Petersburg, may have caused an increase in stand volume growth during the latest few decades.
2. The temperature and precipitation in the growing season do not show an increase in the last 60 years. On the contrary, the summers from 1961 to 1990 were cooler than in the preceding decades.
3. The long-term cyclic variation in tree growth is so pronounced that it can conceal the beginning of abnormal trends.
4. It is very likely that the increase in the total production of Finland's forests during the last 40 years can be explained by changes in silvicultural practices.

References

Alekseev AS (1995) The long-term trends of the tree ring width in Scots pine stands of North-West part of Cola peninsula. The Proceedings of the XX IUFRO World Congress, p 7 (in press)

Arovaara H, Hari P, Kuusela K (1984) Possible effect of changes in atmospheric composition and acid rain on tree growth. An analysis based on the results of Finnish National Forest Inventories. Communicationes Instituti Forestalis Fenniae 122: 1-16

Eronen M, Zetterberg P (1992) Dendrochronology and climatic history in the subarctic area of Fennoscandia. In: Kanninen M, Anttila P (eds) The Finnish research programme on climate change. Progress Report, Publ. of the Academy of Finland, VAPK-Publishing, Helsinki, pp 13-18

Eronen M, Lindholm M, Zetterberg P (1994) Extracting paleoclimatic information from pine tree-rings in Finland. In: Climatic trends and anomalies in Europe 1675-1715. Paläoklimaforschung - Paleoclimate Research 13: 43-50

Gustavsen HG (1980) Talousmetsien kasvupaikkaluokittelu valtapituuden avulla. Abstract: Site index curves for conifer stands in Finland. Folia Forestalia 454: 1-31

Hari P, Arovaara H, Raunemaa T, Hautojärvi A (1984) Forest growth and energy production, a method for detecting trends in growth potential of trees. Canadian Journal of Forest Research 14: 437-440

Hari P, Arovaara H (1988) Detecting CO_2 Induced Enhancement in the Radial Increment of Trees. Evidence from Northern Timber Line. Scandinavian Journal of Forest Research 3: 67-74

Herttonen H (1990) Kuusen rinnankorkeusläpimitan kasvun vaihtelu Etelä-Suomessa. Summary: Variation in the diameter growth of Norway spruce in Southern Finland. University of Helsinki, Department of Forest Mensuration and Management, Research Notes 25: 1-80

Mielikäinen K, Timonen, M (1995) Growth trends of Scots pine in unmanaged and regularly managed stands in Southern and Central Finland. Report in this publication

Nöjd P (1995) Mäntyjen kasvu muuttuvassa ympäristössä. In: Tikkanen E (ed) Kuolan saastepäästöt Lapin metsien rasitteena. Itä-Lapin metsävaurioprojektin loppuraportti, Jyväskylä, pp 183-195

Nöjd P, Reams GA (1995) Growth variation of Scots pine across a pollution gradient on the Kola Peninsula, Russia. A manuscript submitted for publication

Sennov SN (1995) Growth trends of south taiga forests. The Proceedings of the XX IUFRO World Congress (in press)

Sinkevich SM (1995) Short- and long-term natural trends of pine radial growth in north- and mid-taiga forests in Karelia. The Proceedings of the XX IUFRO World Congress (in press)

Siren G (1961) Skogsgränstallen som indikator för klimatfluktuationer I norra Fennoskandien under historisk tid. Commun. Inst. For. Fenn. 54, p 66

Tomppo E, Siitonen M (1991) The National Forest Inventory of Finland. Paperi ja Puu - Paper and Timber 73(2)

Short- and Long-Term Natural Trends of Scots Pine (*Pinus sylvestris*, L.) Radial Growth in North- and Mid-Taiga Forests in Karelia

Serguei M. Sinkevich[1] and Markus Lindholm[2]
[1]Forest Institute. Petrozavodsk, Russia
[2]Centre for Environmental Sciences. FIN-57130 Savonlinna, Finland

Abstract

Tree-ring data of Scots pine were collected in north-western Russia, in the Republic of Karelia, between the 62th and 65th latitudes. Due to the remote location and favourable position in regard to prevailing winds, the region is considered relatively free of airborne pollution.

The analysis of the dynamics of radial growth was based on separating of quasiperiodic components and approximating by a polyharmonic model. Forecasting increment dynamics was done by summing up the reliable harmonic constituents for a defined span of time.

In the north taiga, the occurrence of about 18- and 25-year periods is approximately equal. In the eastern mid-taiga subzone, the spectrum expansion shows a decrease in frequency corresponding roughly to a 22-year period. A temporal shift of approximately 30 years between stands in the north- and mid-taiga was observed. A 50-year fluctuation was also evident which coincided with noticeable increment decrease. The most noticeable increment decrease occurs when low- and middle-frequency cyclic components simultaneously reach the minimum. In the present century, a deep increment minimum took place around the year 1960 in the north taiga, and around the year 1990 in the mid-taiga. The lowest levels of increment variability in the mid-taiga subzone are close to the characteristic levels of variation in the north taiga.

1. Introduction

In the search for evidence of cycles in tree-ring series, the most frequently emerging observations include the so-called quasibiennal-oscillation, cycles around 3 or 4 years, and 5 to 7 years. In addition, in the literature, there is generally some support for the 11-year (sunspot) cycle, and, more importantly, substantial evidence for the 20-year cycle. According to Burroughs (1994), the evidence does not distin-

European Forest Institute Research Report No. 5
Growth Trends in European Forests
Edited by H. Spiecker, K. Mielikäinen, M. Köhl and J. P. Skovsgaard

guish between this being the 18.6-year lunar cycle or the 22-year double sunspot cycle. There exists also strong support for 80- to 90-year and 180- to 200-year cycles.

In the present work, fluctuation of periods between 40-60 years and about 100 years were found most promising for extrapolation. Forecasting the radial dynamics is based on the assumption that if a simple periodic process has taken place in the course of several complete cycles, then it may last for at least the next whole cycle. Apart from this, the link between biological phenomena and periodic processes in planetary movements in space lasting thousands of years and staying almost unchanged, support the idea. After verifying reliable harmonic constituents, the forecasting of increment dynamics is done by summing them up over a defined span of time. Naturally, the longer the series of available observations, the more reliable is its extrapolation by polyharmonic model.

The periodic changes in total increment variation are very complex even in undisturbed forest sites. Consequently, at present no unambiguous conclusions can be made about amelioration or deterioration of forest productivity. The statistical confidence of the amplitude of separated sinusoids is evaluated in this work by mean-square errors of the approximations. In addition to statistical confidence, a plausible physical explanation is needed, a biological sense of each component and consideration of its occurrence in other natural processes.

2. Material

The pine stands where tree-ring material was collected are located between 62° and 65° N in the Republic of Karelia, in north-western Russia. Samples were collected in two main areas, the Kostomuksha Nature Reserve in the north-taiga subzone, and the Vodlozero National Park in the mid-taiga subzone. Due to the remote location and favourable position in regard to prevailing winds, the aerotechnogenic discharges from Western European cities have not been detected in the region.

The investigation covered 100-300-year-old, even- and uneven-aged stands, located on the upland plain. No signs of tree-felling or other kinds of human activities were evident in the sites. The soils are sandy podzols on glacial deposits of varying thickness. Plant community type is Pinetum vaccinioso-myrtillosum. Average height of trees was 22 m, and the volume of the growing stock between 150 and 300 m^3 ha^{-1}.

3. Methods

In order to obtain long-term radial-increment series, living trees of varying ages as well as dead trees were cored at a height of 1.3 m. Sample trees generally formed the upper storey. Measurement series were cross-dated to produce unified time-series. Total ring widths were measured to the nearest 0.01 mm, and early- and late-wood widths estimated. Ring-width series and corresponding series of late wood percentage were obtained. The first ten annual rings were omitted.

The measurement series were indexed by the corridor method developed by Shiyatov (1986). The corridor boundaries are represented by the gently sloping curves which show the maximum (upper) and minimum (lower) width of annual rings. The increment indices for each year were calculated with the following formula:

$$I(t) = [X(t)-A(t)] / [B(t)-A(t)] \times 200\%,$$

where $X(t)$ = ring width in year t,
$A(t)$ = current value for the lower boundary,
$B(t)$ = current value for the upper boundary.

Indexing removes the age-produced changes, noise-related trends and other perturbations in the raw series. In addition, standardisation allows for the estimation of mean-value functions (Cook et al. 1990, Schweingruber 1993). The mean index series, also called site chronologies, were produced by combining individual trees within each stand.

The index series were generalised within the uniform conditions of the site and climate. When generalising for each calendar year, all the trees were checked, and the data were excluded which fell beyond the range of normal distribution at a 10% confidence level. When the current-year indices differed greatly from the normal distribution, the maximum mode was taken as a mean value. The 7 to 25 cores selected from each of the 18 stands, enabled the construction of five generalised mean value functions. Two master chronologies for the main research areas are presented in Fig. 1. The Kostomuksha Nature Reserve represents the north taiga subzone, and the Wodlozero National Park represents the mid-taiga subzone.

Analysis of the dynamic processes present in the tree-ring chronologies was based on determining the possible quasi-periodic constituents and on a successive approximation of the series by polyharmonic models. Firstly, the spectral density of the time series was estimated by the maximum entropy method (Radovsky 1975, Mazepa 1986, 1990). This method allows for the evaluation of the frequency areas according to their relative weight or importance (Fig. 2). Secondly, these areas were filtered with narrow-band filters to dampen fluctuations beyond a chosen band. This enables successively distinguishing the narrower bands. They are then approximated with sinusoids. After separating out a sufficient number of periodic components and summing them up, it is possible to describe the course of the processes with precision.

Generalisations of the series, spectral analysis, expansion and approximation were made with the aid of software developed at the Laboratory of Dendrochronology, Institute of Plant and Animal Ecology, Ekaterinburg, and the University of Arizona, Tucson.

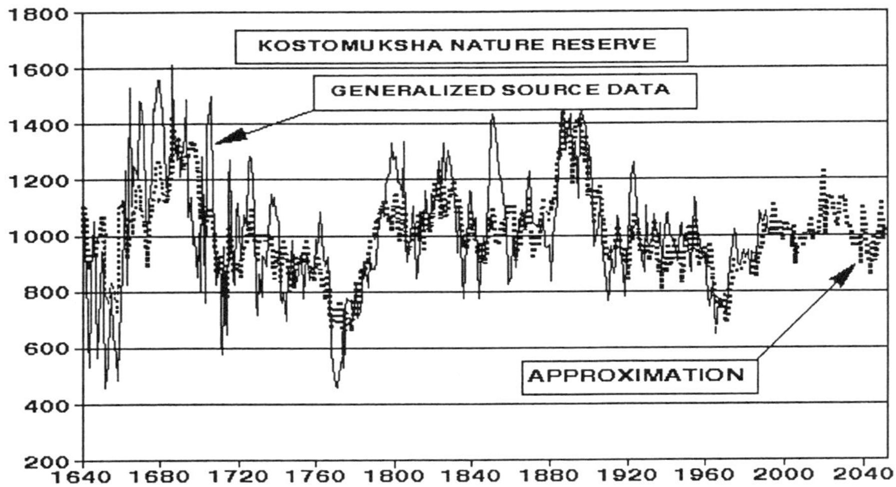

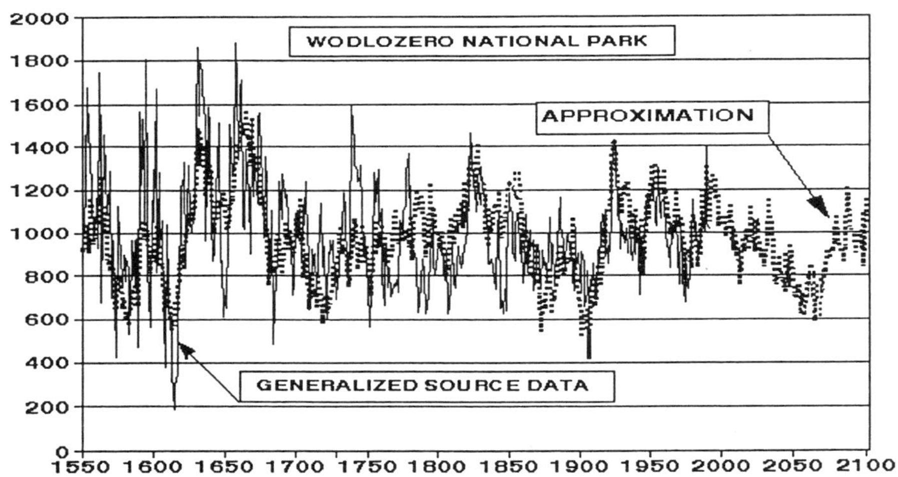

Fig. 1. Graphs of generalised radial increment series from Kostomuksha region (north taiga) and Wodlozero region (mid-taiga) and their approximation by polyharmonic models.

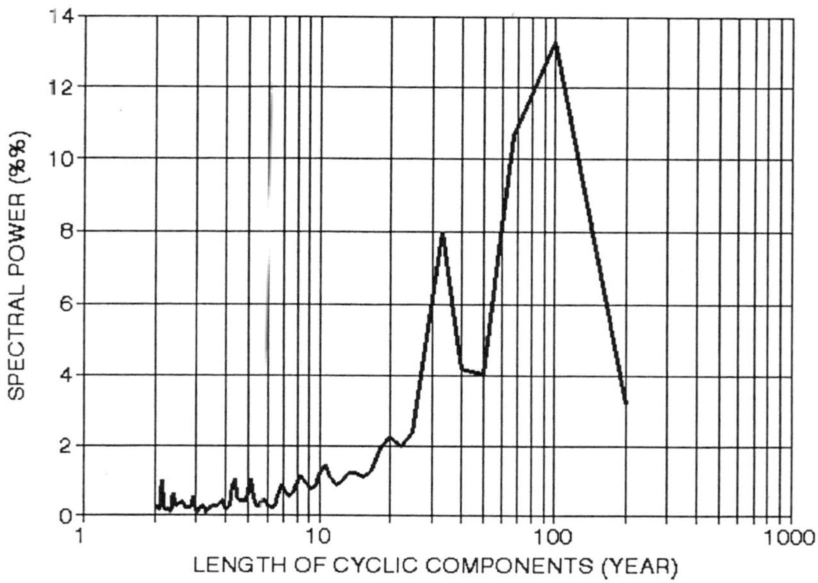

Fig. 2. Distribution of the spectral power between the different temporal components

4. Results

Correlation between tree-ring series of individual trees forming the stand are of great importance in determining the accuracy and reliability of the information obtained. Various correlation-based techniques may be used to quantify the signal strength of tree-ring chronologies (Briffa and Jones 1990). In even-aged closed stands, the correlation coefficients between the increments of individual trees seldom exceed the average of 0.4. Since codominant specimens were chosen for the analysis, the effect of competition between trees can be neglected. Thus, the main reason for differences in variability is due to the diversity in site conditions. On the other hand, it is quite permissible to believe that codominant trees reflect the dynamics of stand productivity over long periods, because it is these trees that accumulate the bulk of wood.

Changes in radial growth depend mostly on variations in temperature, amount of light and humidity. These factors, in combination with various site conditions, determine the variability in tree-ring time series. Complex feedback links in the biogeocenosis also influence and smooth local conditions, which, in their turn, may modify the dynamics of climatic factors common in the region. Although the total

Table 1. Cyclic components of mean increment series in pine stands in the mid-taga subzone.

Amplitude	Period (year)	Mean-square approx error	Start year
15,490	85,246	3,640	1840
8,838	42,857	4,776	1835
14,705	21,040	5,489	1820
9,838	11,454	6,180	1815
10,039	9,495	4,196	1818
7,860	7,423	5,294	1808
6,295	6,059	2,866	1812
4,483	5,169	2,419	1810
5,358	4,672	3,382	1813
4,301	4,236	3,118	1811
3,236	3,908	2,285	1813
4,754	3,544	2,979	1815
5,851	3,373	3,446	1810
2,698	3,073	1,811	1810
5,127	2,827	2,658	1810
2,706	2,638	2,743	1812
4,605	2,471	4,234	1802
3,097	2,330	2,438	1813
3,449	2,164	3,103	1804

Table 2. The main spectral peaks in increment series of the north taiga pine stands.

Period range (year)	Period length (year) on the plots NN		
	I	III	IV
>50	120.00	100.00	200.00
			57.14
25-50	30.00	40.00	35.50
15-25	18.20	23.50	18.20
10-15	13.30	11.80	12.50
	12.50		10.00
	11.10		
5-10	7.40	9.5	7.14
	6.40	6.70	5.40
	5.70	6.10	5.00
	5.00	5.00	
2.5-5	4.20	4.10	4.40
	3.60	3.60	3.60
	2.90	2.90	2.90
	2.70	2.60	2.60
2.5<	2.35	2.33	2.33
	2.25		2.20
	2.11		2.15

variance distribution of radial increment series shows the presence of similar in-length periods in the majority of stands studied (Tables 1 and 2), the temporal changes in these fluctuations make the general pattern seem like a "stochastic process".

The analysis of the bands of indices characterising the increment dynamics at each individual stand resulted in 15 to 23 distinct elementary periodic processes. Statistics for the stands located in the centre of southern Karelia (mid-taiga) are presented in Table 1, and those for the north-taiga subzone in Table 2. The main cyclic components are described by the length of periods, the amplitudes, the mean-square errors of approximation and the starting years. The error values indicate that the approximation of the frequency bands with sinusoids is quite satisfactory.

Correlation between the estimations produced with the sinusoids and the source data range from 0.65 to 0.75 for various stands. In all, the selected sinusoids explain 42-45% of the total series dispersion. It should be borne in mind that no less than 40% of the total spectral power is made up by fluctuations from 2 to 8 years, and that narrow and sharp peaks (both positive and negative) are generally poorly estimated. The technique used here eliminates this drawback. However, biological interpretation of the fragmented cycles of less than 5 years' duration is difficult.

The well-known 11-year period which is often connected with solar activity undergoes such changes in length and phase that it is not apparent when generalising the data on seemingly unified climatic conditions. However, it is clearly evident in the chronologies built for individual stands. The 21-22-year period, also-known from literature sources, is quite distinct and manifests itself with only a 1-3-year shift in central Karelia. In the north taiga, the occurrence of 18- and 25-year periods is approximately equal (Fig. 3), and in the eastern mid-taiga subzone, the spectrum expansion shows a power decrease in frequencies corresponding roughly to the

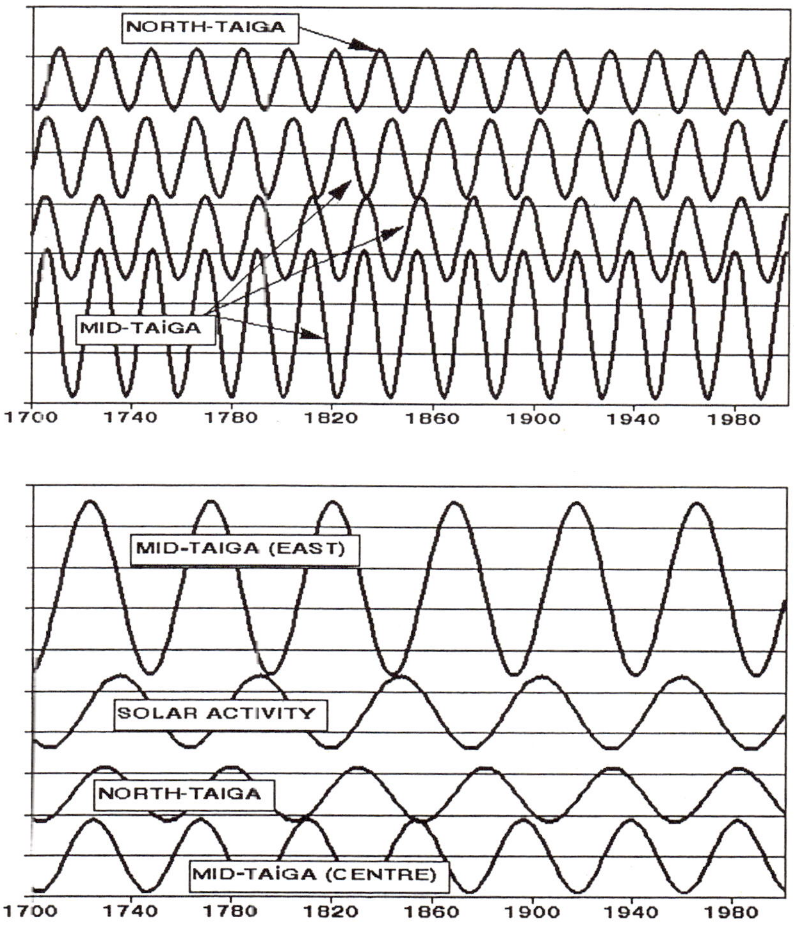

Fig. 3. Temporal shifts in cyclicity of some 22 and 50 years in length in the north- and mid-taiga subzones.

22-year period. For estimations of the dynamics of forest productivity, the increment fluctuation with periods between 40 and 60 years and about 100 years appear most promising. They are least dependent on the growth conditions in individual stands and more influenced by regional climatic peculiarities. Nonetheless, a temporal shift, as long as 30 years, between stands in the north- and mid-taiga is observed, in addition to a 50-year fluctuation (Fig. 3).

In order to investigate cycles of 100 years in length, the length of the chronologies must be no less than 300 years. It is not certain whether or not such fluctuations even occur at all in the dynamics of any natural phenomena. Repeatability, or close similarity, of the main periods in the dynamics of various environmental factors is evidence for regularities in their course. In addition to the information on radial increment series, also other processes were analysed to investigate the possibility of finding corresponding cyclic components. This analysis included averaged data on solar activity in vegetation periods (Wolf's number), average monthly temperatures of the growing season and the amount of rainfall in July. In Table 3, the spectral characteristics of these processes are presented. The forms of these factors and increment dependency are complex.

The 91-year cycle is in the Wolf's number fluctuation for summer. So, the presence of fluctuations similar to it must not be ruled out even from the increment series not longer than 300 years. Preliminary analysis of the data obtained shows a

Table 3. The main spectral peaks in the series of meteofactors and solar activity.

Period range (year)	Period length (year) in the series of					
	mean monthly temperature of				rain in July	Wolf's number
	May	June	July	August		
> 50	100.00		66.70	66.70	100.00	100.00
						66.70
25-50	40.00			33.30	28.60	28.60
15-25		20.00		16.70	22.20	22.20
10-15	13.30		12.50	13.33	10.52	11.10
				10.00		10.50
5-10	8.70	8.70	8.33	7.14	8.00	8.00
		7.14	6.67		6.71	5.88
		6.00	5.13			
2.5-5	3.23	3.39	2.98	2.67	4.25	3.180
	2.98	2.78			3.18	
	2.81				2.63	
2.5<	2.25	2.44	2.35	2.35	2.35	2.27
	2.04		2.25			2.15
			2.15			

noticeable increment decrease coinciding with a 50-year cycle. A most marked decrease occurs when long and middle-length cycle components simultaneously reach their minimum. In the current century, the deep increment minimum took place around 1960 in the north taiga, and around 1990 in the mid-taiga. Comparing the increment dynamics in different stands shows that the site quality is of primary importance in the shift. The lowest levels of increment variability in the mid-taiga subzone are close to the characteristic levels of variation in the north taiga (Fig. 4).

Large variability in increment series makes the search for links difficult. In this respect, the analysis of late-wood content in annual rings seems promising. The late-wood content is affected greatly by hydrothermal conditions and its range of natural variability is much less. The dynamics of the late-wood content is stable to a considerable extent. This is evident in different spots in Karelia and, at the same time, there seems to exist a possible connection with changes in solar activity (Fig. 5). It is reasonable to expect the curves of radial increments and late-wood contents to reflect at least the main inflexion points, since it is common knowledge that a decrease in the annual ring width is necessarily accompanied by an increase in the late-wood content. However, some natural deviations from this relation might take place due to factual combinations of soil and climate conditions.

Forecasting the radial dynamics in Figure 4 is based on the assumption that if simple periodic process has taken place in the course of several complete cycles, then it may last for at least the next whole cycle. Apart from this, the link between biological phenomena and periodic processes in planetary movements in space, lasting thousands of years and staying almost unchanged, support this idea. After having reliable harmonic constituents, the forecasting of increment dynamics is done by summing them up at a defined span of time. Naturally, the longer the series of available observations, the more reliable is its extrapolation by polyharmonic model.

5. Discussion and Conclusions

The problem of changing forest productivity (Spiecker et al. 1994) includes two crucial questions concerning (1) the anthropogenic impact and (2) the natural development. To answer the first question seems impossible without investigating the other. If these changes do have a cyclical character, they may be used in forecasting the main trends in wood increment dynamics.

In the present work, the occurrence of about 18- and 25-year periods was detected in the north taiga. In the eastern mid-taiga subzone, roughly 22-year and 50-year fluctuations were evident. The long-term increment dynamics of stands growing at their distributional limits have been investigated in numerous studies (e.g., Douglass 1919, Briffa and Schweingruber 1992, Graybill and Shiyatov 1992). Dealing with northern European tree-ring data, Briffa and Schweingruber (1992), as well as Siren and Hari (1971), both found concentrations of variance corresponding to cycles

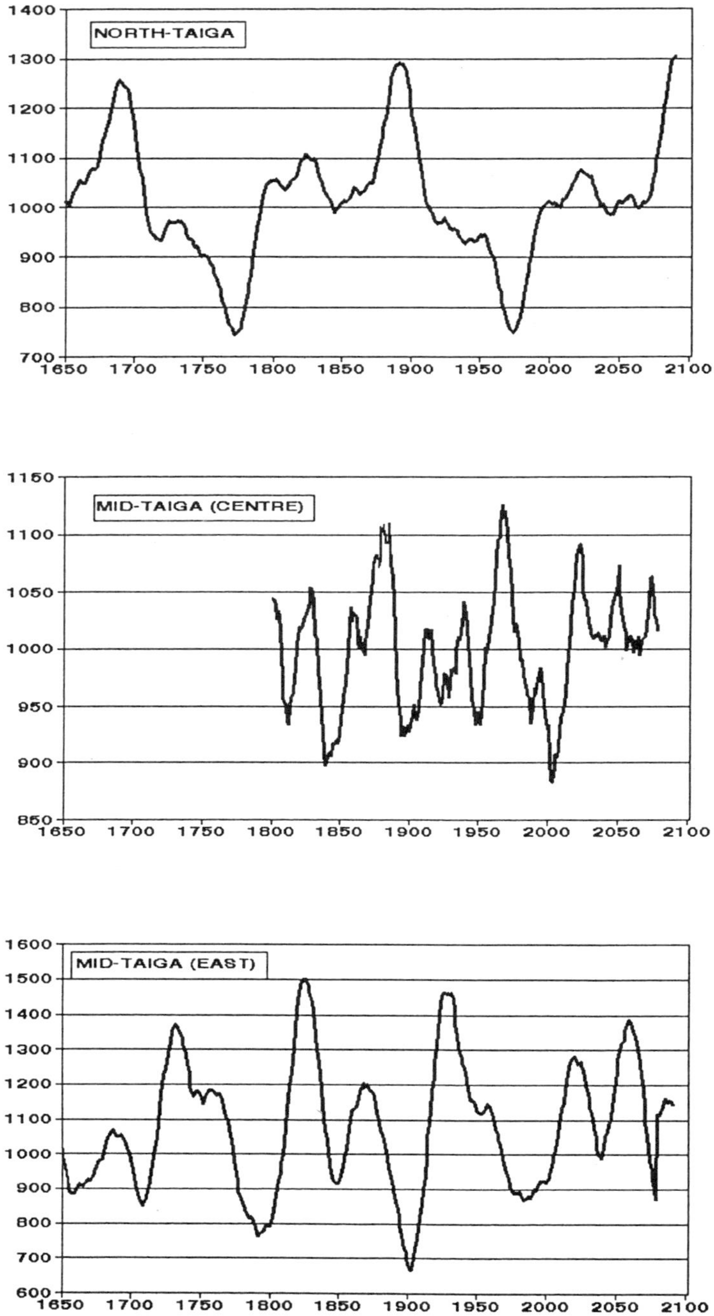

Fig. 4. Smoothed graphs of the mean increment series in the north- and mid-taiga subzones.

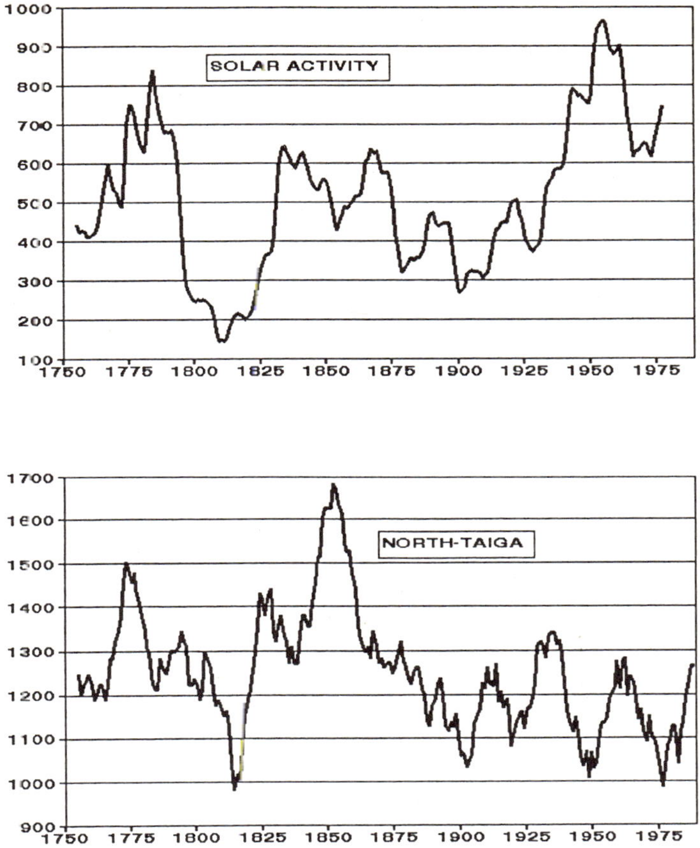

Fig. 5. Smoothed graphs of index series of the solar activity and late-wood contents in north taiga.

with periods of about 3.6-33.0 and 80-96 years. By far the most significant peaks in the temperature reconstructions of Briffa and Schweingruber (1992), however, correspond to a period between 33 to 38 years.

Burroughs (1994) has reviewed dendrochronological literature in a search of evidence of cycles. He concludes that there is generally some support for the 11-year (sunspot) cycle, and, more importantly, substantial evidence for the 20-year cycle. According to the above author, the tree-ring evidence does not distinguish between this being the 18.6-year lunar cycle or the 22-year double sunspot cycle. There exists also strong support for 80- to 90-year and 180- to 200-year cycles.

The present work is an attempt to predict the baseline trend in current tree growth. However, cyclicity in increment series is such a complex phenomenon that no definite conclusions can be made about productivity changes over great areas.

References

Briffa KR, Cook PD (1990) Methods of response function analysis. In: Cook E, Kairiukstis L (eds) Methods of dendrochronology: applications in the environmental science. Kluwer Academic Publishers, Dordrecht, pp 240-246

Briffa KR, Jones PD (1990) Basic chronology statistics and assessment. In: Cook E, Kairiukstis L (eds) Methods of dendrochronology: applications in the environmental science. Kluwer Academic Publishers, Dordrecht, pp137-152

Briffa KR, Schweingruber HF (1992) Recent dendroclimatic evidence of northern and central European summer temperatures. In: Bradley RS, Jones PD (eds) Climate Since A.D. 1500. Routledge, New York, pp 366-392

Burroughs WJ (1994) Weather cycles: real or imaginary? University Press, Cambridge, p 201

Cook E, Briffa K, Shiyatov S, Mazepa V (1990) Tree-ring standardization and growth-trend estimation. In: Cook, E. & Kairiukstis, L. (eds) Methods of dendrochronology: applications in the environmental science. Kluwer Academic Publishers, Dordrecht, pp104-122.

Douglass AE (1919) Climatic cycles and tree-growth: A study of the annual rings of trees in relation to climate and solar activity. Carnagie Institute of Technology, Washington, D.C., p 127

Graybill DA, Shiyatov SG (1992) Dendroclimatic evidence from the northern Soviet Union. In: Bradley BS, Jones PD (eds) Climate since A.D. 1500. Routledge, New York, pp 366-393

Mazepa V (1990) Spectral approach and narrow band filtering for assessment of cyclic components and ecological prognoses. In: Cook E, Kairiukstis L (eds) Methods of dendrochronology: applications in the environmental science. Kluwer Academic Publishers, Dordrecht, pp 302-308

Mazepa V (1986) Ispolzovanie spectralnogo predstavleniya i lineinaya filtraciya stacionarnyh posledovatelnostey prianalize cyclichnostiv dendrochronologicheskikh ryadah. In: Dendrokhronologiya i dendroklimatologiya, pp 49-68

Radovsky HR, Fougere PF, Zawalick EJ (1975) A comparison of power spectral estimates and applications of the maximum entropy method. J.Geophys.Res., pp 619-626

Shiyatov SG (1986) Dendrochronologia verkhney granicy lesa na Urale. Moscow, p 136

Siren G, Hari P (1971) Coinciding periodicity in recent tree rings and glacial clay sediments. Reports Kevo subarctic research station 8, pp 155-157

Spiecker H, Mielikäinen K, Kohl H, Untheim H (1994) Growth trends of European forests - has site productivity changed? European Forest Institute. Working paper 4, p 68

Schweingruber FH (1993) Jahrringe und Umwelt. Dendroökologie. Vologda, p 473

Growth Trends of Scots Pine (*Pinus sylvestris*, L.) in Unmanaged and Regularly Managed Stands in Southern and Central Finland

Kari Mielikäinen[1] and Mauri Timonen[2]
[1]Finnish Forest Research Institute. FIN-01301 Vantaa, Finland
[2]Finnish Forest Research Institute. Rovaniemi Research Station
FIN-96300 Rovaniemi, Finland

Abstract

Annual ring data collected from strict nature reserves and long-term permanent plots of the Finnish Forest Research Institute (METLA) did not reveal any long-term trend in the radial increment of Scots pine (*Pinus sylvestris*) in the southern part of Finland during the last 100 years. This result was supported by the analysis of standwise increment data from permanent thinning experiments.

The height development of dominant sample trees had been faster than that of previous site index curves. This was supposed to be mainly due to differences in the data of this study and of the index curves. According to our investigations, there were some indications of increasing productivity on some sample plots in the southernmost part of the country during the last 40 years.

Possible reasons for non-existing growth trends are as follows:

1. The wide climatic variation of tree growth in northern latitudes could conceal the possible trend. The years 1961-1990 were slightly cooler than the years 1931-1960 in Finland.
2. The SO_2- and N-depositions in Finland during the past few decades have been too low to be able to cause growth trends of such magnitude that could be separated from natural growth variation.
3. The study material consists of untreated stands or stands with known management history. Thus, trends caused by changing silviculture are outside the scope of this study.

On the basis of the non-existing trends in this study and results from eight successive national forest inventories beginning as early as in the 1920s, it can be concluded that the main factors leading to a 44% increase in the total growth of Finnish forests during the last four decades are changes in stand structures (age, density) and silvicultural practices (regeneration, thinnings, drainage, fertilization).

European Forest Institute Research Report No. 5
Growth Trends in European Forests
Edited by H. Spiecker, K. Mielikäinen, M. Köhl and J. P. Skovsgaard

1. Introduction

1.1. Growth Trends in Individual Trees and Forest Stands

The total growth of Finnish forests has increased by more than 40% during the last 40 years. This can be observed in the measurement data collected in the course of national forest inventories (NFIs; Tomppo and Siitonen, 1991). The main cause for the increased productivity is supposed to lie in the intensified silviculture practised for more than four decades, sustainable wood production being the main target. The chief growth-promoting measures carried out were the draining of peatlands, the rejection of selection cutting and higher densities in young and mature stands.

In addition to this man-made increase in stand growth, trees have also shown some deviation from expected growth. Hari and Arovaara have reported evidence of increasing growth trends for Scots pine in the 1900s (1984, 1988) and Henttonen (1990) for Norway spruce from 1970 to 1985. According to the study by Henttonen, the short-term increase in the productivity of Norway spruce correlated with increasing temperature and precipitation during that period.

Most yield tables used in forest management in Finland are based on measurements carried out before 1960. Quite often, the measured growth levels of stands today exceed yield table estimates. The site type, estimated using the species composition of the ground vegetation (Cajander 1909), has in some cases improved by one class during the duration of experiments begun in the early years of this century. The proportion of forests dominated by bilberry (*Vaccinium myrtillus* - fertile site) has also increased and the proportion of cowberry (*Vaccinium vitis idea* - poor site) has decreased according to the inventory results from the past 40 years. It is not yet clear whether the abundance of bilberry is due to changes in stand structures (density and age) or to improving soil fertility.

Clear decline in tree growth and even increasing mortality have been observed around industrial plants for decades. The negative effect of pollutants on tree growth has, in extreme cases in Russia, been observed as far as at distances of 30 to 40 km from pollution sources in Russia (Nöjd and Reams 1995). The said damage is possibly caused by gaseous compounds acting on needles or by the indirect effects of sulphur and heavy metals on the soil.

1.2. Environmental Changes

Chemical compounds emitted into the atmosphere as the result of human activity have markedly altered the properties of the air, the soil and the water. The concentration of atmospheric SO_2 in Finland is clearly at a lower level than in Central Europe. In the 1980s, the emissions of SO_2 in Finland decreased by 60% (Fig. 1).

During the last few decades, nitrogen deposition has been at the level of 10 kg $ha^{-1}year^{-1}$ in southern Finland (Kauppi et al., 1990). This is about 50% of the recommended dose for forest fertilisation. Half of the amount is in the form of nitrates

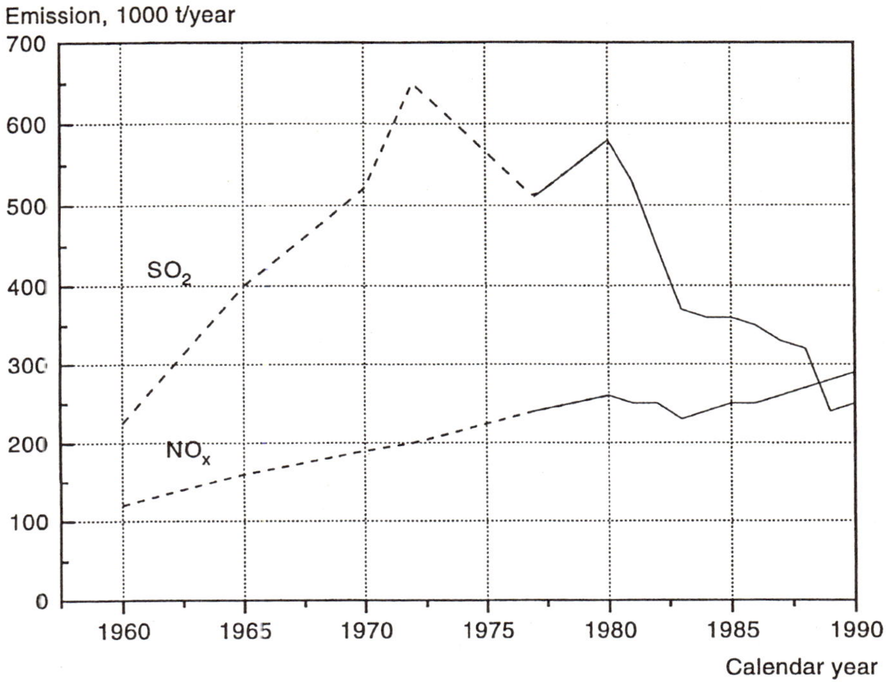

Fig. 1. The emissions of sulphur dioxide (SO_2) and nitrogen oxides (NO_x) in Finland (Savolainen and Tähtinen 1989).

emitted mainly by traffic - the other half is made up of ammonium originating from agriculture. In Central and Northern Finland, nitrogen deposition is much lower (1-5 kg $ha^{-1}year^{-1}$).

The third gas affecting tree growth is CO_2, the increase of which is supposed to promote global warming in future decades. Both temperature and CO_2 influence tree growth. These two growth factors are expected to enhance the biomass production of Finnish forests in the future.

On the basis of the above facts, it can be assumed that part of the increase in forest growth in Finland during the last 40 years may be attributed to environmental changes. Contrary to the greenhouse theory, the mean temperature of the growing season in Finland during the last 30 years has been lower than in the period 1931-1960. This may partly counteract the expected increase in forest productivity.

The aim of this study was to find out if Scots pine trees in Finland presently grow at the same rate as pines of the same size or age in earlier times. The analyses were done using increment cores obtained from strict nature reserves (Nature Parks) or from old permanent sample plots established by the Finnish Forest Research Institute in the southern part of the country.

2. Material and Methods

2.1. Old Trees Measured in Strict Nature Reserves

To get an overview of past long-term trends, the first study focused on untreated Scots pine stands on relatively poor sites in Southern and Central Finland. Scots pine was selected as the study object because most old pine stands were established after forest fires on open land. Consequently, the growth history of the studied trees has been quite regular without any suppression phases caused by the overstory in youth.

In order to estimate the mean growth level of trees prior to the era of atmospheric depositions, pine stands, 100 to 200 years of age, were sampled. The sampling was quite similar to that done by Hari and Arovaara (1984). The biggest problem with the material is the unknown stand history. Competition between trees in their youth causing growth decrease and natural mortality, and frequent forest fires bring uncertainty into the interpretation of the results. Trees surviving partial forest fires can provide us with long growth trends, thanks to decreasing competition and the irregular healing of scars made by fire. Stands with visible traces of forest fire were rejected as study objects.

The material measured in strict nature reserves consisted of fourteen stands on poor sites in Southern and Central Finland. From each subjectively selected research plot, ten dominant or co-dominant trees closest to the plot centre were bored to the pith from their south and north sides. The size of this material was 207 cores. The measured tree age at breast height was between 98 and 200 years (mean 151 years); the corresponding diameter was between 17.4 and 43.3 cm and the total number of annual rings was 30545.

The material, based only on increment borings without any information on stand history, is suitable for simple analyses. The only characteristics known throughout the entire lifetime of the trees are the age, diameter, and radial increment for each year. Analysing this type of data can provide some indication of the possibility of the existence of growth trends, but not prove it.

2.2. Long-Term Experiments

Based on Stand Increment. The known management history of stands makes the results of analyses much more reliable because of the possibility of modelling the effects of competition. In this respect, the best possible data can be found on the permanent experiments, the first of which were established in the 1920s by the Finnish Forest Research Institute. Increment data of individual trees combined with stand characteristics, measured every 5 to 10 years, markedly improves the reliability of the trend estimation.

The standwise material consists of experiments begun in the 1950s and 1960s in order to study the growth reactions of whole stands after low thinning and fertiliza-

Table 1. Stand characteristics on permanent experiments used for tree analyses. Plots marked with an asterisk refer to unmanaged control plots.

Experiment	Plot	Site index	Number of assessm.	Year of first assessm.	Age, years	Mean height m	Basal area m^2ha^{-1}	Year of last assessm.	Mean height m	Basal area m^2ha^{-1}
Punkaharju 1	3	30	12	1924	32	10.9	36.8	1993	30.1	29.3
Punkaharju 18	2	30	12	1924	26	9.5	22.5	1993	31.6	32.1
Punkaharju 18	3*	30	12	1924	26	9.3	35.5	1993	27.7	61.2
Punkaharju 36	1	30	8	1948	18	6.0	22.3	1992	25.2	24.4
Punkaharju 37	1	30	9	1948	23	7.5	31.5	1992	26.5	22.6
Punkaharju 39	1	24	6	1962	58	13.7	18.6	1993	22.6	23.5
Mustila 13	1	28	8	1949	25	8.6	17.0	1993	23.1	22.2
Mustila 17	1	30	8	1949	24	7.9	23.8	1993	19.4	28.9
Ruotsinkylä 5	1	24	10	1931	8	0.5	2.0	1993	19.0	28.9
Ruotsinkylä 13	1	27	10	1931	8	2.4	2.7	1993	23.1	25.5
Parkano 2	2*	27	5	1963	34	9.3	28.7	1993	13.5	34.2
Parkano 2	3	27	5	1963	34	10.0	26.2	1993	19.9	25.5
Parkano 4	2*	24	5	1963	34	8.3	7.9	1993	18.2	27.6
Parkano 4	3	24	5	1963	34	7.8	7.4	1993	16.6	21.3
Parkano 6	3	16	5	1963	78	11.1	21.1	1993	15.9	22.8
Koli 7	1*	19	7	1924	94	16.7	31.7	1993	22.7	38.4
Koli 7	2	19	7	1924	94	17.0	30.6	1993	24.6	21.1
Koli 8	1	20	9	1924	53	9.1	14.6	1993	21.9	25.1
Koli 8	2	20	9	1924	53	9.8	19.3	1993	22.5	18.8
Evo 2	1	24	8	1925	48	13.1	19.6	1993	28.5	6.3
Evo 3	1*	24	10	1925	45	9.7	18.7	1993	21.5	31.2
Evo 3	2	22	10	1925	45	9.5	19.9	1993	23.4	22.0
Vesijako 38	1	30	9	1930	14	4.3	10.1	1993	26.8	33.9
Vesijako 38	4*	30	9	1930	14	4.0	9.1	1993	22.0	54.6
Vesijako 44	1*	29	7	1931	59	19.0	45.8	1993	29.8	64.7
Vesijako 44	2	29	7	1931	59	19.6	43.9	1993	30.4	52.7
Vesijako 46	1*	28	7	1948	31	9.2	29.6	1993	20.2	36.5
Vesijako 46	2	28	7	1948	31	9.7	31.5	1993	24.2	31.4
Solböle 2	1*	30	7	1930	28	10.0	35.4	1993	24.3	46.6

tion. Thinned control plots without fertilisation were used in this study. The total number of stand plots measured was 98.

Based on Individual Trees. The most reliable picture about growth trends and causes for trends can be obtained by conducting single-tree analyses in stands whose history is known exactly. This type of material can be obtained by taking increment cores or discs from trees growing on successively measured permanent plots.

Taking increment cores from standing trees damages them and makes them susceptible to rot. In this study, the sample trees were cored only in the study stands which were no longer objects of active research. In order to guarantee a balanced

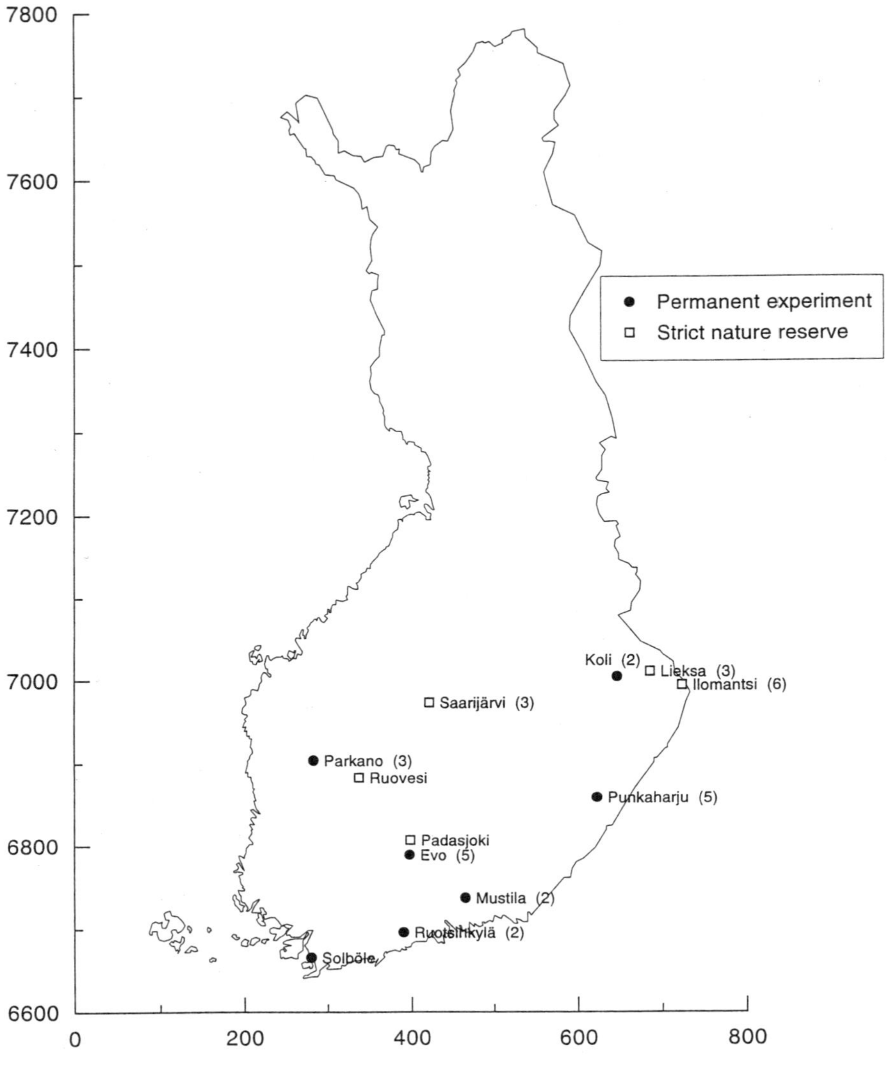

Fig. 2. The location of the study stands. Number of experiments or plots in parenthesis.

age structure for individual calendar years, cores were also taken from trees on sample plots of new experiments. In these stands, the sample trees were selected from within buffer zones between study plots. This was possible because the buffer zones had been managed in the same way as the experimental plots.

The data for individual-tree studies consisted of 20 experiments (Fig. 2). For most of the experiments, one unthinned and one thinned plot were selected. The total number of sample plots was 29. Successive low thinnings guarantee that there is no growth trend due to changes in silviculture.

Each selected plot was first measured for stand and tree characteristics. Ten dominant and co-dominant trees closest to the plot centre were then selected for increment analysis. Cores extending to the pith were taken at the height of 1.3 metres from the direction of the plot center. The age and diameter structure of the annual ring material by calendar years is shown in Fig. 3.

In order to reconstruct the dominant height development of the study stands, three stems on each plot were felled and measured to serve as stem analysis trees. The trees were selected using the PPS (Probability Proportional to Size) sampling method. The trees do not exactly represent the dominant height of the stands, but are close to it. The annual rings of the trees were counted at height intervals of one meter. Along the first 3 and the last 5 m, the annual rings were counted at intervals of 0.5 m.

All increment cores taken on the temporary and permanent plots of this study were crossdated using the Dynaclim software (van Deusen, 1992). Peak years with exceptional annual rings were used in the crossdating.

3. Results

3.1. Growth at a Stand Level

The trends in stand increment were studied using the growth model presented by Nyyssönen and Mielikäinen (1978), which was based mainly on data from the 1940s and 1950s. The model predicts the volume increment of a stand for the following 5-year period as function of forest site type, stand age, volume and mean diameter.

The mean diameter as a predictor variable is important for detecting possible growth trends. The correlation between stand age and mean diameter can conceal a trend if the growth model is based only on stand age and volume. Rejecting the mean diameter as one predictor variable decreases the significance of forest site type as a dummy variable. In the worst case, this can change the coefficient of this variable negative.

The growth model used was

$$\begin{aligned} \ln Iv5+ = & -0.5885 \\ & - 0.09667 \times (\ln T)^{**}2 \\ & + 1.2503 \times V^{**}(1/V)^{**}0.3 \\ & + 0.1796 \times (\ln D)^{**}8/10^{**}4, \end{aligned}$$

where T is stand age, years,
V is stem volume, $m^3\ ha^{-1}$,
D is mean diameter (median of basal area), cm.

Constant = -0.7702, if the forest site type is Calluna Type.

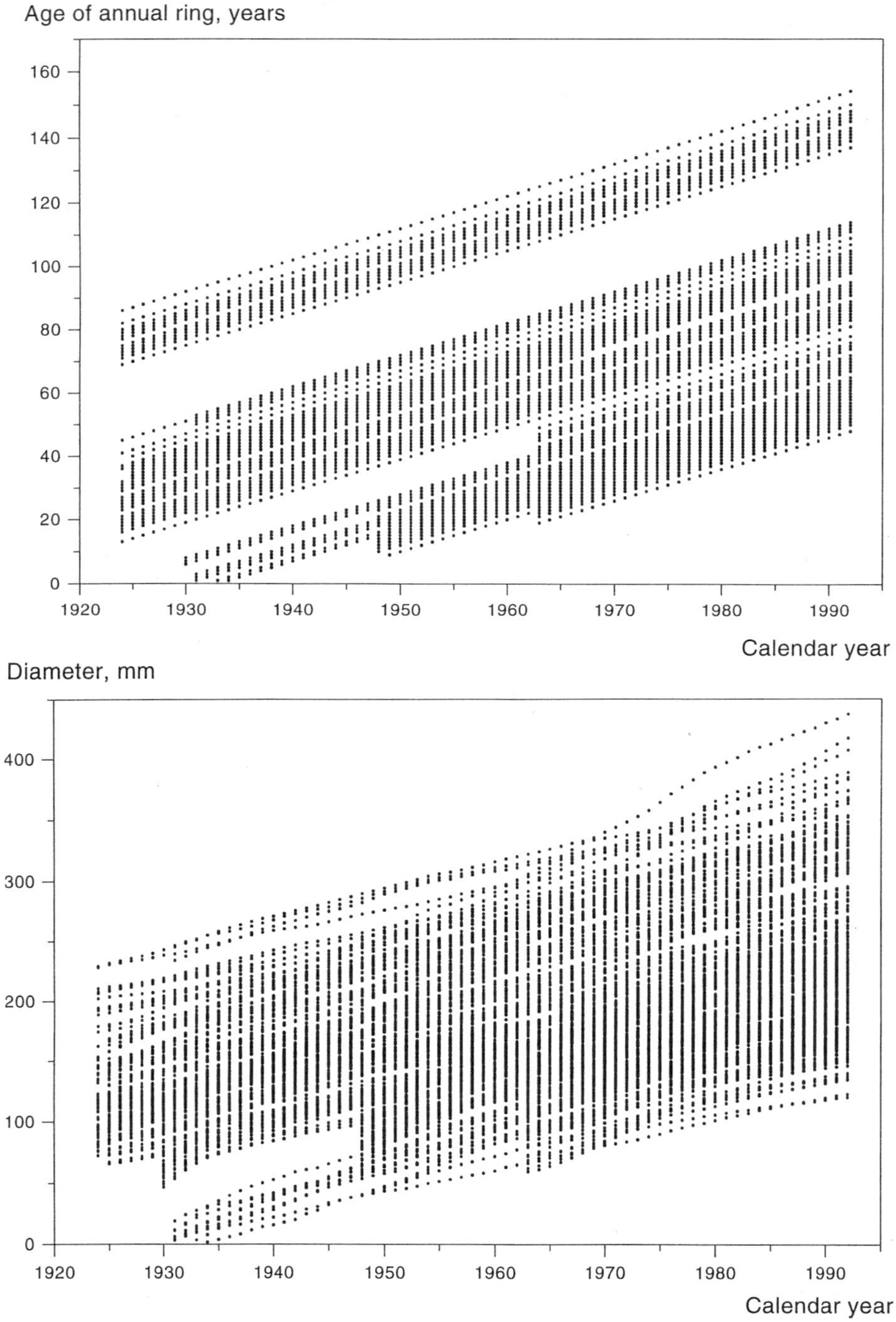

Fig. 3. The age and diameter structure of the annual ring data for growth modelling against calendar year on the permanent experiments.

The existence of a trend was investigated by comparing the measured volume increment of the experimental stands to the model estimates. First, the mean of the model estimates was calibrated to correspond to the growth level of the data. The presence of a trend could then be revealed by examining the residual deviations from model estimates against the calendar year.

The residual variation of stand volume increment around the model estimate (Fig. 4) does not show any trend in stand growth from 1950 to 1990. This result does not guarantee the absence of a long-term growth trend. Climatic growth variation can interfere with the comparisons during the relatively short observation period of 40 years. On the other hand, the use of periodic 5-year increments smoothes annual variation.

3.2. Growth of Individual Trees in Strict Nature Reserves

The radial increment of sample trees in different age classes for each calendar year does not show the presence of a long-term trend during the last 50 years. The favourable years (1957, 1967 and 1990) can be observed in all age classes in the same way as the years of low growth (1942, 1956 and 1985, Fig. 5).

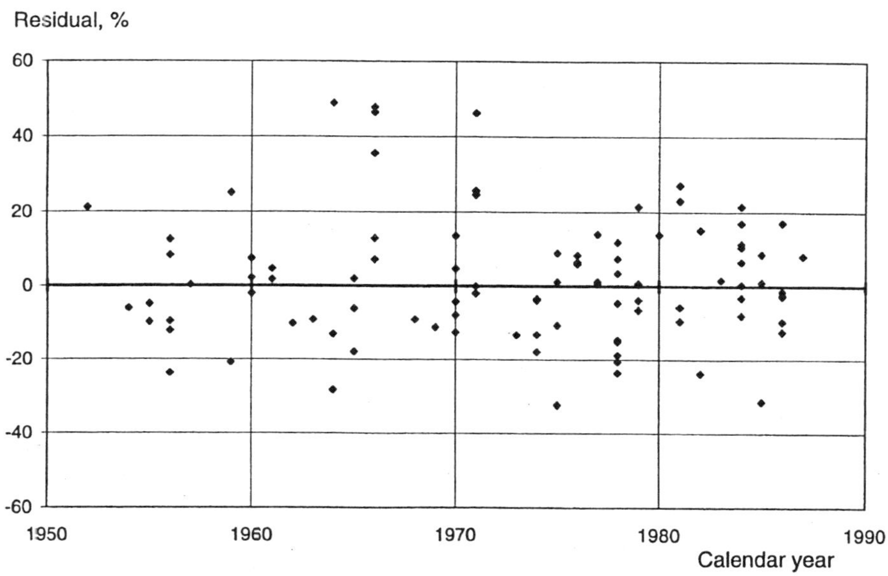

Fig. 4. The relative difference (residual variation) of the measured stand volume increment on permanent thinning experiments from the model estimate of Nyyssönen and Mielikäinen (1978) against the calendar year.

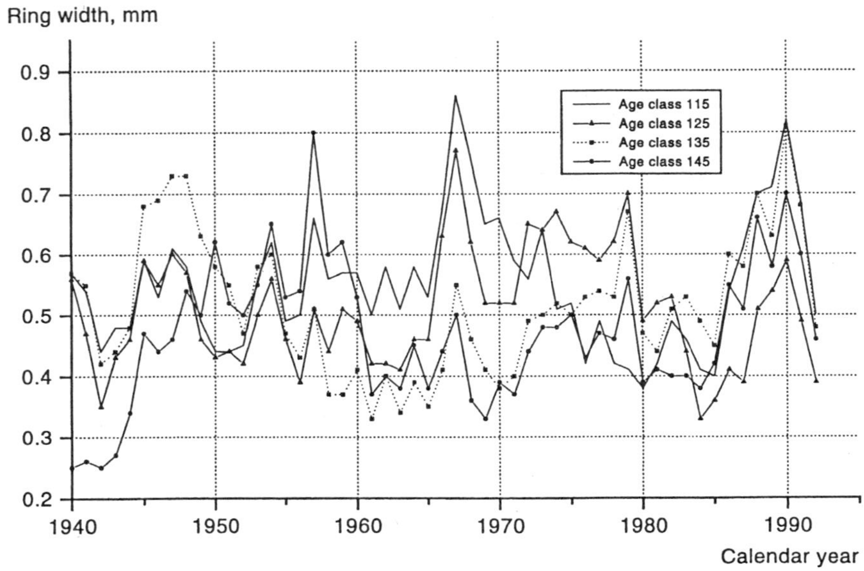

Fig. 5. The radial increment of Scots pine in different age classes in the material from strict nature reserves. Age classes 115, 125, 135 and 145 years.

A more detailed analysis was made by trying to model the radial increment of sample trees and studying the residual variation of the data against the calendar year. In the model, the width of a single annual ring was predicted by tree diameter as the only predictor variable. Tree age was not used because it could, when used together with diameter, conceal a possible trend (Elfving and Tegnhammar 1995). This is because the size of a tree at a certain age is a measure of soil productivity itself.

In order to make the age and diameter distribution of the material as homogenous as possible against the calendar year, the data was first restricted to contain only annual rings in age classes from 90 to 120 years, diameter classes 16 to 28 cm and the calendar years 1893 to 1992 (Fig. 6). Without this, the data may have been skewed (correlation between diameter and calendar year), causing problems in growth modelling and interpreting of the results.

The restriction of the data resulted in a growth model with a zero coefficient of determination (R^2). This means that the width of annual rings in the age class 90-120 years has been constant in unmanaged stands for the last 100 years.

The deviation of the radial increment from the mean (Fig. 7) along the calendar year shows both short- and long-term growth variation during this century. A slight descending trend ($p = 0.98$, $t = -2.38$) could be detected for the whole study period. This trend may be caused by the lower mean age of annual rings in the data before 1920 (see Fig. 6). From 1943 to 1992, no trend exists. In this respect, the results dif-

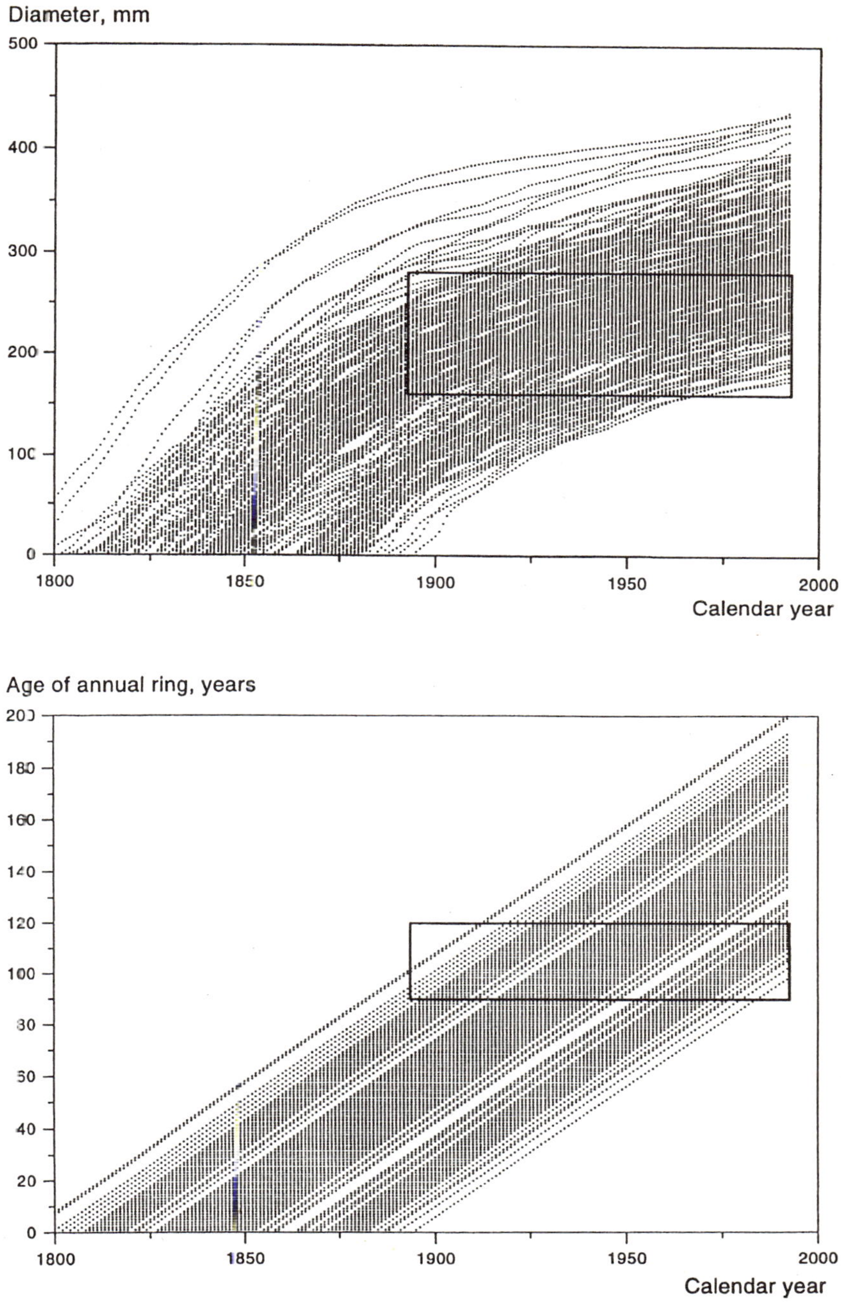

Fig. 6. The homogenisation of the annual ring material from strict nature reserves for modelling tree growth.

fer clearly from those published by Hari and Arovaara (1984), who used the same kind of data from untouched stands. The method used by Hari was based on the increment curve against age from the 19th century as reference. Some uncertainty in both studies is caused by possible human influences in the past. Most of the strict nature reserves were set aside as late as the 1950s, and consequently their management history is not always exactly known. So some of the annual rings in the first few decades of this century may have been influenced by selection cuttings commonly used in those times.

3.3. Individual Trees on Permanent Experiments

Regular measurements of trees on permanent experiments offered a good basis for modelling the effects of competition on tree growth. The method was roughly the same as was used with the material from strict nature reserves. Prior to modelling, the material was limited to 30-60 years in age. This limitation decreased the variation in the material and thus resulted in a growth model with a very low degree of determination. The study could also have been made on the basis of existing growth models. This alternative was rejected because the models contained tree age and diameter as predictor variables (see Elfving and Tegnhammar in this volume).

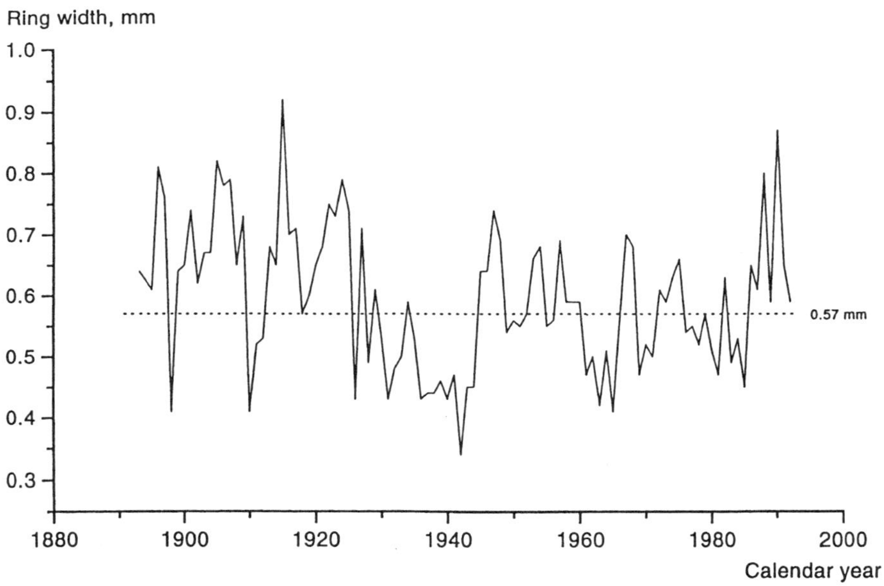

Fig. 7. The variation of radial increment against the calendar year in the material from strict nature reserves after homogenisation. The mean of annual rings = 0.57 mm.

The growth model developed was as follows:

Dependent variable: Width of annual ring, mm

Independent variable	Coefficient	t-value
Constant	1.804	23.6
d	0.005307	21.6
N	-0.00007018	5.8
G	0.05301	7.7
G^2	0.0009618	8.3
H_g	-0.1043	21.6
D_g	0.04071	8.5

where d = tree diameter (cm), N = number of stems per ha, G = stand basal area (m^2 ha^{-1}), H_g = mean height (m), D_g = mean diameter (cm). Height and diameter weighed with basal area.

Constant = 1.804 if site type CT (H100 = 21)
Constant = 1.987 if site type VT (H100 = 24)
Constant = 2.327 if site type MT (H100 = 27)

Number of observations = 5354
Degree of determination $R^2 = 0.22$.

The deviations from the model do not show any long-term trends from 1924 to 1992 ($p = 0.1653$, $t = -0.21$). The small differences between the results from strict nature reserves and from permanent experiments may be caused by the relatively small number of samples in both materials (Fig. 8).

3.4. Height Development of Dominant Trees

The development of dominant height is regarded as one of the best stand characteristics for showing changes in soil productivity because dominant height is considered to be insensitive to the management history (density) of the stand. In this study, the site index of every stand was calculated at different dates on the basis of felled sample trees. The model, based on data from national forest inventories from the beginning of the 1950s (Gustavsen 1980), is well suited for this study because it cannot contain the possible growth trend during the past 40 years.

In Fig. 9, the relative site indices (year 1950 = 1) for felled sample trees calculated using the model presented by Gustavsen (1980) show an increasing trend of about 10 per cent from 1950 to 1990. Gustavsen's model is based on temporary sample plot measurements conducted in connection with national forest inventories.

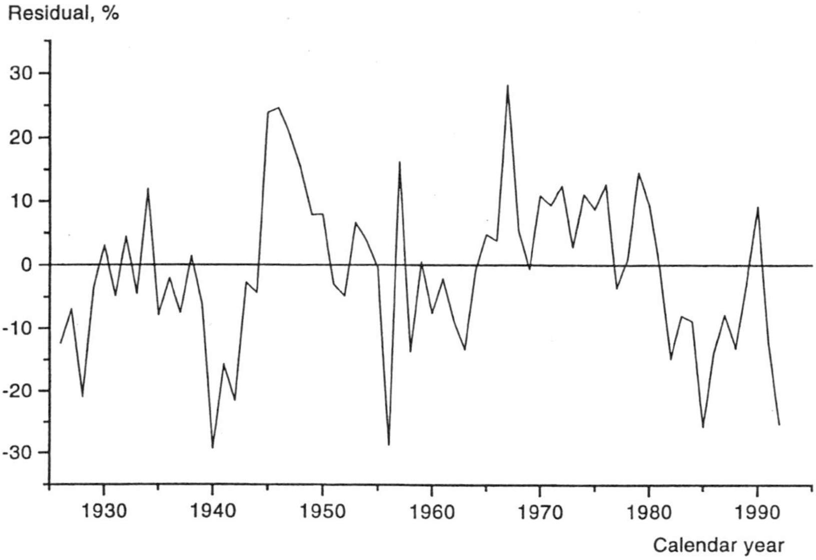

Fig. 8. The residual variation of the radial increment from the individual tree growth model against the calendar year in the data from permanent experiments.

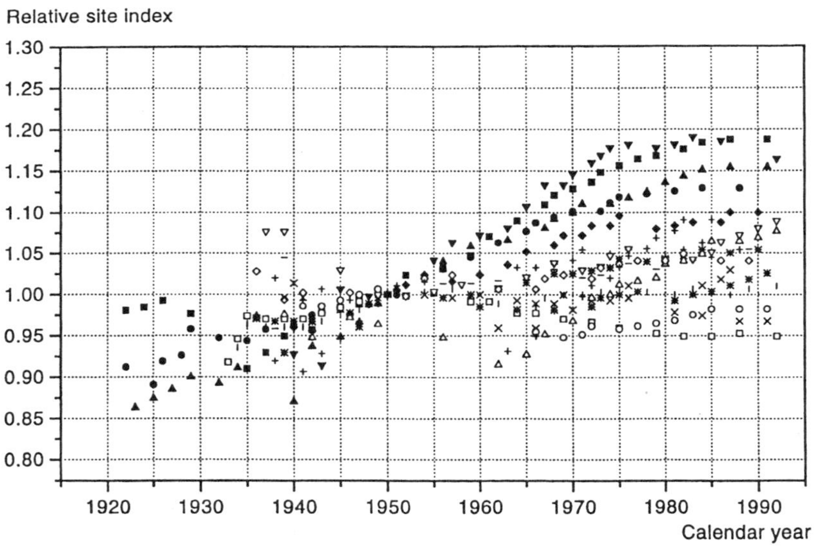

Fig. 9. The change in the relative site index of Scots pine (site index in the year 1950 = 1) on relatively poor soils (site index H100 = 24-27 m) as estimated using the model presented by Gustavsen (1980). Felled sample trees on permanent experiments.

The different management history of young and old stands in this type of material can cause problems when using the model estimates as representing the real height development of stands managed by low thinnings. Thinning from above was commonly used in Finland in the first part of this century, which must have had an effect on Gustavsen's site index curves (see Tamminen 1993). A trend lasting throughout the entire century suggests that the trend is mainly due to differences in these two data materials.

Results deviating from the average were obtained in study stands #5 and #13, close to the Helsinki-Vantaa airport. An acceleration in the height development of dominant trees (3 trees per stand) began in these stands in the 1940s (Fig. 10). The current volume increment also shows the same atypical trend (Fig. 11). It is possible that the increase in wood production in these stands is at least partly due to the increased deposition of nitrogen in this area.

4. Discussion and Conclusions

The results obtained contain many uncertainties. The long-term climatic variation of tree growth is one of them. The acute effect of exceptional climatic events can lead to long-lasting growth decline as a result of root damage and defoliation (Jalkanen et al. 1995). The recovery of and decreasing competition between surviving trees after catastrophes or management measures can result in very long-lasting trend-like growth reactions.

The major weakness of the data from old stands of strict nature reserves is the lack of information on stand history. Forest fires have occurred in pine stands in southern Finland at intervals of 60 to 100 years (Parviainen 1994). The signs of fires, storms and old cuttings may have disappeared elsewhere, but not from annual rings. Fire has presumably caused long-lasting trends in the annual rings of some old trees in this material. The initial density of the study stands has also played a decisive role in the competition between trees, natural mortality and diameter growth of the stands in their youth.

The problems in connection with permanent experiments are the relatively small size of the data sample and the uneven age structure of the data. The material is concentrated at the beginning of this century in young and at the end of the century in old age-classes. The study method presupposes that the age structure of the annual ring material remains constant throughout the century. To solve this problem, the material was limited to age classes from 30 to 60 years. In addition to old experiments, also experiments established during the last few decades were included in the material. This in turn results in variation caused by annual rings originating from different stands in each calendar year.

A problem common to all materials is that of site classification. If there is a growth trend, this trend is included in the site class, regardless of whether the clas-

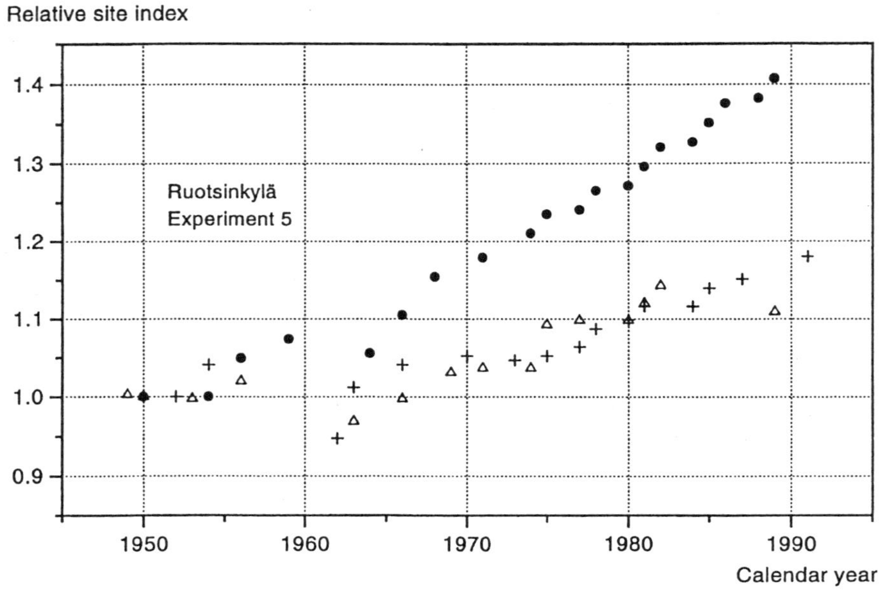

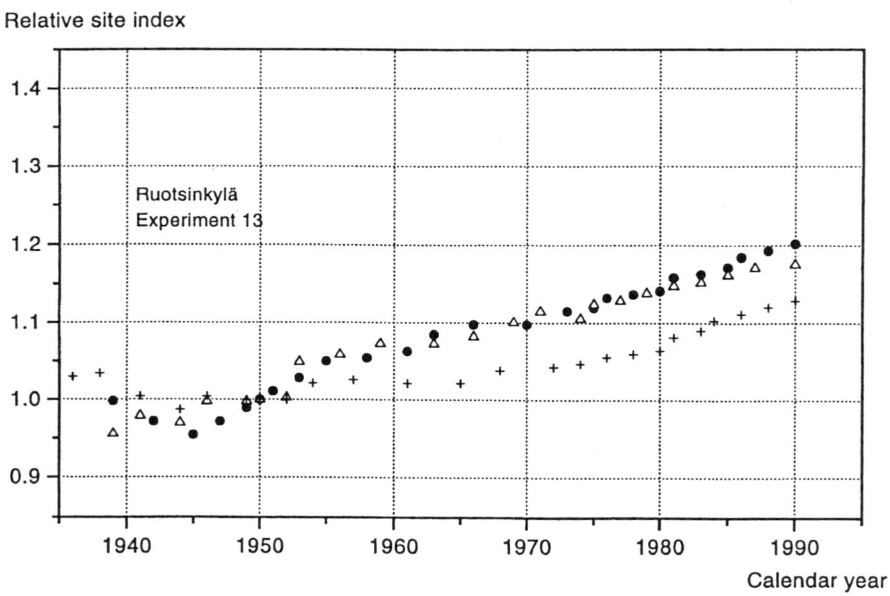

Fig. 10. The change in the relative site index (see Fig. 9) on two permanent experiments exposed to above-average nitrogen deposition in southern Finland. Three dominant trees in experimental stands #5 and #13 in Helsinki, Vantaa.

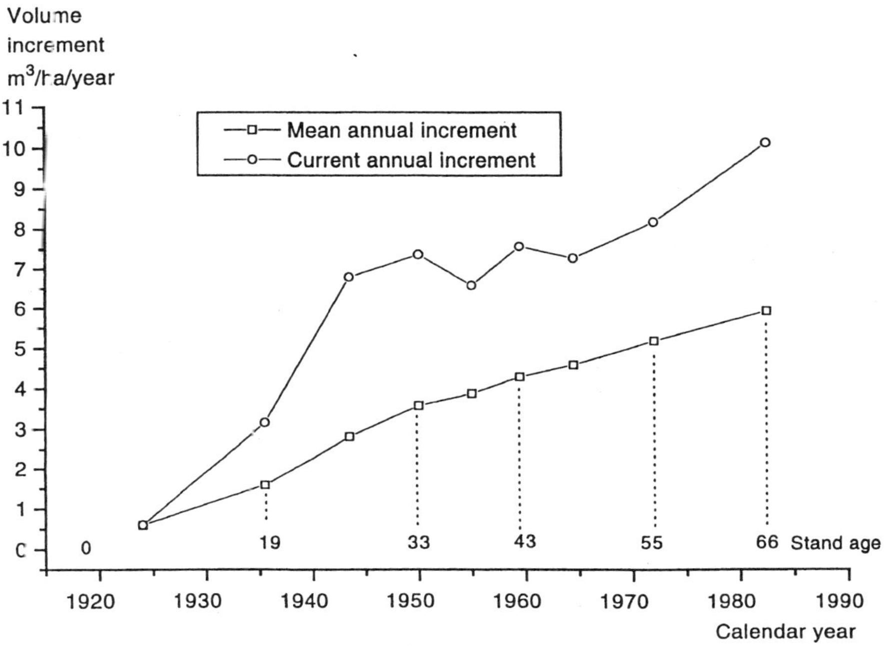

Fig. 11. Current and mean annual volume increment of experimental stand #13 in Helsinki, Vantaa.

sification is based on the development of dominant height or ground vegetation. The changing site quality conceals the existing growth trend. Without information on the change rate of the site, the comparison of experiments established in different decades in different locations involves some uncertainty (Keller 1992).

Our data, collected from natural forests and from old experiments, do not reveal trend-like changes in tree growth. Only few indications about trends could be found on some plots in the southernmost part of the country. The following are some of the reasons for the lack of trends:

1. In northern latitudes, the typically marked climatic variation in tree growth can conceal small emerging trends.
2. The temperature and precipitation of the growing season do not show a clear trend during this century. The summers of 1961-1990 were slightly cooler than those during the preceding 30 years. Only the temperature of May is showing an increasing trend during the past few decades.
3. The fairly low nitrogen deposition rate (5 kg $ha^{-1}a^{-1}$) in Central Finland during the past 20 years is unlikely to cause growth trends that could be separated from long-term growth variation yet.

4. The study material consists of unmanaged natural stands and experimental stands that have been subjected to low thinnings. Thus, trends caused by changing silviculture are outside the scope of this study.

On the basis of the non-existing growth trends of this study and the results obtained from eight successive national forest inventories, it can be concluded that the main factors leading to a high increase in the total growth of Finnish forests during the last 40 years are the result of changes in stand structures (age, density etc.) and silvicultural practices (regeneration, thinnings, drainage, fertilization etc.).

Acknowledgements

We would like to thank Merja Arola and Hilkka Ollikainen for field measurements and preliminary data processing, Pekka Nöjd for valuable discussions concerning study methods and Erkki Pekkinen for the English correction of the manuscript.

References

Cajander AK (1909) Über Waldtypen. Acta Forestalia Fennica 1(1): 1-175

Elfving B, Tegnhammar L (1995) Trends of tree growth in Swedish forests 1953-1992: an analysis based on sample trees from the National Forest Inventory. Scandinavian Journal of Forest Research 10 (in press)

Gustavsen HG (1980) Talousmetsien kasvupaikkaluokittelu valtapituuden avulla. Abstract: Site index curves for conifer stands in Finland. Folia Forestalia 454: 1-31

Hari P, Arovaara H, Raunemaa T, Hautojärvi A (1984) Forest growth and energy production, a method for detecting trends in growth potential of trees. Canadian Journal of Forest Research 14: 437-440

Hari P, Arovaara H (1988) Detecting CO_2 Induced Enhancement in the Radial Increment of Trees. Evidence from Northern Timber Line. Scandinavian Journal of Forest Research 3: 67-74

Henttonen H (1990) Kuusen rinnankorkeusläpimitan kasvun vaihtelu Etelä-Suomessa. Summary: Variation in the diameter growth of Norway spruce in Southern Finland. University of Helsinki, Department of Forest Mensuration and Management, Research Notes 25: 1-80

Jalkanen R, Aalto T, Derome K, Niska K, Ritari A (1995) Männyn neulaskatoon 1987 johtaneet tekijät Pohjois-Suomessa. Loppuraportti. Summary: Needle loss in Lapland. Factors leading to needle loss of Scots pine in 1987 in northern Finland. Final report. Metsäntutkimuslaitoksen tiedonantoja 544: 1-75

Keller W (1992) Bonität in Fichten-Folgebeständen ehemaliger Fichten-Versuchsflächen der WSL. DVFFA, Sektion Ertragskunde, Tagungsber., pp 123-129

Nöjd P, Reams GA (1995) Growth variation of Scots pine across a pollution gradient on the Kola Peninsula, Russia. Environmental Pollution (in press)

Nyyssönen A, Mielikäinen K (1978) Metsikön kasvun arviointi. Summary: Estimation of stand increment. Acta Forestalia Fennica 163: 1-40

Parviainen J (1994) Natural forests as models for forest management. Nordic Workshop on Sustainable Forest Management in Rolighed, Denmark, 9-11 November 1994 (in press)

Savolainen I, Tähtinen M (1989) Emissions of SO2, NO2 and CO2 in Finland -Future Control Scenarios. In: Kraftwerk und Umwelt 1989, VGB-Kraftwerkstechnik GMBH, Essen, p 51-55

Tamminen P (1993) Pituusboniteetin ennustaminen kasvupaikan ominaisuuksien avulla Etelä-Suomen kangasmetsissä. Summary: Estimation of site index for Scots pine and Norway spruce stands in South Finland using site properties. Folia Forestalia 819: 1-26

Tomppo E, Siitonen M (1991) The National Forest Inventory of Finland. Paperi ja Puu - Paper and Timber 73 (2)

Van Deusen PC (1992) DYNACLIM Version 3.2. Institute of Quantitative Studies, Southern Forest Experiment Station, USDA Forest Service, New Orleans

Studies on Growth Trends of Forests in Sweden and Norway

Björn Elfving[1], Lars Tegnhammar[2] and Björn Tveite[3]
[1]Department of Silviculture. Swedish University of Agricultural Sciences
S-901 83 Umeå, Sweden
[2]Department of Forest Soils. Swedish University of Agricultural Sciences
S-750 07 Uppsala, Sweden
[3]Norwegian Institute of Forest Research. N-1432 Ås, Norway

Abstract

In the Swedish National Forest Inventory (NFI), a steady increase in the estimated productivity of forest land has been noticed since inventory was begun in 1923. Young stands generally indicate higher site indices than old stands at equal site conditions. For spruce, this rise of site index has been estimated at 0.05-0.11 m·year^{-1}, with the highest value in the south. In a study based on sample tree data from the NFI, a highly significant annual increase of both height and basal area growth of the magnitude 0.5-0.8% was found for the period 1953-1992. At the same time, top height seems to develop according to the site curves on permanent plots both in Sweden and Norway.

There are two alternative explanations to these observations:

1. Altered silviculture has changed the population structure.
2. A long-term rise of natural site productivity is taking place.

In the first instance, an altered way of cutting, from selective cutting to clearfelling and thinning from below, has probably had the biggest impact so far. Genetic improvement, better species-site adaption and different site treatments performed during the last decades will enhance this trend.

In the second case, the increasing atmospheric deposition of nitrogen is suspected to have the biggest influence. A long-term trend of this type may have been built into our prediction models. Comparison of consecutive generations on the same site should then reveal the trend. A spruce study in southern Sweden indicated a rise, while comparison of pine plantations established in the same way on similar sites during different decades gave a more dubious result.

European Forest Institute Research Report No. 5
Growth Trends in European Forests
Edited by H. Spiecker, K. Mielikäinen, M. Köhl and J. P. Skovsgaard

1. Introduction

The potential productivity of a site has often been regarded as given by nature and stable in a long-term perspective. However, in the last decades, with human impact on the environment increasing, this stability may be questioned. In the Swedish National Forest Inventory (NFI) a steady increase in estimated potential productivity (maximum mean annual increment, m^3ha^{-1}) of forest land has been noticed since inventory was begun in 1923 (Fig. 1). The rise in productivity from 3.2 m^3 in the 1920s to 5.2 m^3 in 1990 means a relative increase of about 60 %. The cause has been considered to be changing silvicultural and measurement methods. The estimate was based on the same site curves, using basal area weighed mean height and age, up until 1982, but its practice in mixed, uneven, and open stands is thought to have changed. Since 1982 the estimate has been based on height and age of top height trees (defined in text to Fig. 2). The observed growth has increased in line with increasing growing stock, while the forest area has been fairly constant.

A related observation is that young stands often indicate higher site indices than old stands at comparable sites. This has caused much frustration both in yield research and practical forestry. The interpretation of the trend is of importance for the choice of silvicultural methods and cutting levels. The aim of this chapter is to summarise some trend studies in Sweden and Norway in order to cast more light on the problem. The tree species concerned in the paper are Scots pine (*Pinus sylvestris* L.) and Norway spruce (*Picea abies* [L.] Karst.).

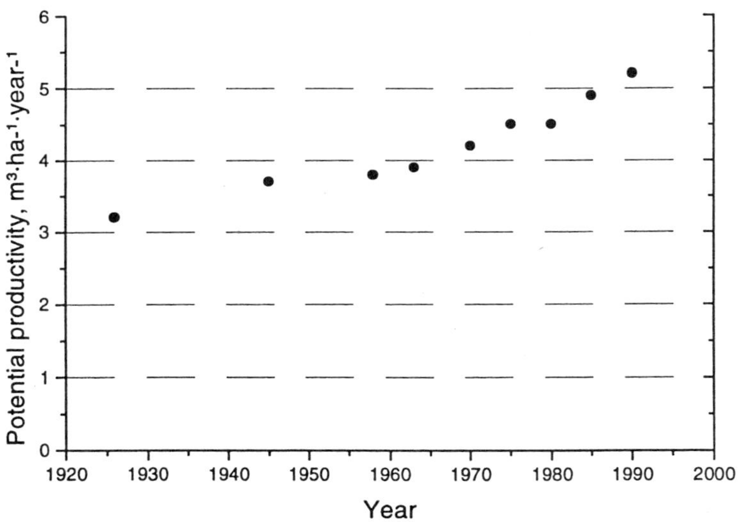

Fig. 1. Average potential productivity of forest land in Sweden according to estimations at the National Forest Inventory in the period 1923-1992. Redrawn from official statistics.

2. Site Index in Young and Old Stands

Tegnhammar (1992) examined the age trend (decreasing site index over the age scale) on NFI plots in spruce stands measured 1983-1987 (Fig. 2). The obvious age trend depends partly on a sorting effect. Stands on better sites are harvested at lower ages than stands on poor sites. The trend can be divided into two parts, one before lowest harvest age (50 years at breast height) and one after this age. After correction for site differences, the decrease of site index before age 50 was estimated to 5-11 cm per year, with the highest value in the south. The basic aim of the study was to stabilize site index as a response variable in functions for site index estimation with site factors.

Functions for site index estimation with site factors were presented by Hägglund and Lundmark (1977). They did not take the age trend into account, so the estimated site indices were connected to the average age in the data. For spruce plantations in northern Sweden, the difference between site indices estimated with that function and with site curves was very large. Elfving and Nyström (1996a) investigated 91 operational and experimental spruce plantations in the age interval 27-46 years, where site curves (existing top height trees) indicated 4.6 m higher site index than estimated with the site factor function. One important reason for the difference was thought to be early suppression of top height trees in the data behind the site factor function. In Sweden, most of the old spruce stands have been subjected to diameter limit fellings. Some of the plots were remeasured after 10 years, and no significant change of the site index estimated with site index curves was detected.

3. Reliability of the Site Index Curves

One reason for the age trend could be bias in the site index curves. The Swedish curves in use were presented by Hägglund (1972, 1973, 1974). They were based on data from sectioned top height trees that were felled in the period 1941-1965. A check against data from remeasured permanent plots showed good agreement. In a later check by Elfving (1990), observed and predicted top height increment in the period 1970-1985 were compared on 235 permanent plots with good distribution on age classes and site types. The top height increment was about 5% higher than predicted, and this was assumed to be an effect of better than average growing conditions in the actual period. It was stated that the deviation was marginal and that the site index curves seemed to be reliable.

The Norwegian site index curves in use were constructed by Tveite in 1976 (pine, see Braastad 1980) and 1977 (spruce, Tveite 1977). These curves were based mainly on data from remeasured permanent plots. Tveite (1994) checked the curves against data from permanent plots in northern Norway. Data comprised 33 pine

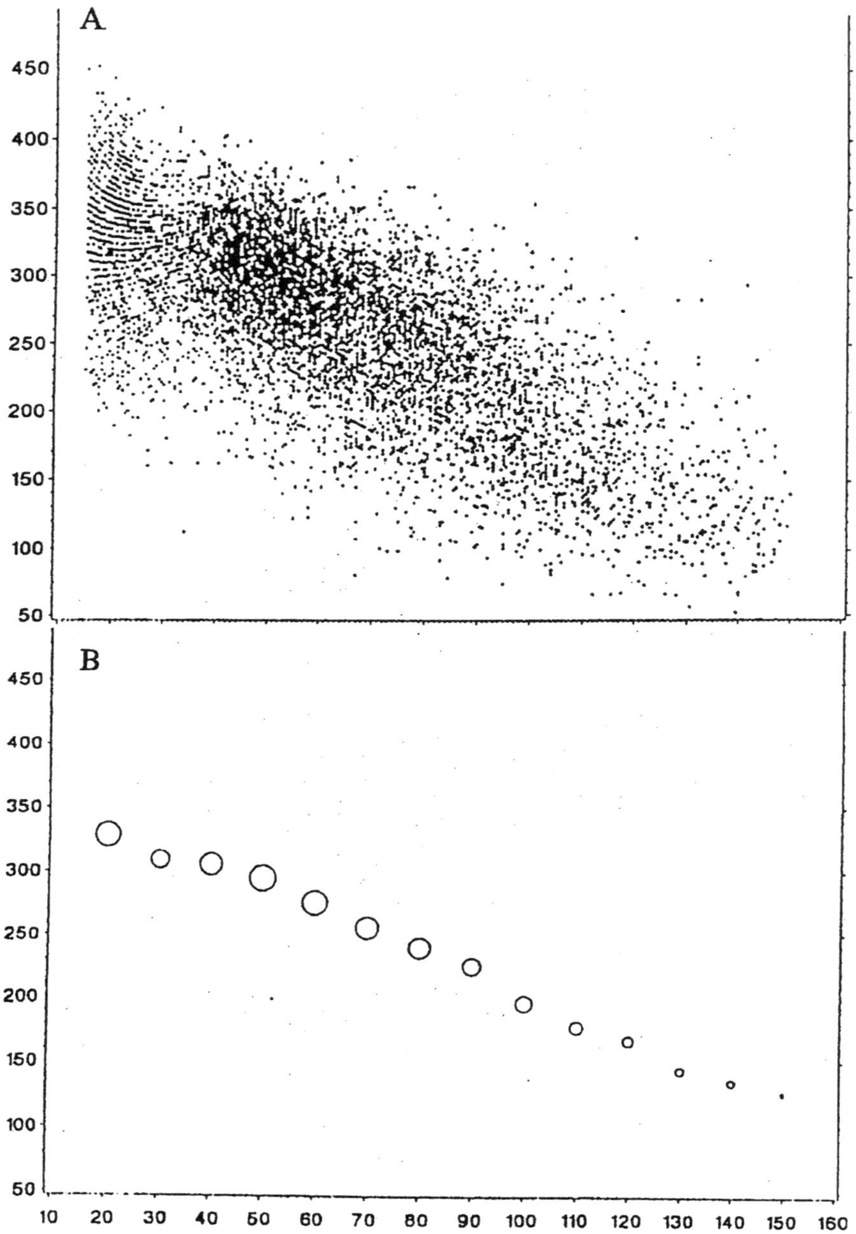

Fig. 2. Site index over age on 6561 NFI-plots in spruce stands. Site index means top height at total age 100 and is estimated with site index curves from measured height and age of the two by diameter largest trees on a plot with 10 m radius. A. Single observations. B. Averages per age class. From Tegnhammar (1992).

plots and 25 spruce plots, most of which had been followed more than 30 years with repeated measurements. Early measurements on some of these plots were included in the data used for the curve construction.

Pine plots covered the age interval 16-165 year and spruce plots 16-72 year. On an average, site index did not change significantly during the observation period. Furthermore, a grouping of the site index changes over calendar years did not show any significant trend for pine (Fig. 3). For spruce, site index was stable during the period 1965-1985. Increasing values before 1965 and after 1985 were suspected to depend on specific weather conditions. Also, the volume growth was checked against simple volume growth functions that were based mainly on data from more southern plots. In that study, residuals showed a decreasing growth trend. The reason was suspected to be deviating growing conditions at these northern latitudes.

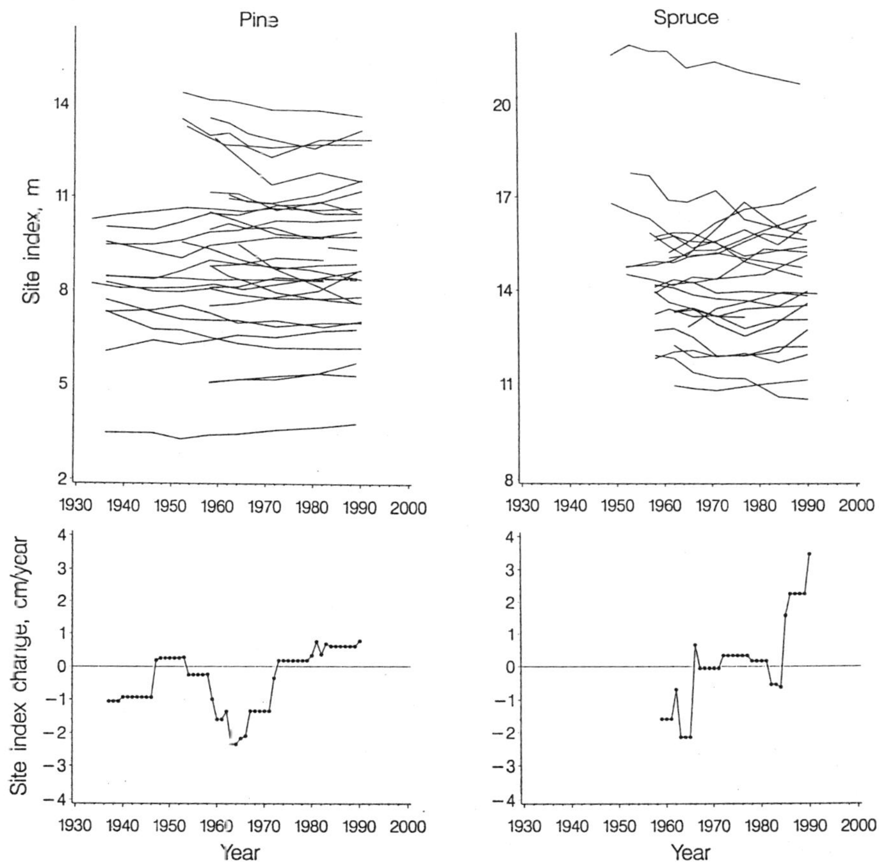

Fig. 3. Site index changes on permanent plots in northern Norway. From Tveite (1994).

4. Productivity in Consecutive Generations

Eriksson and Johansson (1993) studied the total volume yield to age 40 in two consecutive generations of spruce plantations at the Tönnersjöheden experimental forest in southwest Sweden. First and second generation growth occurred mainly during the periods 1880-1919 and 1950-1989, respectively. The volume yield was found to be 40 % higher in the second generation. Seedling material and stand treatment were similar in the two generations and the authors attributed the increased nitrogen deposition to be the most important factor for the increasing yield.

Elfving and Nyström (1996b) studied site index over planting year on permanent plots in 197 pine plantations that were planted in the period 1886-1977. Evaluation with the current site index equation (Hägglund 1974) did not reveal any certain trend, but with a new site index equation, based on repeated measurements of the permanent plots, a significant annual increase of site index of 2.4 cm was estimated in an ordinary linear regression (Fig. 4). However, the main difference was between stands established before and after 1940. This difference was significant, but not the residual, continuous trend. The sudden increase after 1940 could depend on differences in establishment and tending of the stands.

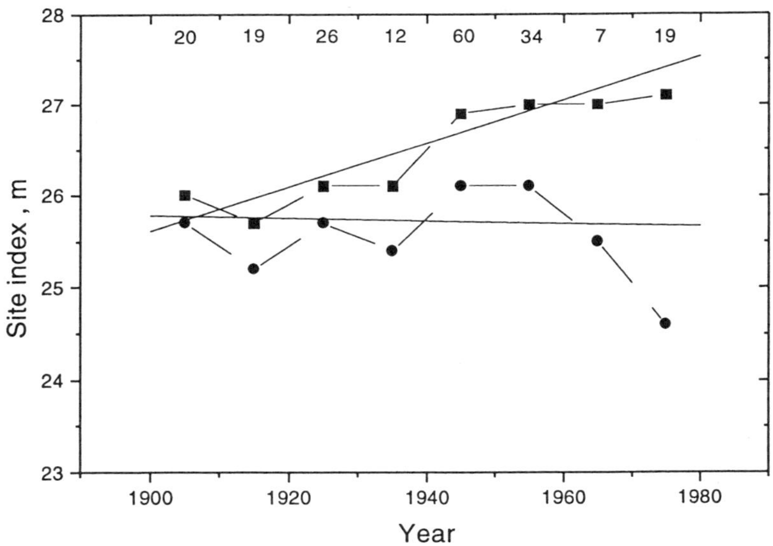

Fig. 4. Site indices in Scots pine plantations over year of planting. Site index according to Hägglund 1974 (●—●) and to a new site index equation based on repeated measurements of permanent plots (■—■). Residuals per decade from average trends (—) are plotted and the number of observations are given at the upper frame. From Elfving and Nyström 1996b.

5. Trends of Tree Growth 1953-1992

Elfving & Tegnhammar (1996) examined the growth trend on the basis of data from sample trees of pine and spruce that were measured by the NFI in the period 1953-1992. Over 100 000 trees of each species were available. Average tree height and basal area growth were studied as functions of species, age, stand density, latitude, altitude, and year of inventory in the period 1953-1992. A highly significant increase at given tree age of both height and growth was found in the 40-year period, as illustrated for pine in Fig 5. The trend was of the magnitude 0.5-0.8 % per year for both traits and species. This means, for example, that trees of age 50 at breast height were on an average about 12 m in 1953 and 15 m in 1992.

Inclusion of the tree diameter as an explanatory variable in the growth model raised the multiple correlation considerably and reduced the growth trend to zero. In other words, trees with the same age and diameter had the same growth in 1953 as in 1992. The combination of age and tree size as explanatory variables must be avoided in studies of growth trends.

The trend in tree height was significantly stronger for spruce than for pine, but the trend in basal area growth was equal for both species. Furthermore, the trend in basal area growth was somewhat higher than that of tree height. In separate studies for different age classes, geographical regions, and ownership categories, only minor deviations from the main trend were revealed. Spruce below age 40 showed a weaker trend in both height and growth. The trend in basal area growth was stronger in state and company forests than in private ones. This discrepancy was stronger for pine than for spruce, and it did not correspond with the height trend.

Possible causes of the trends were discussed. Silvicultural and land use effects were distinguished from effects by environmental changes. Ditching and nitrogen fertilisation have certainly had some effect. A rough calculation based on statistics on treated areas and estimated effects indicates that these treatments can explain up to 25 % of the growth trend. But the silvicultural influence of largest importance was thought to be the shift around 1950 from various forms of selective cutting and dimension felling to clearcutting and thinning from below. Among environmental changes, the increasing N deposition was thought to be the most important.

6. Discussion and Conclusions

The main impression from the studies cited is that trees are growing faster today than 40 years ago, but they are still growing along the site index curves. The homogenous growth response in different geographical regions and age classes indicates that the acting force is universal.

The study of pine plantations from different periods of establishment was hard to interpret. A trend in increasing site index was found, but reasons other than in-

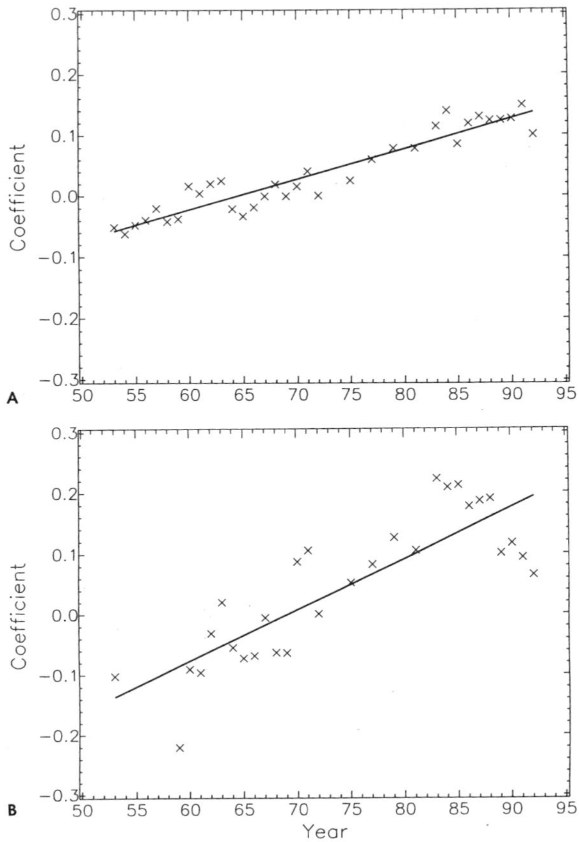

Fig. 5. Estimated height (A) and basal area growth (B) trends for Scots pine in the period 1953-1992 on basis of sample tree data from the Swedish National Forest Inventory. Crosses are estimated coefficients to dummy variables for years of inventory in the model: tree height or basal area growth = f (tree age,stand density, site location, year of inventory). Solid lines are estimated trends for (A) height (0.0050 ± 0.0003) and (B) basal area growth (0.0084 ± 0.0011). From Elfving and Tegnhammar (1996).

creased site productivity were possible. However, planted spruce in southwestern Sweden really grew better in the latter than in the former generation. The reason was suspected to be increasing nitrogen deposition. That deposition is largest in the southwestern part of Sweden. It is much less in the northern parts, so it can hardly cause a homogenous response all over the country. There has been some discussion whether trees in southern Sweden are N- saturated or not. Pettersson (1995) found that the growth response to N fertilization has not changed during the last 25-30 years, even in southern Sweden. This result indicates that N is still a deficient nutrient, despite the high deposition.

The increasing content of carbon dioxide in the air could have a more universal impact. Luxmore et al. (1993) pointed out that increasing CO_2 is expected to enhance growth only under dry conditions, when the gain in CO_2 -intake per quantum of evaporated water is crucial. However, Yoder et al. (1994) found through studies of carbon isotope distribution that the water stress in a tree probably increases with tree height and can be severe even in a more humid climate.

Ultimately, an altered silviculture is certainly responsible for the main part of the 0.5-0.8 % annual increase of tree growth found by Elfving and Tegnhammar (1996). The altered way of cutting, from selective cutting to clearfelling and thinning from below, has probably had the biggest impact so far. Andrén (1954) postulated on the basis of the bad forest conditions in northern Sweden around 1950 and observations of the good growth in older plantations that it should be possible to double the yield. In the coming years, the effects of genetic improvement, adaptation of species selection to site conditions and different site treatments performed in the last decades will certainly enhance the growth trend.

References

Andrén T (1954) Produktionsmöjligheter på medelgod norrländsk skogsmark. Norrlands Skogsvårdsförbunds Tidskrift 1954: 653-669

Braastad H (1980) Growth model computer program for Pinus sylvestris. Norwegian Forest Research Institute, Report 35: 265-359 (in Norwegian with English summary)

Elfving B (1990) Bonitet och tillväxt i den nya skogen. In: Albrektsson A (ed) Aktuella skogsskötselfrågor. Sveriges Lantbruksuniversitet, inst. för skogsskötsel, Arbetsrapporter nr 47: 65-76

Elfving B, Nyström K (1996a) Yield capacity of planted Norway spruce in northern Sweden. Scand. J. For. Res. 11: 38-49

Elfving B, Nyström K (1996b) Stability of site index in Scots pine (Pinus Sylvestris, L.) plantations over year of planting in the period 1900-1977. Spiecker et al. (ed.) Growth Trends of European Forests. European Forest Institute Research Report No. 5. (in this volume)

Elfving B, Tegnhammar L (1996) Trends of tree growth in Swedish forests 1953-1992: an analysis based on sample trees from the National Forest Inventory. Scand. J. For.Res. 11: 26-37

Eriksson H, Johansson U (1993) Yields of Norway spruce (Picea abies [L.] Karst.) in two consecutive rotations in southwestern Sweden. Plant and Soil 154: 239-247

Hägglund B (1972) Site index curves for Norway spruce in norhtern Sweden. Royal College of Forestry, Dept of Forest Yield Research, Research Notes 21, p 298 (in Swedish with English summary)

Hägglund B (1973) Site index curves for Norway spruce in southern Sweden. Royal College of Forestry, Dept of Forest Yield Research, Research Notes 24, p 49 (in Swedish with English summary)

Hägglund B (1974) Site index curves for Scots pine in Sweden. Royal College of Forestry, Dept of Forest Yield Research, Research Notes 31, p 54 (in Swedish with English summary)

Hägglund B, Lundmark J-E (1977) Site index estimation by means of site properties. Scots pine and Norway spruce in Sweden. Stud. For. Suec. 138, p 38

Luxmore RJ, Wullschleger SD, Hanson PJ (1993) Forest responses to CO_2 enrichment and climate warming. Water, Air and Soil Pullution 70: 309-323

Pettersson F (1994) Tillväxtökningen efter kvävegödsling densamma idag som för 30 år sedan. SkogForsk, Uppsala. Resultat nr 9. ISSN 1103-4173

Tegnhammar L (1992) On the estimation of site index for Norway spruce. Swed. Univ. of Agri. Sciences, Dept of Forest Survey, Report 53, ISSN 0348-0496, p 259 (in Swedish with English summary)

Tveite B (1977) Site index curves for Norway spruce. Norwegian Forest Research Institute, Report 33:1-84 (in Norwegian with English summary)

Tveite B (1994) Høgdeutvikling og volumproduksjon i gran- og furuskog nord for Saltfjellet. In: Bergan J (ed) Faglige emner innen primærproduksjonen i skogbruket i Nord-Norge. Norsk Institutt for Skogforskning, Ås, ISBN 82-7169-655-6, pp 93-103

Yoder BJ, Ryan MG, Waring RH, Schoettle AW, Kaufmann MR (1994) Evidence of reduced photosynthetic rates in old trees. Forest Science 40: 513-527

Stability of Site Index in Scots Pine (*Pinus sylvestris*, L.) Plantations over Year of Planting in the Period 1900-1977 in Sweden

Björn Elfving[1] and Kenneth Nyström[2]
[1]Swedish University of Agricultural Sciences. Department of Silviculture. S-901 83 Umeå
[2]Swedish University of Agricultural Sciences.
Department of Forest Resource Management and Geomatics. S-901 83 Umeå

Abstract

The hypothesis that environmental changes have affected site productivity was tested through examination of site index over year of planting. Data included 197 experimental plots established 1886-1977 and 617 survey plots established 1952-1968, all with Scots pine (*Pinus sylvestris* L.). Effects of site differences were balanced in a multiple regression analysis. Evaluation with the current site index equation did not reveal any certain trend, but with a new site index equation based on data from repeated measurements of the experimental plots, a significant annual increase of site index of 2.4 cm was estimated for the period 1900-1977. However, the main difference was between stands established before and after 1940. Segmentation indicated a significant difference between these groups but no significant, continuous trend. The difference between the two groups could be related to differences in establishment and tending of the stands.

1. Introduction

In a study of pine and spruce in Sweden, it was found that average tree height and basal area growth of single trees at a given age have increased at a rate of 0.5-0.8% per annum in the period 1953 - 1992 (Elfving and Tegnhammar 1996). Silvicultural improvements were assumed to be the main reasons, but the effects of environmental changes such as increased natural nitrogen deposition could not be excluded. One way to explore the nature of the growth trend is to compare site indices in plantations established in different years, cf. Spiecker et al. (1994). Selection of data from stands with known history will facilitate control of silvicultural effects like ditching, nitrogen fertilization and a changed cutting system.

European Forest Institute Research Report No. 5
Growth Trends in European Forests
Edited by H. Spiecker, K. Mielikäinen, M. Köhl and J. P. Skovsgaard

Hägglund (1974) showed that even those plantations established at a later point in time increase their top heights according to site curves based on old, sectioned trees. However, if there is a long-term, increasing trend in the growth level due to a common site amelioration, the trend has been incorporated in the site curves. Comparison of site indices in plantations established in a standardized manner for different decades should reveal the trend. The aim of this study is to make such a comparison based on available data from Scots pine plantations in Sweden. The study concentrates on Scots Pine (*Pinus sylvestris* L.), since the data for this species are the most extensive and homogenous available.

2. Data and Methods

Data consisted of two separate sets, in the following text referred to as set 1 and set 2.

Set 1. Data from pine plantations in the Hugin young stand survey, cf. Elfving (1982). Five circular plots per stand in 180 pine plantations of age 18-36 years were measured from 1981-1984. The stands were selected in a stratified, random manner from a large data base of regenerations that had been checked in a regeneration survey. Detailed data on stand history were available. Height and age at breast height of top height trees were recorded, as well as site factors according to Hägglund and Lundmark (1977). The two trees with the largest breast height diameters within a 10-m radius from plot centre were selected as top height trees. Average height of the two largest trees in a 314 m² plot corresponds with the average height of the ten largest trees within a 1000 m² plot, according to empirical studies (Fries 1974). Only those plots with two pines as top height trees were used in the study, a total of 617 plots. Tree ages were registered through whorl counting, totally independent of the known stand history. In several cases the top height trees seemed to originate from advance growth.

Set 2. Permanent yield plots in artificially regenerated pine stands (sown or planted), generally emerging from seed of local origin. Most plots were established as experiments on regeneration methods, provenances, spacing or thinning. The plot data in this study are often based on values from several subplots. Top height was calculated as the height corresponding to the 100 largest diameter trees per hectare, according to the height curve. Increment coring was avoided on these plots, and age at breast height was generally unknown. However, in younger stands, the age at breast height of top height trees was registered through whorl counting. The only available site descriptions in common for these plots were latitude, altitude and soil moisture. Only plots on mesic sites were included in the analysis, 197 stands established in 1886-1977. Data from the last measurement were used. This was generally performed in the period 1977-1990.

Table 1. Statistics on the Scots pine plots used in the site index study. Mean values and range (minimum and maximum) are given for key variables

Variable	Data set	
	1. Survey plots	2. Permanent plots
No. of plots	617	197
Site index, m (SIH)	23.7 (14.6-33.3)	24.6 (17.0-31.2)
" (SIEK)	-	25.4 (17.5-31.7)
Latitude, °N	61.5 (55.6-67.7)	62.3 (56.7-67.7)
Altitude, m a.s.l.	248 (10-800)	245 (5-635)

Data in both sets had a wide distribution over latitude and altitude (Table 1). For each plot, site index (SIH, m) was determined with the site index equation presented by Hägglund (1974). The independent variables are height and age at breast height of the top height trees, and the dependent variable is site index, defined as expected top height at total age 100. Number of years from the seed's emerging to its reaching breast height was expressed as a function of site index. In the present study, this function was used to estimate age at breast height on those plots in data set 2 where only total age was known.

For data set 2 an alternative site index (SIEK) was also determined, with a site index equation based on data from repeated measurements on most of the plots in this data set (Elfving and Kiviste in prep.). The equation is based on the Hossfeld II model : $Y = b_0/(1+b_1/x^{b_2})$, adapted to the data with the difference equation method after transformation in the way proposed by Cieszewski and Bella (1989) and smoothed in a non-linear regression. The result is a polymorphic, base-age invariant equation with site-specific asymptotes. The equation uses total age as independent variable, and gives steeper site index curves than the Hägglund equation. SIEK also gives expected height at total age 100. The reason for not using this equation with data set 1 was that some top height trees on these plots seemed to originate from advance growth. Total age was thus a more uncertain variable.

For each data set, site index was expressed as a function of latitude (LAT, °N), altitude (ALT, m a.s.l.) and year of planting (YR). For data set 1, the ground vegetation type was also included in the function. Classification of ground vegetation types was based on coverage of different species in the field and bottom layers according to Hägglund and Lundmark (1977). Four classes were distinguished: herb type (20), grass type (40), blueberry type (9) and poor type (31). Figures within parentheses give the proportion of all plots in %. Indicator variables (value 0 or 1) were inserted for the classes herb, grass and poor, and the blueberry type was chosen as reference class. The procedure REG in the SAS computer program was used in the multiple regression analysis.

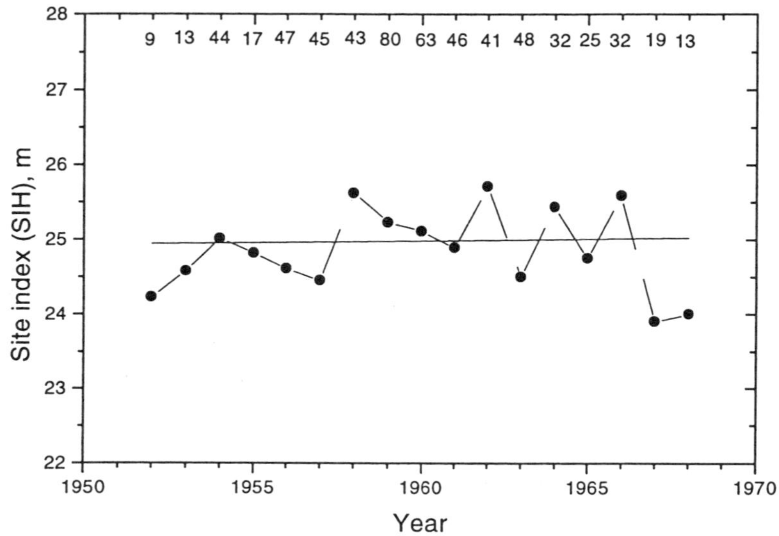

Fig. 1. Residuals in site index from function F1 over year of planting. Residuals (●—●) are presented as deviations from the predicted (—) site index at latitude 61.5 °N, altitude 250 m a.s.l., veg. type grass, in different years. The number of plots in different years are indicated below the upper frame.

3. Results

In data set 1 there was no significant trend over time in site index for plantations established in the period 1952-1968 (Fig. 1 and Table 2, function F1: $\beta(YR) = 0.0052$, $p = 0.81$).

In data set 2 the trend in SIH over year of establishment was also insignificant (Fig. 2 and Table 2, function F2: $\beta(YR) = -0.001$, $p = 0.78$). Stands established after 1970 deviate negatively from the other observations. Their breast height age was only 5-10 years. If these stands are excluded, the residual observations form a weak positive trend ($\beta(YR) = 0.013$, $p = 0.034$).

However, SIEK in data set 2 had a clear positive trend over time (Fig. 2 and Table 2, function F3: $\beta(YR) = 0.024$, $p < 0.0001$). This could be expected, since the new site index curves are steeper than the former. The residual standard deviation was smaller for function F3 than F2, mainly because the youngest stands indicated higher site indices over total age than over breast height age.

Estimated site indices (SIH) with function F1 and F2 differ about 0.5 m at comparable sites, cf. Figs. 1 and 2. Experimental plots had the highest SIH-level. In the comparison, the vegetation type was set to "grass type" in function F1, since this vegetation type approximately represents the average on the experimental plots.

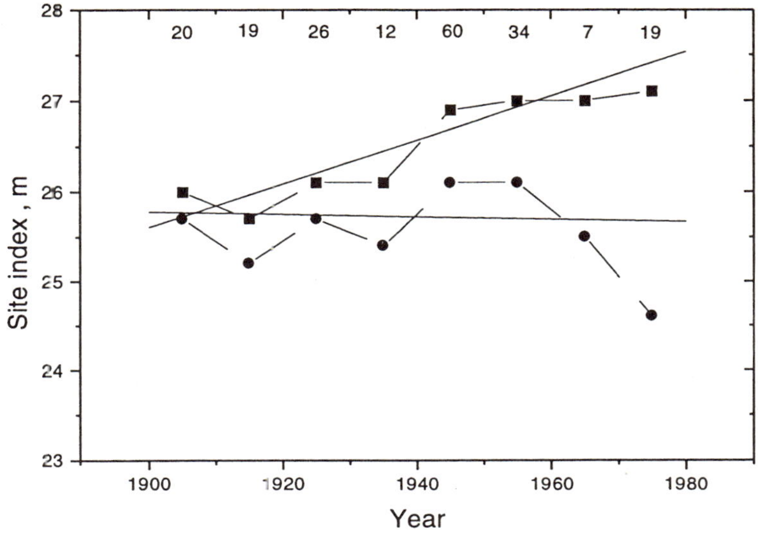

Fig. 2. Residuals in site index from functions F2 (SIH, ●—●) and F3 (SIEK, ■—■). Mean values for regenerations from different decades are plotted as deviations from calculated (—) site index at latitude 61.5 °N, altitude 250 m a.s.l., in different decades. Two plots established before 1900 were shifted to the group 1900-1909. The number of plots in each group are indicated below the upper frame.

Table 2. Regression functions for site index in Scots pine plantations. Estimated coefficients (ß) and probabilities for non-deviation from zero (p) are given. For variable description see text

	Data set and function number (dependent variable)					
Variable	Set 1, F1 (SIH)		Set 2, F2 (SIH)		Set 2, F3 (SIEK)	
	ß	p	ß	p	ß	p
Intercept	25.35	<0.0001	28.67	<0.0001	28.52	<0.0001
$(LAT-54)^2$	- 0.0333	<0.0001	- 0.0459	<0.0001	- 0.0453	<0.0001
ALT/100	0.7055	<0.0001	-	-	-	-
$(ALT/100)^2$	- 0.1980	<0.0001	- 0.2693	<0.0001	- 0.2316	<0.0001
(LAT-54)(ALT/100)	-	-	0.0732	0.0120	0.0578	0.0329
Herb type	1.36	<0.0001	-	-	-	-
Grass type	0.67	0.0341	-	-	-	-
Poor type	- 0.86	0.0074	-	-	-	-
YR-1900	0.0052	0.8159	-0.0014	0.7850	0.024	<0.0001
Adjusted R-square	0.572		0.679		0.722	
Root MSE	2.05		1.47		1.36	

4. Discussion

Different site index curves gave different results in this study. The studies of SIH did not indicate any significant time trend. However, after exclusion of the youngest stands, a weak positive trend was registered in data set 2. It is possible that the equation for SIH has a minor bias at low ages. This was indicated in the residual study of the equation (Hägglund 1974). Such a bias can also hide a positive time trend in data set 1.

Site index according to the new equation (SIEK) had a clear time trend in data set 2. The trend was estimated at +2.4 cm y^{-1}. However, the trend is not smooth. There seems to be a step between stands planted before and after 1940. In a regression with an indicator variable for "planted after 1940" included, the coefficient for this variable was estimated at 0.84 m ($p= 0.031$). The coefficient for the continuous variable "year of planting" diminished to 0.006 ($p= 0.53$).

The difference in data set 2 between stands established before and after 1940 could be a treatment effect. Most stands established before 1940 had been thinned, whereas those stands established after that time seldom were. As height growth seems to decrease for a period after thinning in stands of normal density (Valinger 1992), this could be one explanation. Furthermore, most stands established after 1940 were well tended regeneration experiments, whereas many of the older stands were selected at a top height of about 13 m for thinning experiments, and less is known about their establishment. The regeneration experiments were established on uniform sites without shelter trees after intensive site preparation. Dominating seedlings in sown patches were released early, and naturally regenerated seedlings were systematically removed. It is thus possible that a difference in stand establishment practice before and after 1940 could be another explanation for the sudden increase in site index.

The difference in SIH-level of approximately 0.5 m between the two data sets probably depends on the bias of the site index equation at low ages mentioned earlier and on selection effects. Permanent plots were generally established on sites more homogenous than average, and the included thinning experiments avoided stands with gaps, low survival etc.

In conclusion, plantations established after 1940 indicated 0.84 m higher site indices than older plantations. Besides this sharp increase, no significant trend could be detected. The sudden increase could be a treatment effect.

References

Cieszewski CJ, Bella IE (1989) Polymorphic height and site index curves for lodgepole pine in Alberta. Can. J. For. Res. 19: 1151-1160

Elfving B (1982) Hugins ungskogstaxering 1976-1979. Sveriges lantbruksuniversitet, projekt Hugin. Rapport 27, p 87

Elfving B, Tegnhammar L (1996) Trends of tree growth in Swedish forests 1953-1992: an analysis based on sample trees from the National Forest Inventory. Scand. J. For.Res. 11: 26-37

Fries J (1974) Practical site classification. Sveriges Skogsvårdsförbunds Tidskrift 72: 551-5--63 (in Swedish with English summary)

Hägglund B (1974) Site index curves for Scots pine in Sweden. Royal College of Forestry, Dept of Forest Yield Research, Research Notes 31, p 54 (in Swedish with English summary)

Hägglund B, Lundmark J-E (1977) Site index estimation by means of site properties. Scots pine and Norway spruce in Sweden. Stud. For. Suec. 138, p 38

Spiecker H, Mielikäinen K, Köhl M, Untheim H (1994) Growth trends of European forests. Has site productivity changed ? European Forest Institute Working Paper No. 4, ISSN 1237-5136, p 67

Valinger E (1992) Effects of thinning and nitrogen fertilization on stem growth and stem form of Pinus sylvestris trees. Scand. J. For. Res. 7: 219-228

Long-Term Changes in Site Index in Growth and Yield Experiments with Norway Spruce (*Picea abies*, [L.] Karst) and Scots Pine (*Pinus sylvestris*, L.) in Sweden

Harry Eriksson and Kjell Karlsson
Department of Forest Yield Research. Swedish University of Agricultural Sciences
S-776 98 Garpenberg, Sweden

Abstract

Based on records from long-term yield experiments in pure Scots pine and Norway spruce stands in Sweden, the development of top height was analysed. The top height of a sample plot is defined as the average height of the largest 100 trees (in terms of diameter) per hectare. Site indices, i.e. top height in 100-year-old stands, were estimated using site-index functions commonly applied in Sweden on two different occasions for each sample plot. The observation periods varied between 27 and 39 years in one study group and between 15 and 26 years in a second group.

In most regions, the site indices have increased significantly in both Scots pine and Norway spruce stands during the last decades. The increase of site index has been highest in southwestern Sweden where the atmospheric deposition of nitrogenous compounds is high, i.e. 15-30 kg N $ha^{-1}y^{-1}$.

1. Introduction

The question as to whether or not the productivity potential of Swedish forests has changed during the last decades has stirred debate among forestry researchers as well as politicians. Earlier, it was more or less assumed that site productivity changes very slowly owing to the great stability of factors influencing growth, such as mean temperature, mean precipitation, nutrient release and the genetic origin of the stands. This resulted in the paradigm that, for example, height growth functions and yield tables based on retrospective studies could be used for predicting stand development without any corrections. However, in the light of findings from various kinds of recent investigations, more and more researchers have become convinced that a general increase in site fertility has been taking place during the past few decades (e.g. Becker 1989; Becker et al. 1990; Kenk et al. 1991; Eriksson and Johans-

European Forest Institute Research Report No. 5
Growth Trends in European Forests
Edited by H. Spiecker, K. Mielikäinen, M. Köhl and J. P. Skovsgaard

son 1993; Spiecker 1995; Sverdrup and Warfvinge 1993). Spiecker et al. (1994) ascribed these apparent changes in potential production to the following factors:

- climatic changes
- changes in management practices
- change of target population
- changes in methods used for assessing and estimating (volumes etc.)
- new genotypes produced through plant breeding
- genetic sources
- use of impropriate or faulty references (e.g. growth and yield tables)
- increased deposition levels
- occurrence of pests and diseases
- other biotic and abiotic impacts.

In their review, aimed at addressing the question of whether long-term increases in forest productivity have occurred, Spiecker et al. (1995) discussed different techniques and compiled data from various sources. They distinguished between the use of (1) tree analysis, (2) permanent research plot data and (3) forest inventory data.

In the present study, materials from permanent sample plots at the Department of Forest Yield Research, Swedish University of Agricultural Sciences, were used as a data base for analysing possible growth long-term trends for *Picea abies* and *Pinus sylvestris.*

2. Materials and Methods

2.1. Materials

Our department has continually used records from permanent sample plots for developing prediction tools, such as yield tables and height growth functions, used in forestry planning. Permanent sample plots were first established in Sweden in 1902 and have been treated according to specified routines in order to develop a data base for analysing the effects on various management practices on forest yields. Management factors assessed in these long-term studies include thinning intensity, thinning intervals, thinning form, initial spacing, fertilization, liming and pruning. Today, data from permanent sample plots are available for different species, localities and sites, for various thinning methods and for stands of different age.

For the analyses performed here subsamples from the total data base were used. Two main groups of data were selected which satisfied the following criteria:

- They were collected from relatively pure Scots pine and Norway spruce stands
- The age at last assessment was less than 150 years
- Top height was at least 10 m at the first assessment used

- The plots had been subjected to low thinning or crown thinning. In other words, plots subjected to thinning from above or selection thinning were excluded
- The plots had not been fertilized or limed.

Groups 1 and 2 consist of plots inspected during the period 1956-1965 and 1971-1980, respectively, which had also in most cases been inspected both before and after these respective periods. Among the selected plots, the first assessment available for each plot represents the beginning and the last assessment of the same plot represents the end of each observation period. In a few cases, data from the inspections during the period 1956-1965 and 1971-1980, respectively, were used as the first or the last assessment in the study.

The choice of the period 1956-1965 for group 1 was based on the following assumptions:

- The site index functions used (Hägglund 1972, 1973, 1974) were derived from stem-analysis data obtained from temporary sample plots collected during the period 1941-1965. Residual analyses performed by Hägglund (op.cit.) indicated that these functions are free from systematic trends over stand age. Thus, the growth functions should describe top-height development of the permanent sample plots up to 1965 without systematic errors.
- Other investigations (e.g. Eriksson and Johansson 1993) have demonstrated increases in productivity that took place mainly after around 1950.

Group no. 2 was chosen in order to obtain more material in the geographical groups which depend on the laying out of a large number of new permanent sample plots during the 1960s and 1970s. The number of plots used in groups 1 and 2 are given in Tables 1 and 2, respectively.

Sample plots vary between 0.1 and 0.5 ha in size. There were approximately 40-70 sample trees before 1970 and 25-30 afterwards. Throughout this period the ten trees with the largest diameter on each plot were selected as sample trees. The same sample trees were used from one inspection to the next. When a sample tree was removed in a thinning a new sample tree was systematically selected from among undamaged remaining trees.

The source material was divided into four regions, named A - D, as follows:

Group A. Northern Sweden (Norrland), north of latitude 60°30'N
Group B. "Middle Sweden" (Svealand), between latitudes 59° and 60°30'N
Group C. Southeastern Sweden (Götaland), south of latitude 59°N and east of longitude 14°E.
Group D. Southwestern Sweden (Götaland), south of latitude 59°N and west of longitude 14°E.

The division of the country into these regions is based on the pattern of the atmospheric deposition (Brodin 1990; Björkdahl and Eriksson 1989). Götaland was divided into two parts mainly owing to large differences in precipitation between the western and the eastern regions (Eriksson 1981).

2.2. Methods

In both groups 1 and 2, data from the first and last inspections of each sample plot, with tree heights estimated according to the description given above, were used to calculate (1) top heights and (2) site indices.

The top height for each plot and for each inspection was calculated according to the following definitions of the term top height:

The current top height of a sample plot refers to the mean height of the 100 trees with the largest diameter per hectare. Since it is generally impractical to measure the heights of all such trees on the plots the top height is calculated using a static height function according to Näslund (1936). This function is used for each plot and each assessment based on all sample trees for which height records are available, and has the following analytical form:

$h - 1.3 = [d/(a + bd)]^n$, where

d is diameter at breast height
h is tree height over ground
n is 2 for Scots pine and 3 for Norway spruce
a, b are regression parameters.

Top height calculated according to this height/diameter function will systematically be higher than the mean height of the 100 largest trees due to the concavity of the height/diameter curve. This bias is of very little importance here and does not affect the reliability of later interpretations of the results.

Site index refers to the top height of the 100 trees with the largest diameter per hectare at a reference age of 100 years (total age from seed). To estimate site index, the top height growth functions published by Hägglund are used. The site index function for northern Sweden (Hägglund 1972) was used for Norway spruce plots located north of approximately latitude 60°30'N and the function for south Sweden (Hägglund 1973) for the other spruce plots. Site index for Scots pine was estimated according to Hägglund (1974).

3. Results

Table 1 shows the results of the calculations for group 1 for each species and geographical region.

In Scots pine there was a significant [P(t)>0.005] increase in site index during the observation period in all regions except southeastern Sweden. Observation periods were, on average, 37.9, 39.1, 34.8 and 36.4 years long in regions A, B, C and D, respectively. Over the past 40 years, the site indices increased by 0.29, 0.62, 0.38 and 2.11 m in regions A, B, C and D, respectively. Region D included only 21 plots.

Table 1. Characteristics of plots included in group 1, i.e. permanent sample plots inspected during the period 1956-1965. N = number of plots; A1 = average total age at first assessment used; A2 = average total age at last assessment used; TH1 = average top height, m, at first assessment used; TH2 = average top height, m, at last assessment used; SI1 = average site index, m, at first assessment used according to Hägglund (1972, 1973, 1974); SI2 = average site index, m, at last assessment used according to Hägglund (1972,1973, 1974); DIFF = mean difference between site indices, weighted with length of observation period for each plot, cm yr^{-1}; t = t-value according to Student's t-test

	Assessments before 1965				Assessments after 1956				
Group	N	A1	TH1	SI1	A2	TH2	SI2	DIFF	t
Scots pine									
A	147	71.2	16.1	20.39	109.1	20.8	20.72	0.74	2.58
B	144	73.5	17.2	21.58	112.3	22.6	22.21	1.56	5.97
C	52	53.2	15.5	23.57	88.0	22.2	24.11	0.96	1.50
D	21	52.6	15.1	22.66	89.0	23.2	24.52	5.28	5.05
Norway spruce									
A	21	67.8	16.9	23.17	100.3	22.4	23.34	0.37	0.30
B	30	69.3	18.8	24.81	96.7	24.4	25.91	3.71	2.80
C	31	48.8	18.2	29.83	78.9	26.5	30.37	1.76	3.06
D	91	48.7	18.7	29.82	79.8	27.9	31.53	4.43	6.34

In Norway spruce the observation periods were, on average, 32.5, 27.4, 30.1 and 31.1 years long for regions A, B, C and D, respectively, and the number of sample plots was limited in regions A to C (21, 30 and 31). In Norway spruce stands significant increases in site indices were demonstrated for regions B to D. There was a tendency for site index to increase in northern Sweden but it was not significant. As can be seen in Table 1, site indices for Norway spruce during the last 40 years increased by 0.15, 1.48, 0.70 and 1.77 m in regions A, B, C and D, respectively.

The relative increase in site index during the period 1955-1995 was 1.5, 2.9, 1.6 and 9.3% in Scots pine stands and 0.6, 6.0, 2.4 and 5.9% in Norway spruce stands for north Sweden, middle Sweden, southeast Sweden and southwest Sweden, respectively (Table 1).

In Table 2 the corresponding figures are given for the larger set obtained in the inspections of plots inspected during the period 1971-1980.

The observation periods for this group were shorter than those in group 1. For Scots pine the length of the observation periods was 22.5, 26.2, 21.9 and 18.6 years in regions A, B, C and D, respectively. For all regions site indices for Scots pine tended to increase, and the t-values are high, indicating a strongly significant trend. Based on the data analysed we cannot say whether this tendency will persist in the future. Over the past 20 years, increases in site indices were 0.75, 0.64, 0.70 and 1.45 m for regions A, B, C and D, respectively.

Table 2. Characteristics of plots included in group 2, i.e. permanent sample plots inspected during the period 1971-1980. For symbols in table heading see explanations in Table 1

	Assessments before 1980				Assessments after 1971				
Group	N	A1	TH1	SI1	A2	TH2	SI2	DIFF	t
Scots pine									
A	337	51.0	13.7	22.63	73.5	18.4	23.23	3.76	9.62
B	268	55.4	14.6	22.88	81.6	19.8	23.46	3.22	7.64
C	145	39.8	13.5	25.30	61.7	19.4	25.92	3.52	6.93
D	46	38.7	13.4	25.29	57.3	18.6	26.34	7.26	7.44
Norway spruce									
A	50	53.4	14.2	24.09	74.0	19.7	25.02	5.96	5.23
B	121	47.8	15.5	29.08	62.5	21.2	30.10	7.59	6.44
C	85	31.0	13.8	32.54	51.7	22.5	33.50	5.38	4.42
D	203	40.0	17.3	31.76	55.7	23.3	32.60	5.77	7.86

Site indices for Norway spruce plots in all regions of Sweden increased during the observation periods. Despite the results for group 1, spruce plots in north Sweden also showed an increase that was faster than the one predicted by Hägglund's growth functions for dominant height (1972). During the past 20 years, site indices have increased in Norway spruce by 1.19, 1.52, 1.08 and 1.15 m in regions A, B, C and D, respectively.

4. Discussion and Conclusions

The results presented here support the conception that forest productivity has changed. Thus, it would not be accurate to use "historical" data for predictive purposes in forest production research or in forestry planning processes in the future.

Results obtained while analysing permanent sample plot data demonstrate that site indices have tended to increase. This increase has been most pronounced in southwestern Sweden, for both Scots pine and Norway spruce. Although there are certainly several factors involved, no conclusions as to their relative importance could be made based on the data analysed here. Nevertheless, based on results from other investigations, some of the factors that could have changed and their effects on the growth patterns can be discussed.

Measurement errors when collecting tree height data are often of importance in analyses like those made here. In this case, height over ground of the sample trees was measured using a tripod hypsometer. According to Eriksson (1970), the measurement error associated with the hypsometer used averages 2.1 dm (standard deviation of random error). Since the top height determination of each plot was based on at least ten sample trees, the error in measuring top height was small (<0.7 dm) and is therefore regarded to be of negligible importance.

Potential biases in Hägglund's site inex curves (1972, 1973, 1974) are of great importance for the interpretation of results obtained in the study. The possibility that some of the change in site indices could be an artefact due to problems in the methods used by Hägglund must be considered. However, in analyses made by Hägglund (op.cit.) there was no indication that the model used, the so-called Chapman-Richards' equation (Bertalanffy 1941; Richards 1959), is less exact than other growth functions used for fitting stem analysis data from Scots pine and Norway spruce stands. Based on this assumption, we concluded that the method of comparing site index estimates obtained on different occasions of each sample plot is an appropriate technique for use in analysing changes in productivity.

Other techniques - for example, measuring the annual height growth in top height between two consecutive inspections - proved to have disadvantages due to changes in population structure that occurred over long periods of time. Thus, these techniques were not considered suitable for our purposes.

Initial tests of Hägglund's site index curves for Norway spruce and Scots pine were performed using height records from inspections of permanent sample plots made before 1965. There were no systematic long-term patterns of change in site index for spruce or pine in the regions studied except in southwestern Sweden, where a signifiant increase in site index was demonstrated.

Results from both tests made by Hägglund for systematic changes in site index over time and our own intitial checks of biases in Hägglund's site index curves further confirmed that Hägglund's site index functions could be used for permanent sample plots assessed before 1965 without any risk of introducing systematic errors into site index determinations.

Changes in important climatic parameters in Sweden such as temperature and precipitation were analysed by Eriksson (1982, 1983). Compared with the period 1931-60 the mean temperature during the period 1951-80 was 0.2-0.7°C lower (especially in the northern part of Sweden). The mean precipitation during the period 1951-80 was nearly the same as that for 1931-60. Thus, the changes in site productivity cannot be explained by more favourable climatic conditions during the past three decades.

Management practices used on the plots studied over the past few decades are basically the same as those used earlier.

It should be emphasised that in the present investigation measurements were made repeatedly on the same plots, and the observation period was relatively long (27-39 years in group 1 and 15-26 years in group 2). Thus, problems associated with the use of different populations were avoided.

The methods used for assessing top height on different occasions are standardized and the same methods for field measurements and selection of sample trees were used throughout the observation period.

Because the same sample plots were studied at two different dates, none of the variation observed could have been due to changes in plant genetic sources or to plant breeding.

The permanent sample plots studied did not show abnormal levels of damage due to pests or diseases, and plots subjected to heavy windthrough were excluded from the study.

Several investigations have shown that the atmospheric deposition of sulphur and nitrogen compounds has increased since around 1950 (Odén 1968), as have the effects of the acidification of the forest soils (Grennfelt and Hultberg 1986; Hallbäcken and Tamm 1986). Furthermore, the effects of the atmospheric deposition of different compounds have been analysed in many investigations (e.g. Anon. 1984; Anon. 1988; Nilsson 1991).

Many reports published during the past few years deal with the hypothesis that the increased concentrations of CO_2 in the atmosphere should have influenced the growth rate of forest stands. At present, there are different opinions among researchers regarding the magnitude of the effects that increasing levels of CO_2 have had on forest yield. CO_2 effects were considered to be of only minor importance when interpreting the causes of the demonstrated increases in site productivity.

Deposits of nitrogen are largest in the southwestern parts of Sweden (e.g. Brodin 1990). Since the increase of site indices for both Scots pine and Norway spruce have also been highest in this region it is probable that the increased availability of nitrogen can largely explain why top-height growth was higher than normal there.

According to yield tables, there is a close relationship between site index and productivity in terms of the mean annual increment of stem wood. Based on this relationship, it can be concluded that site productivity has increased in most parts of Sweden during the past 30-40 years. One important factor contributing to the increase in site productivity, at least in southwestern Sweden, is assumed to be the increase in nitrogen deposition. This conclusion corresponds with other hypotheses regarding the most probable factors responsible for increases in forest yield during the past decades (Kenk 1990; Eriksson and Johansson 1993).

References

Anon. (1984) Impacts de l'Homme sur la Forêt. Symposium international. Strasbourg (France) 17-22 septembre 1984

Anon. (1988) Critical loads for sulphur and nitrogen. In: Nilsson J, Grennfelt P (eds) Report from a workshop held at Skokloster, Sweden. 19-24 March,1988

Becker M (1989) The role of climate on present and past vitality of silver fir forests in the Vosges mountains of northwestern France. Canadian Journal of Forest Research. 19(9): 1110-1117

Becker M, Bräker O-U, Kenk G, Schneider O, Schweingruber F-H (1990) Kronenzustand und Waldwachstum von Waldbäumen an Driländereck Deutschland-Frankreich-Schweitz in den letzten Jahnzehnten. Allgemeine Forstzeitung 45: pp 263-266, pp 272-274

von Bertalanffy L (1941) Stoffwechseltypen und Wachstumstypen. Biol. Zentralbl. 61: 510-532

Björkdahl G, Eriksson H (1989) Effects of crown decline on increment in Norway spruce [Picea abies (L.) Karst.] in southern Sweden. In: Brække F, Bjor K, Halvorsen B (eds) Air pollution as stress factor in the Nordic forests. Meddelser fra Norsk institutt for skogforskning 42: 19-36

Brodin Y-W (1990) Belastning av svavel och kväve. In: Brodin Y-W (ed) Effekter av svavel- och kvävebelastning på skogsmark, yt- och grundvatten. Swedish National Protection Board Report, Stockholm. 3762, pp 17-33

Eriksson B (1981) The "potential" evapotranspiration in Sweden. Swedish meteorological and hydrological institute. Report 28 (in Swedish)

Eriksson B (1982) Data concerning the air temperature climate of Sweden. Normal values for the period 1951-80. Swedish meteorogical and hydrological institute, Report 39 (in Swedish)

Eriksson B (1983) Data concerning the precipitation climate of Sweden. Normal values for the period 1951-80. Swedish meteorological and hydrological institute, Report 28 (in Swedish)

Eriksson H (1970) On measuring errors in tree height determination with different altimeters. Royal college of forestry, Department of forest yield research, Research Notes 19 (in Swedish)

Eriksson H, Johansson U (1993) Yields of Norway spruce (Picea abies [L.] Karst.) in two consecutive rotations in southwestern Sweden. Plant and Soil 154: 239-247

Grennfelt P, Hultberg H (1986) Effects of nitrogen deposition on the acidification of terrestrial and aquatic ecosystems. Water, Air, Soil Pollut. 31: 945-964

Hallbäcken L, Tamm O (1986) Changes in soil acidity from 1927 to 1982-84 in a forest area of south-west Sweden. Scandinavian Journal of Forest Research 2: 219-232

Hägglund B (1972) Site index curves for Norway spruce in northern Sweden. Royal college of forestry, Department of forest yield research, Report Notes 21 (in Swedish)

Hägglund B (1973) Site index curves for Norway spruce in southern Sweden. Royal college of forestry, Department of forest yield research, Research Notes 24 (in Swedish)

Hägglund B (1974) Site index curves for Scots pine in Sweden. Royal college of forestry, Department of forest yield research, Research Notes 31 (in Swedish)

Kenk G (1990) Effects of air pollution on forest growth in southwestern Germany - hunting for a phantom? Proceedings, Division 2, XIX IUFRO World Congress, Montreal, Canada

Kenk G, Spiecker H, Diener G (1991) Referenzdaten zum Waldwachstum. KfK-PEF Forschungsbericht 82

Näslund M (1936) Die Durchforstungsversuche der Forstlichen Versuchsanstalt Schwedens in Kiefernwald. Primärbearbeitung, Meddelanden från statens skogsförsöksanstalt 29(1) (in Swedish)

Nilsson L-O (1991) Investigations of impact of air pollution on tree vitality, forest damage and forest production. The Skogaby project. Biomass performance after three years of treatment. IUFRO and ICP-forest workshop on Monitoring Air Pollution Impact on Permanent Sample Plots, Data Processing and Results Interpretation. Prachatice, Czechoslovakia, September 1991

Odén S (1968) The acidification of air and precipitation and its consequences for the natural environment. Bull.Ecol.Comm. Stockholm No.1: 1-86 (in Swedish, English translation)
Richards FJ (1959) A flexible growth function for empirical use. Journal of Experimental Botany 10(29): 290-300
Spiecker H (1995) Growth dynamics in a changing environment - long-term observations. In: Nilsson LO, Hüttl RF, Johansson UT (eds) Nutrient Uptake and Cycling in Forest Ecosystems. Proceedings of the CEC/IUFRO Symposium. Halmstad, Sweden, June, 7-10, 1993. Plant and Soil, Volumes 168-169: 555-561
Spiecker H, Mielikäinen K, Köhl M, Untheim H (1994) Growth trends of European forests. Has site productivity changed? European Forest Institute, Working paper 4
Sverdrup H, Warfvinge P (1993) Effect of soil acidification on growth of trees, grass and herbs, as expressed by the (Ca+Mg+K)/Al ratio. Lund Institute of Technology, Department of Chemical Engineering II, Internal report

Increasing Site Productivity During Consecutive Generations of Naturally Regenerated and Planted Beech (*Fagus sylvatica* L.) in Denmark

J.P. Skovsgaard and H.A. Henriksen
Danish Forest and Landscape Research Institute
Hørsholm Kongevej 11. DK-2970 Hørsholm

Abstract

Two large-scale silvicultural experiments on natural regeneration of beech in Denmark demonstrated a significant increase in site productivity as measured by height growth. In both cases there was an increase from the previous to the present generation and, in one case, also during the development of the present generation. The increase was approx. 3.6 m (at a reference age of 100 years) during the period 1920s-1990. The conclusion is sustained by comparisons with the height growth of beech planted after beech.

1. Introduction

One of the major concerns in sustainable forestry is to maintain or even improve site productivity. Because it is largely dependent on climate, soil and topography, site productivity is an inherent characteristic of the location. For practical purposes, forest site productivity may be defined broadly as the timber production potential of a site for a particular tree species or forest type (see e.g. Vanclay 1994). In consequence, forest site productivity is usually considered in terms of present or potential tree growth rather than in terms of abiotic factors that account for the production potential.

The concern for site productivity relates to the long-term yield capacity of the soil and more specifically involves issues such as the effects of species and silvicultural practices on soil and other site factors, the effect of consecutive generations of the same tree species or forest type, and the direct or indirect effects of air pollutants and climate changes. With the high forest silviculture system, consecutive rotations of the same species or forest type seem potentially to affect site productivity in different directions. For Norway spruce (*Picea abies* [L.] Karst.), for example, there are examples of declining (Wiedemann 1925), more or less constant (Holmsgaard et al. 1961) and increasing site productivity (Eriksson and Johansson 1993; Elfving and

European Forest Institute Research Report No. 5
Growth Trends in European Forests
Edited by H. Spiecker, K. Mielikäinen, M. Köhl and J. P. Skovsgaard

Nyström 1996). The temporal trend evident in these examples may or may not be real, but emphasizes the complexity of the issue in question.

For beech (*Fagus sylvatica* L.) in Denmark, there are indications that site productivity has increased during this century (Andersen 1982; Holstener-Jørgensen 1958; Sabroe 1958a). This conclusion has been based mainly on comparisons between management-level inventory data and the standard site index curves of Møller (1933). Critics may argue that the possible increase is an artifact due to unreliable height measurements (Møller 1958 *vs.* Sabroe 1958b), or due to biases in the site classification system [these are evident both from the authors' practical experience, and in Møller and Nielsen's (1953) evaluation even though they, by misjudgement, explained away the bias]. In addition, it has been debated whether the method of regeneration or stand establishment may influence the attainable site class, i.e. whether a naturally regenerated beech stand is generally of a lower site class than a planted stand (Jakobsen 1989, 1990; *vs.* Billeschou et al. 1990).

Since beech is of major commercial and environmental importance in Denmark as well as on the European continent, its site productivity is an essential issue for forest management, particularly in relation to long-term planning and silviculture. The detection of changes or developments in forest site productivity requires long-term observations, but those at hand often originate from sources that have not been established with this type of problem as the primary objective. Like many others, the present case study is not suited to reveal causal relations or to filter out any individual effects. Despite methodological controversies, case studies are nevertheless relevant, as they may indicate possible trends and help identify favourable paths for further research.

2. Material

The material in this investigation originates from two silvicultural experiments on natural regeneration of beech in the growth region of East Jutland, Denmark. The experiments are located in Stagsrode Forest and Als Nørreskov (Fig. 1). Results for the naturally regenerated stands are compared with observations on planted beech stands of similar age and growth conditions and located nearby in the same forest or, in some cases, in other forests on the island of Als. All stands are even-aged and of local provenance (except one planted stand on Als, originating from the central Carpathians, but equal in growth development to nearby planted stands of local origin).

Growth conditions in this region are optimal for beech (current annual increment of approx. 12 m^3 ha^{-1} yr^{-1} for a 100-120-year rotation). Mean annual temperature is 8 °C with an average minimum of 0 °C and an average maximum of 16 °C. The precipitation is approx. 700 mm y^{-1}, with 300 mm in the growth period (May-September). The soils are sandy loam to sandy clay loam till. The topography is moderately undulating.

Fig. 1. Location of Stagsrode Forest (S) and Als Nørreskov (A).

The areas of Stagsrode Forest and Als Nørreskov have probably been forested for several centuries (cf. maps by the Royal Danish Academy of Science). The experiments were started in the mid 1920s at or shortly after the initiation of regeneration. The stands were regenerated under the uniform shelterwood system (see e.g. Matthews 1991 or Röhrig and Gussone 1990). The silvicultural regime in the present stands is characterized by thinnings predominantly from below, favouring high quality trees, with a target residual basal area of approx. 24 m^2 ha^{-1}.

The Stagsrode experiment covers approx. 30 ha. The regeneration was initiated when the age of the previous stand was approx. 100 years from seed, and the regeneration period lasted 46 years (between appearance of the first seedlings and removal of the last seed trees). The thinning grade just prior to regeneration was at an intermediate level. The regeneration established itself spontaneously without site preparation.

The Als experiment covers approx. 8 ha. The regeneration was initiated when the age of the previous stand was 130 years from seed, and the regeneration period lasted 18 years. The thinning grade just prior to regeneration was light. The regeneration was planned in advance and included site preparation.

For this study, the Stagsrode material comprises 4 naturally regenerated plots and 5 observations of planted beech. The Als material comprises 5 naturally regenerated plots and 4 observations of planted beech.[1] Further details and results for the Stagsrode material is given by Bornebusch (1946) and Henriksen and Bryndum (1996), and for the Als material by Krarup (1946) and Henriksen (1996).

3. Methods

Due to lack of information on the younger stages of the previous generation of beech, the methodological approach in this study is limited to comparisons of height growth and height growth patterns. Stand height is expressed as the height corresponding to the tree of mean basal area, H_g. Data are based on direct height measurements on site, supplemented and verified by measurements of felled trees and by stem analyses.

Height developments are compared with two very similar growth models for beech in this region, one established by Oppermann (1914) and based mainly on material from the 19th century, the other established by Henriksen and Bryndum (1996) and based mainly on material from the 20th century, but including older material as well. Site index refers to stand height, realized or predicted, at the reference age 100 years from seed. The experiments in this study were not used for model construction.

Stands and measurements are chosen, as far as possible, so as to cover approximately the same span of ages, and in any case representing the same silvicultural era, around the turn of the century characterized by perfectionistic regenerations. Since all stands are not of the same age, in particular the planted stands, the height-based site index is used as a means for comparison.

To avoid dependence between consecutive observations, temporal changes in site index within locations and methods of regeneration are statistically evaluated by the paired t-test (i.e. testing whether observed changes in site index differ from zero). For the naturally regenerated stands, parts of the original 1920s field notes have been lost and height measurements cannot be traced to individual plots. In these cases, a t-test could rely on an external estimate of the variance, but it has not been possible to provide reliable estimates. This does not, however, detract from the clarity of the results.

---

[1]For the Als material, 1920s-measurements in one planted stand are excluded due to poor silviculture for the previous generation (exploitation thinnings, waterlogging etc.). Likewise, 1962-measurements in another planted stand are excluded because the stand age - height combination is outside the valid range of the growth model.

4. Results

Table 1 shows a comparison of the height development for the present generation of naturally regenerated beech. The Als experiment had a slower start than Stagsrode, at the age of 35 they were approximately equal, but in later years height growth in the Als experiment has been exceptionally good. The experiment in Stagsrode followed very closely the height development described by the growth models. The fact that the 1914-model and the 1996-model almost coincide serves as an indication that there probably is no significant built-in trend in the present height growth of beech in Denmark. So. at least height growth is hardly affected by changes in environmental factors such as air pollution or climate. The late occurrence of difference in site index contradicts the general experience that site class differences (by height) originate at an early stage. Also, unexpectedly, there seems to be no negative shelter effect from the previous generation of beech on the early height development of the present generation.

Table 2 shows the site index developments for the present generation of naturally regenerated and planted beech during the period 1962-90. For Stagsrode, the site index is constant over time and unaffected by regeneration method. The variance is considerably larger for the planted stands, reflecting their less homogeneous origin and development. For Als, there is a 2.6 m increase in site index for both regeneration methods. In this case the variance for naturally regenerated beech is comparatively large, reflecting site heterogeneity due to a relatively hilly topography.

Table 3 shows the site index development during two consecutive rotations of beech. During the period 1920s-1990 there was an increase at approx. 3.5-3.7 m for the naturally regenerated beech, and approx. 3.9 m for planted beech on Als. The increase for planted stands is not statistically significant, but in all cases amounted to at least +2.0 m. The variation within the naturally regenerated experiments remains unknown, but for planted beech on Als it is quite large compared to the period 1962-90. This probably reflects the large variation in local silvicultural conditions in the 1920s following German administration during 1864-1920 (which implies "good" silviculture) and the First World War (which resulted in exploitation-

Table 1. Height development for the present generation of naturally regenerated beech (site index 30 according to the 1996-model).

	Height (m)		
Age (yrs)	Als	Stagsrode	SI 30
8	0.4	1.8	-
18	3.8	6.0	6.5
35	12.4	12.9	13.3
67	25.7	22.8	23.7

Table 2. Site index development for the present generation of beech (site index according to the 1996-model).

	Obs.	Age period (averages)	Average site index		Change[1]	
			1962	1990	Mean	Std.dev.
Stagsrode						
Natural regeneration	4	38-67	29.4	29.5	0.10^{NS}	0.356
Planted	5	44-73	29.6	29.6	0.00^{NS}	1.214
Als						
Natural regeneration	5	35-67	29.5	32.1	2.66***	0.658
Planted[2]	3	27-55	31.1	33.7	2.63*	0.666

[1] Average change in site index, calculated and tested as described in text.
*, ** or *** Significant at 5 %-, 1 %- or 0.1 %-level, respectively.
^{NS} Not significant.
[2] Two planted stands measured 1966 and 1991.

Table 3. Site index development during two consecutive rotations of beech (site index according to the 1996 model).

	Obs.	Average site index		Change[1]	
		Previous gen. 1920s (age 100+)	Present gen. 1990 (age c. 70)	Mean	Std.dev.
Stagsrode					
Natural regeneration	4	c. 26	29.5	3.5	-
Als					
Natural regeneration	5	28.4	32.1	3.7	-
Planted	3	29.3	33.2	3.90^{NS}	4.057

[1] Average change in site index, calculated and tested as described in text.
^{NS} Not significant (P=0.170).

like conditions in forests near towns). For example, the planted stand that was excluded from this part of the analysis, located close to the major town on Als, suffered from exploitation thinnings and waterlogging during the 1920s, resulting in an extremely large increase in site index at approx. 10 m until the present generation in 1990. This example clearly demonstrates the overall importance of silviculture for making the most of forest growth potential, and in the interpretation of growth trend analyses.

5. Conclusion

In conclusion, these two large-scale silvicultural experiments on natural regeneration of beech in Denmark demonstrate a significant increase in site productivity as measured by height growth. In both cases there was an increase from the previous to the present generation and, in one case, also during the development of the present generation. The increase was approx. 3.6 m (at a reference age of 100 years) during the period 1920s-1990. The conclusion is sustained by comparisons with the height growth of beech planted after beech.

6. Discussion

Detecting changes or developments in forest site productivity requires long-term observations and may imply some controversy because potentially influential and confounding factors mix and often exceed a reasonable number in relation to quality as well as quantity of observations. This is further stressed by the fact that such observations often originate from sources that have not been established with this type of problem as a primary objective. The discussion focuses briefly on examples of these issues relating to the present case study.

In case studies with scarce data, the interpretation of growth trend results relies heavily on adequate and representative models of growth or other reference standards. Using a growth model as a standard may smooth irregularities in individual stands but, on the other hand, whether stated explicitly or not, growth models rely on certain assumptions that may either imply or remove a real trend. This stresses the need for an awareness of historical silviculture and other relevant past conditions for growth, to better comprehend what the reference standard represents. Despite great care in the selection of material, the historical development of natural and management related conditions for silviculture no doubt influences observed changes in site productivity, as exemplified by observations left out for this particular reason. More generally, this concerns factors such as forest structure, its influence on the physical environment and on ecosystem function, tree species composition, utilization of the forest for other purposes than timber production (e.g. cattle grazing, litter removal), soil conditions and administrative relations.

Other important issues are: - How typical is the growth pattern used for comparison? - Could there possibly be changes in growth pattern over time? This further relates to possible temporal changes in height growth-volume growth correlations, indicating that conclusions may depend on choice of variables.

The time or age at which site class changes occur is another unresolved matter. When site class differences occur in the early stages followed by more or less parallel developments in later stages, better tools or indicators of site productivity than available at present for young stands and/or better monitoring of young stands are needed.

Finally, overlaying these effects and mechanisms determined by natural growth conditions and forest management, is the possible long-term role of air pollution and of climatic changes and fluctuations. Initiated in part by the global change issue, it is nevertheless impossible to detect or quantify these most likely influential factors in a study like this.

References

Andersen KF (1982) Er der sket en økologisk forringelse af skovene? En gennemgang af 3 nordsjællandske skoves nyere dyrkningshistorie. Skovreguleringen, internt papir. p 49

Billeschou AE, Holten-Andersen P, Madsen P (1990) Bøgeselvforyngelser contra bøgeplantninger. Skoven 22: 211-214

Bornebusch CH (1946) Bøgeskovens Behandling på Boller Skovdistrikt. Det forstlige Forsøgsvæsen i Danmark 19: 1-80

Elfving B, Nyström K (1996) Yield capacity of planted *Picea abies* in northern Sweden. Scandinavian Journal of Forest Research 11: 38-49

Eriksson H, Johannson U (1993) Yields of Norway spruce (*Picea abies* [L.] Karst.) in two consecutive rotations in southwestern Sweden. Plant and Soil 154: 239-247

Henriksen HA (1996) Bøgeforyngelser i skove på Als. Forskningscentret for Skov & Landskab, Forskningsserien 13: 163-210

Henriksen HA, Bryndum H (1996) Bøgeforyngelser i Stagsrode skov. Forskningscentret for Skov & Landskab, Forskningsserien 13: 5-162

Holmsgaard E, Holstener-Jørgensen H, Yde-Andersen A (1961) Bodenbildung, Zuwachs und Gesundheitszustand von Fichtenbeständen erster und zweiter Generation. 1. Nord-Seeland. Det forstlige Forsøgsvæsen i Danmark 27: 1-167

Holstener-Jørgensen H (1958) Lidt om bøgedyrkningens fremtidsmuligheder bedømt på grundlag af udviklingen i Rude skov. Dansk Skovforenings Tidsskrift 43: 400-408

Jakobsen B (1989) Bøgeforyngelser i dansk skovbrug i de sidste 200 år. Det forstlige Forsøgsvæsen i Danmark 42: 233-265

Jakobsen B (1990) Svar ang. bøgedyrkning. Skoven 22: 215

Krarup F (1946) Langsom Bøgeselvforyngelse. Det forstlige Forsøgsvæsen i Danmark 19: 81-104

Matthews JD (1991) Silvicultural systems. Clarendon Press, Oxford, p 284

Møller CM (1933) Boniteringstabeller og bonitetsvise Tilvækstoversigter for Bøg, Eg og Rødgran i Danmark. Dansk Skovforenings Tidsskrift 18: 457-513; 537-623

Møller CM (1958) Bøgeboniteterne paa Boller. Dansk Skovforenings Tidsskrift 43: 409-417

Møller CM, Nielsen J (1953) Afprøvning af de bonitetsvise tilvækstoversigter af 1933 for bøg, eg og rødgran i Danmark. Dansk Skovforenings Tidsskrift 38: 1-176

Oppermann A (1914) Elleve Prøveflader i Bøgeskov. Det forstlige Forsøgsvæsen i Danmark 4: 189-268

Röhrig E, Gussone HA (1990) Waldbau auf ökologischer Grundlage: Baumartenwahl, Bestandesbegründung und Bestandespflege. Verlag Paul Parey; Hamburg, Berlin. 6. Aufl., p 314

Sabroe AS (1958a) Bøgeboniteter. Dansk Skovforenings Tidsskrift 43: 248-257

Sabroe AS (1958b) Bøgeboniteterne paa Boller. Dansk Skovforenings Tidsskrift 43: 469-471

Vanclay JK (1994) Modelling forest growth and yield. CAB International, p 312

Wiedemann E (1925) Zuwachsrückgang und Wuchsstockungen der Fichte in den mittleren und unteren Höhenlagen der sächsischen Staatsforsten. W. Laux, Tharandt. 2. Aufl.

Height Growth Investigations of Norway Spruce (*Picea abies,* [L.] Karst.) in the Eastern Part of Germany During the Last Century

Günter Wenk and Michael Vogel
Institut für Waldwachstum und Forstliche Informatik, TU Dresden, Germany

Abstract

From 1970 to 1974, an inventory including cutting of sample trees took place in the low mountain ranges of Eastern Germany. Altogether the height growth of 2277 spruces from 448 selected stands was analyzed.

The results show a change in height growth. The young stands appear to have a better height growth, the growth pattern having changed. This process began at the turn of the century and continues to the present day. The young stands have an early culmination with higher increment values than had the older reference samples some decades ago. When aged 60, however, height growth begins to diminish and the height growth curves of young and old stands begin to converge. In contrast to the height development of young stands, old stands had a slow phase of juvenile growth with small but nearly constant height increment over a long time period. A trend to site class changes of old stands within one rotation has not been established by these investigations.

1. Introduction

Norway Spruce is the most important tree species in Saxony and Thuringia, pure spruce stands may be found in the region.

It is known from inventory data (e.g. Kurth and Landbeck 1964, Trebing 1951) and height shoot measurements of trees on temporary experimental plots (e.g. Hofmann 1959) that young stands have better site classes than old stands. Possible causes of site class changes are discussed for example by Wenk et al. (1990).

The question: do the present-day young stands grow better than did young stands 50...100 years ago? has been answered in part by our investigations. An inventory with cutting of sample trees took place in the low mountain ranges of Eastern Germany from 1970 to 1974. The inventory data were supplemented by previous data (archival material of experimental plots and stem analyses).

European Forest Institute Research Report No. 5
Growth Trends in European Forests
Edited by H. Spiecker, K. Mielikäinen, M. Köhl and J. P. Skovsgaard

2. Methods

Altogether 2277 sample trees from 448 selected stands were analyzed:

Table 1. Number of sample plots and trees

Method	District	Number of plots	Number of trees
Height analysis	Thüringer Wald	123	885
	Erzgebirge	63	436
Height shoot measurement	Thüringer Wald	151	425
	Erzgebirge	34	164
	Lausitz	46	147
Stem analysis	Erzgebirge	30	196
	Lausitz	1	24
Total		448	2277

The stands for height analysis were selected from the forest management plans, using random sampling with the following conditions:

- Pure spruce stands under regular management of a minimum area of 0.5 ha without extreme damage (air pollution, snow, wind) and
- Stands with uniform site conditions are preferred.
- If possible uniform distribution among site classes is intended.
- All age classes (except very young or old) are included.

The number of sample trees for height analyses depends on the stand area:

0.5 - 1.0 ha	6 to 8 sample trees
1.0 - 3.0 ha	8 to 10 sample trees
> 3.0 ha	12 sample trees

The sample points were uniformly distributed within the stands and the stand was mapped. Every sample point was approached by means of compass and paces. Each ten trees in the immediate vicinity of such a point were measured. The tree greatest in diameter at breast height was cut and analyzed as sample tree. The growth curves of each height analysis of dominant tree in one stand were averaged. These dominant height growth curves were used for further investigations. Since about 150 years it has been assumed that dominant trees in old stands have had this social class ever within their life. It is not easy to prove this hypothesis by experiments: it would be necessary to have observations about many decades (≥80 years) in long-term experimental plots. Then the known tree-growth of different social classes can be compared with growth pattern of stem analyses.

The results of other investigations tend to support this assumption (e.g. Franz, 1971; Wenk, 1973).

In order to answer the question, whether or not a single stand changes its site class in one rotation (see Fig. 1), a site class assessment independent of yield tables is proposed.

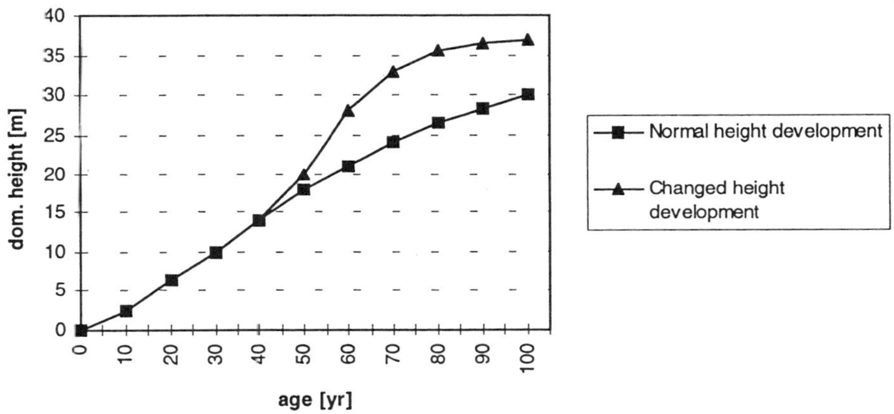

Fig. 1. Comparison of schematic curves showing normal height development and height growth with a "jump"

We have gained experience with respect to several functions (e.g. Korf, Mitscherlich, Richards, Sloboda, Weibull). However, the aptitude of these functions should not be discussed here. The Weibull function with two parameters (Wenk et al. 1990; Wenk 1991) is used because the parameters are easy to interpret and the application of this function is simple.

Within the function $$B = h_{max} * \left(1 - \left(1 - \frac{h}{h_{max}}\right)^{(100/t)^{c}}\right)$$

where the symbols mean

B is site class (mean or dominant height at an age of 100 years),

h_{max} is final height for $t \rightarrow \infty$,

h is mean or dominant height at tree age t,

t is age at measurement,

c is exponent.

The exponent value has remained constant as shown for the yield table collection by Ertelt (1963) and for data from long-term experimental plots. A good approximation is reached with c = 1.2 in more than 80% of investigated data. Consequently, the final height h_{max} is the only unknown parameter. First of all, an arbitrary, but real starting value is selected. By means of function (1), the site class is calculated for each age for which measuring values are available. The value of h_{max} is a suitable approximation as to the final height. If it is impossible to obtain a constant site class (e.g. upper curve in Fig. 1), this may be an indication of growth trend changes due to environmental impacts.

3. Results

The following tendency becomes obvious from the investigations:

The height growth curves of older stands have smaller increment values in the juvenile phase than present-day young stands. (Examples could be found for an inversion of this growth sequence, yet the majority of analysed stands corresponds to this pattern.) Therefore, the young stands have better site classes and their height growth curves lie above those of old stands at the same age. At the age of 60, the height growth of present-day stands decreases when compared to that of present-day old stands at a comparable age. Using a common time axis, the height growth curves of young and old stands have the greatest differences between the ages of 40 and 80. Then the curves draw closer together. The differences between stand heights at an age of 50 are greater than those at an age of 100 years (Fig. 2).

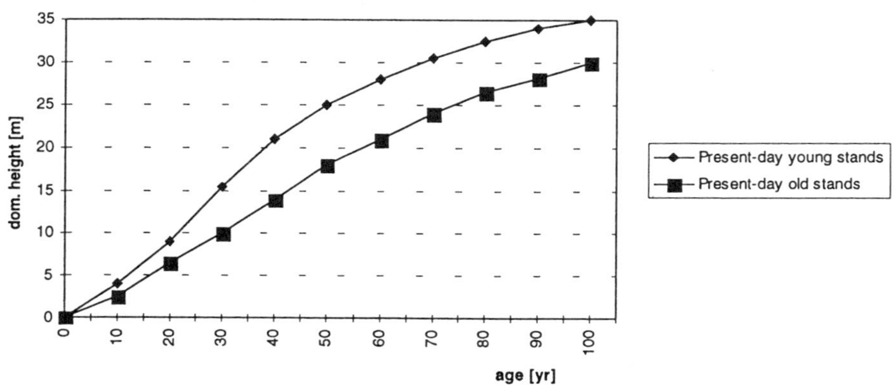

Fig. 2. Schematic height growth comparison of present-day old and young stands at the same age

Today, young stands grow faster, with an early culmination and large increment over the first decades. In contrast to this, old stands had a slow phase of juvenile growth with small but nearly constant height increment over a long time period.

The growth rate as a function of age has changed. This tendency is illustrated by Figs. 3, 4 and 5. Two typical forest districts (Heinzebank in Saxony and Bermbach in Thuringia) have been selected from the multitude of data; however, the same result may be obtained using stands from different forest districts within the research area.

Figure 3 shows the results from the forest district Heinzebank up until the year of analysis in 1974. In this figure, stands have been selected which lie in three neighbouring parcels (17, 18 and 20).

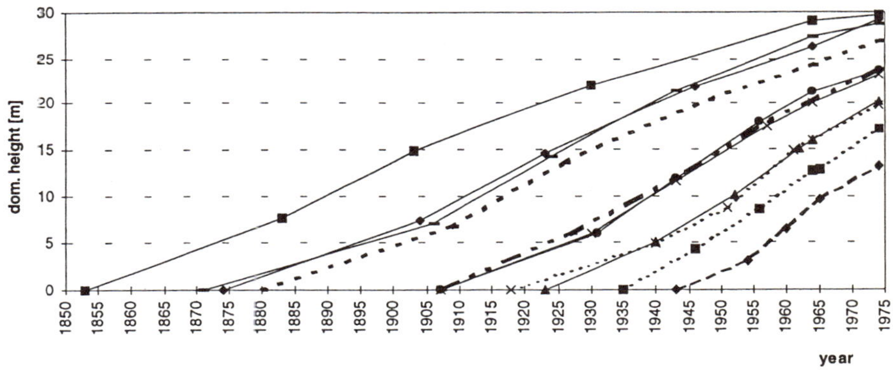

Fig. 3. Height growth curves of spruce stands sampled in 1974 (forest district Heinzebank, parcels 17, 18 and 20)

Younger stand curves show a steeper increase than older stands. This may be seen more clearly in Fig. 4, where the curves from Fig. 3 are traced back to the common age of zero. A schematic summary of these results is given in Fig. 5.

Meanwhile, dominant heights of 20 to 25 meters are measured in 50-year-old long-term experimental plots. This observation bases on measurements gathered from about 220 long-term experimental plots (for example plot Sosa 51a[4] , published by Wenk 1995).

While differences in height growth between young and old stands become discernible, growth rate changes within one rotation of one stand are rarely noted.

The site class assessment independent of yield tables described under point 2 with the Weibull function (1) is represented on two plots:

Both examples seem to be constant in site classes (site classes are given in integers); therefore height developments for these plots thus remain unchanged.

This result has been obtained for the institute's medium-aged plots as well as for its older experimental plots.

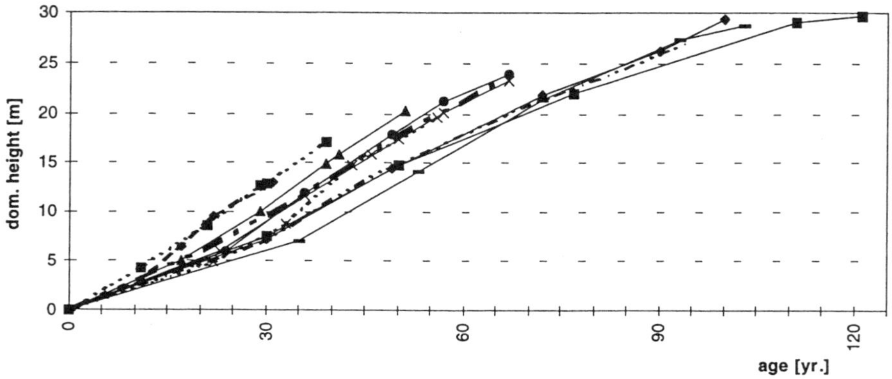

Fig. 4. Presentation of height curves from Figure 1 with a common initial point

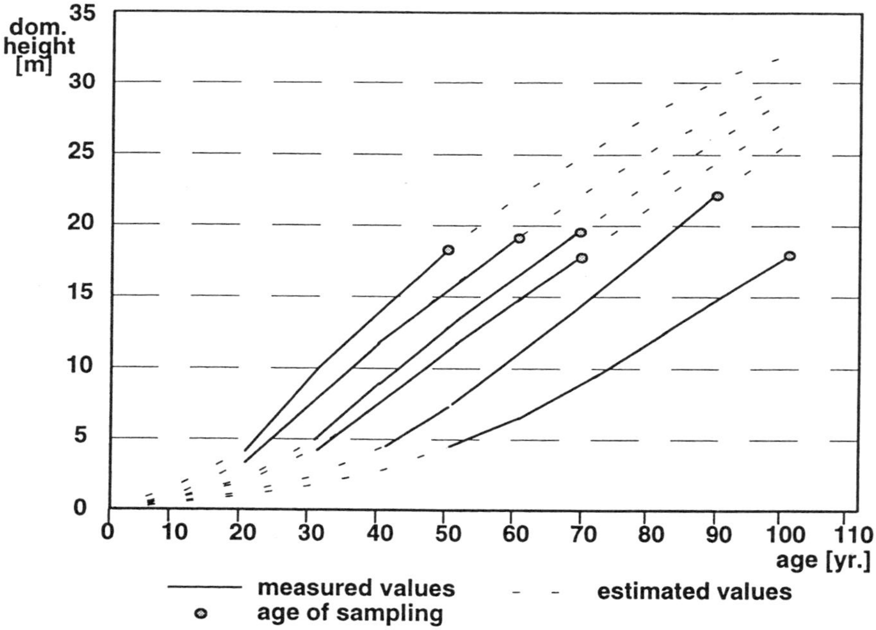

Fig. 5. Height analyses of pure spruce stands with different ages (forest district Bermbach)

Table 2. Site class assessment using the two-parametrical Weibull function

Spruce permanent experimental plot Sachsenried 2/II (Assmann, 1961, p. 150) c = 1.3 hmax = 44m			Spruce temporary experimental plot Heinzebank 17b[1] (see Fig. 2 and 3) c = 1.4 hmax = 46m		
age [years]	dom. height [m]	site class[2] [m]	age[3] [years]	dom. height[3] [m]	site class[2] [m]
32	15.9	38	30.2	7.3	28
37	19.2	39	49.3	14.5	29
42	22.8	39	72.1	21.9	29
47	24.5	39	90.1	26.3	29
52	25.9	38	100.1	29.3	28
58	27.1	38			
69	30.7	38			
74	32.5	38			
79	34.1	38			
84	35.6	38			
95	37.6	38			
101	38.8	38			
107	39.6	39			

[2] Dom. height at an age of 100 y.
[3] Means of 11 trees

4. Conclusions

The changes in height growth trends describe a process which had begun at the turn of the century and continues until today. Determining "bad" or "good" height growth is not the point, rather is it necessary to interpret better height growth as a new growth pattern or a change in growth rate. Earlier yield tables did not take such changes into consideration. Therefore, growth over long time periods should not be prognosticated for young stands using older yield tables. In the future, suitable growth models will be used more and more for prognostic tasks.

Possible causes for growth trend changes are mentioned in Wenk et al. (1990):

- change from seed to plantation,
- wide spacing of plants,
- intensive tending of stands,
- deposition of nitrogen,
- possible long-term change of temperature or
- possible increase of CO_2 in atmosphere.

Furthermore, Mitscherlich (1955) referred to the fact that in the last century the restructuring of broadleaved wood stands to conifer stands has been carried out firstly on poor sites and later on better sites.

If growth changes are induced by global changes (e.g. climate, CO_2) reference stands will not be available. Methods must therefore be developed that will enable us to deduce growth changes from the growth development itself; the authors have proposed to use the Weibull-function.

However, traditional methods of yield science seem to be insufficient to explain causes of trend changes. Only cooperation among the various sciences will enable us to answer the question of why young stands have changed their growth pattern.

References

Assmann E (1961) Waldertragskunde. BLV Verlagsgesellschaft, München-Bonn-Wien, p 490

Ertelt W (1963) Ertragstafelauszüge für den Gebrauch in der Praxis. Neumann Verlag, Radebeul, p 170

Franz F (1971) Grundlagen und Verfahren standortsbezogener Leistungsschätzung. Forstliche Forschungsanstalt, Forschungsbericht, München

Hoffmann J (1959) Untersuchungen über das Wachstum der Fichte auf einigen Standortsformen des süd-östlichen Thüringer Waldes. Archiv für Forstwesen, Bd. 8 - Heft 6/7, pp 536 - 591

Kurth H, Landbeck H (1964) Verbreitung, Ertags- und Altersklassen der Fichte im Mittelgebirge und Hügelland der Deutschen Demokratischen Republik. Wiss. Zeitschr. TU Dresden, 13. Jg. - Heft 5, pp 1477-1488

Mitscherlich G (1955) Zur Frage der dynamischen Bonitierung. Allg. Forstz., 10. Jg. - Heft 14, pp 175-177

Spiecker H, Mielikäinen K, Köhl M, Untheim H (1994) Growth Trends of European Forests - has site productivity changed? - European Forest Institute, Working Paper No. 4, p 68

Trebing R (1951) Ertragstafeln in der Bewährungsprobe. Forstwirtschaft-Holzwirtschaft, pp 80 -83

Wenk G (1973) Ertragskundliche Grundlagen der Bestandesbehandlung. TU Dresden, Sektion Forstwirtschaft, Lehrbrief

Wenk G (1991) Ertragstafelunabhängige Bonitierung mit der Weibullfunktion. In: Sloboda B, Smelko S (eds) Biometrische Beiträge zu statischen und dynamischen Modellansätzen in den Forstwissenschaften und der Praxis. Deutscher Verband Forstlicher Forschungsanstalten, Sektion Forstliche Biometrie und Informatik, Martin/Zvolen 16.-20.9.1991, pp 110 - 116

Wenk G (1995) Der Eibenstocker Standraumversuch. Der Wald 45: 54-57

Wenk G, Antanaitis V, Smelko S (1990) Waldertragslehre. Deutscher Landwirtschaftsverlag, p 448

Wenk G, Römisch K, Gerold D (1985) DDR-Fichtenertragstafel 1984. Agrarwissenschaftliche Gesellschaft der DDR, p 64

Growth Trends of Forests in Southern Germany

Hans Pretzsch
Chair of Forest Yield Science. University of Munich
Hohenbachernstraße 22, D - 85354 Freising, Germany

Abstract

Results from leaf or needle loss inventories and from investigations on growth trends in German forests have been found to contradict one another: while surveys of leaf loss indicate that tree vitality is declining since the 1980s, growth developments of the main species show an upward tendency since the 1950s, even in forest stands with high leaf loss and lie, at present, considerably above yield table level. Remarkable growth trends are apparent in the network of long-term experimental plots, as evident from results from forest inventories in 1970/71 and 1987 and in growth data based on increment cores and from the stem analyses of some 10 000 trees. While the causes underlying growth trends have already been successfully investigated and subsequently much discussed there is as yet no comprehensive representation of increment levels and increment trends for German forest stands. With the present chapter, an attempt is made to make up for this shortcoming. In spruce stands (*Picea abies* [L.] Karst.) diminishing height and volume growth at higher elevations in the Middle Mountains and the Alps are in sharp contrast to improving growth trends in the lowlands, where annual height increment and volume growth reach about 250% of yield table levels. On pure soils, in particular, pine stands (*Pinus sylvestris* L.) show site improvement manifest in an increase in basal area and volume increment of, again, up to 250% of yield tables levels. Growth decline below the level of commonly used growth models occurs only when leaf loss exceeds 60%, but was found to be rather rare. Beech and oak (*Fagus sylvatica* L. and *Quercus petrea* Liebl.) show a similar increase in height growth, volume increment and standing volume since the 1950s, with maximum values far exceeding the range of yield tables. Both species seem to compensate for leaf loss by additional productivity in the remaining leaves and show no definite growth reactions even in heavily damaged stands (60-99 % leaf loss). Dieback in beech stands sets in only after long phases of disease, whereas it tends to occur rather abruptly in oak stands. For the past two decades repeated forest inventories in Bavaria have diagnosed an increase of 10 - 20% in standing volume and show that species-specific annual volume in-

European Forest Institute Research Report No. 5
Growth Trends in European Forests
Edited by H. Spiecker, K. Mielikäinen, M. Köhl and J. P. Skovsgaard

crement exceeds yield tables by 10-40%. Similar production levels and growth trends in other parts of Germany were recorded through the nationwide network of long-term experimental plots.

1. Growth Reduction and Stand Devastation in Smaller Forest Areas

Extensive studies on growth trends in German forest stands were conducted mainly on account of forest damage in regions subjected to strong immissions in the 1970s. A striking loss of foliage was diagnosed in the Harz Mountains, the Black Forest, the Alps and in the border mountains of North-East Bavaria. Growth and yield investigations concentrated on regions with the highest damage rate. There was, literally, a run on severely damaged stands to examine their specific growth development and the correlation between growth and foliage loss at different degrees of damage. These investigations on growth in damaged stands yielded an informative view on the reactions of different tree types to the effects of immissions (Athari and Kramer 1983; Kenk 1984; Pretzsch 1989; Röhle 1987; Schöpfer and Hradetzky 1984 and 1986; Utschig 1989).

1.1. Spruce Stands

Figure 1 shows the characteristic growth behaviour of spruce stands in North and East Bavaria related to varying degrees of damage. Depicted is the percentage basal area increment at a given degree of needle loss between 30 and 99% - in comparison to that of unaffected trees (0-level). The increment curves of the collectives are not represented as absolute figures but in relation to the average increment within a reference period (1959-1968); hence initial differences in increment level of the collectives can be eliminated. In a second step, the increment development of damaged trees (standardized with the help of the reference period) are compared to the standardized increment trend of undamaged trees. The result is the percental deviation in the increment behaviour of damaged trees from that to be expected under undisturbed conditions (Pretzsch and Utschig 1989). As Fig.1 shows, the real increment collapse starts in the second half of the 1970s. Increment losses amount to between 50 and 60% of the reference increment and, as expected, rise steeply with increasing needle loss.

This comparison of collectives, recommended by the German Association of Forestry Research Stations for the diagnosis of increment trends (DVFFA 1988), serves as a detailed analysis of the increment behaviour in damaged stands. Forest management, in particular, has a keen interest in the study of the deviation in increment behaviour of damaged stands from the growth behaviour depicted in yield tables. Figure 2 shows the growth reduction in damaged spruce stands of the Bavar-

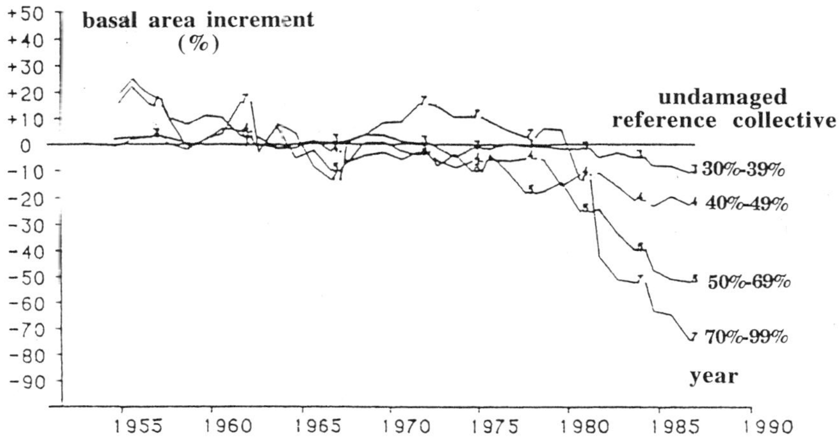

Fig. 1. Trend of annual basal area increment of damaged spruces in North and East Bavaria compared to the unaffected reference collective (0-level). In comparison to the reference collective (n=463), the increments have considerably decreased with increasing foliage loss since the 1970s. The collectives with needle losses of 30-39, 40-49, 50-69 and 70-99% comprise n=257, 75, 34 and 12 trees respectively.

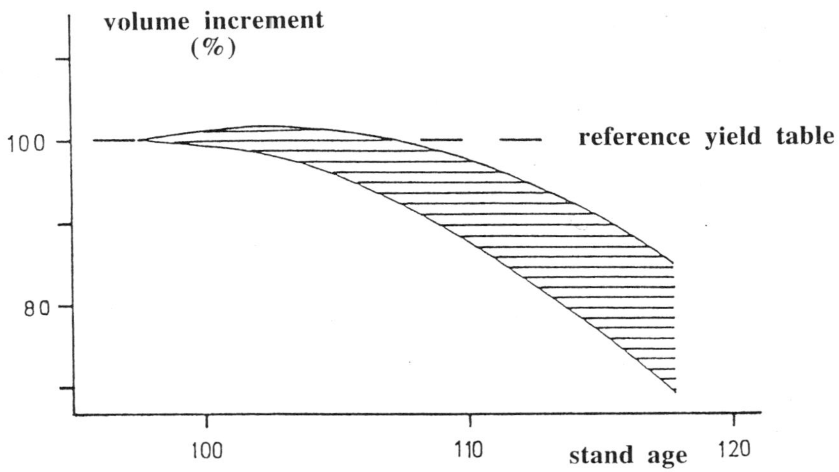

Fig. 2. Development of annual volume increment of damaged spruce stands in the Bavarian forest district Bodenmais compared to the yield table (according to Röhle 1987). Given is the volume increment (100% level). The upper border line of the hatched sector indicates increment loss calculated according to the table by von Guttenberg (1915); the lower border line indicates loss calculated according to the table by Assmann/Franz (1963).

ian forest district Bodenmais according to investigations conducted by Röhle (1987). The increment of stand volume on experimental areas in the border mountains of North-East Bavaria is compared with the growth curves of yield tables (100% level), with tables by Assmann/Franz (1963) and von Guttenberg (1915) used as reference. A comparison between the real increment trends on the experimental areas and yield table values shows that growth rates reached the level of the yield tables at the beginning of the observed growth period, i.e. in the 1960s. In the following years growth rates were significantly reduced. Present growth losses caused by forest damage amount to between 15 and 30% - according to the reference table used. This kind of growth reduction (mostly caused by immission) and the above negative deviations from values to be expected from yield tables cast doubt on whether yield tables are really effective as basic instruments in forest planning. It is well known that similar increment depressions are no occasional phenomena in single spruce stands. At high elevations in the German Middle Mountains, with a high influx of air pollution, all main tree species have, with regional and species-specific differences of course, shown similar reactions during the past 10 to 20 years (Schöpfer and Hradetzky 1984).

1.2. Pine Stands

The growth behaviour of damaged pine stands is comparable to that of damaged spruce: Increment collapse since the 1970s, moderate or strong damage including basal area increment loss up to 50%, and, with heavy needle loss as the prerequisite,

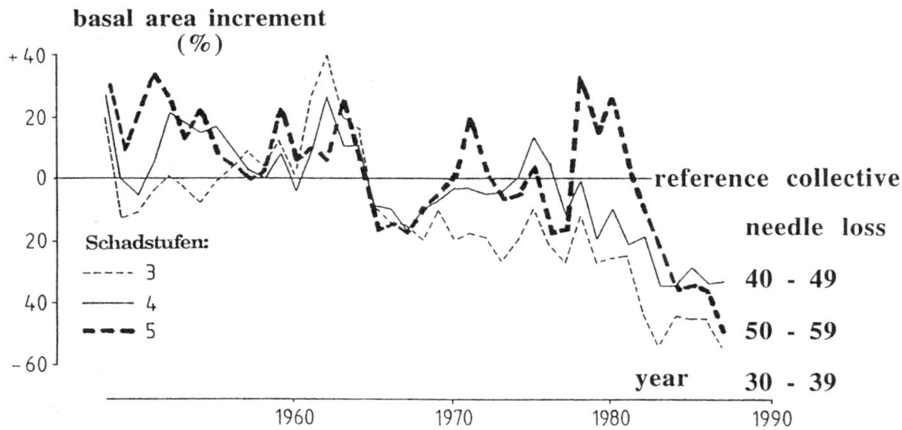

Fig. 3. Development of the current annual basal area increment of damaged pine stands in the forest district Selb in the Fichtelgebirge in relation to growth behaviour of undamaged trees (0-level). Since the mid-1970s, trees of all damage degrees (needle loss between 30 and 59 %) have suffered considerable increment loss (Pretzsch and Utschig 1989).

a correlation between needle density of crowns and increment behaviour (Fig. 3). The development of the basal area increment of damaged pines in the forest district Selb/ Fichtelgebirge shows a particularly serious increment depression and comparatively high increment loss. For pines, there appears to be a particularly high risk of confusing location-related phenotype with damage characteristics. Not in all instances does a close connection exist between estimated needle loss and measured increment development (Franz and Pretzsch 1988).

The comparison of increment trends in damaged pine stands with yield tables leads to some striking results: even damaged stands show growth behaviour that clearly exceeds values given in yield tables; only stands suffering from extreme damage do not come up to yield table expectations. In a schematic representation, Fig. 4 shows the increment behaviour of pine stands as observed on many locations found to suffer from a lack of nutrients in the past such as East German diluvium, Keuper and Jurassic soil and chalk sand soils in Bavaria and Baden-Württemberg. Since the 1950s many stands show an extremely high increment increase. The situation in healthy and damaged stands seems to be the following: in the 1950s and 1960s, i.e. before sustaining any damage, the growth level of stands today diagnosed as having been damaged is, in many cases, between 50 and 100% higher than yield table values. Even after damage began to take effect, growth rates were still higher than those indicated in the yield tables. Also, intensely damaged pine stands suffering from increment decreases between 20 and 30% (compared with unaffected stands) often show growth rates above the values of the yield table; it is only advanced disease that causes growth to drop below yield table level.

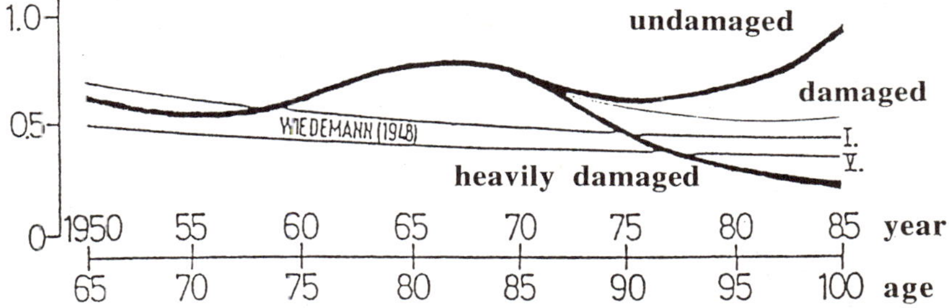

Fig. 4. Development of basal area increment of unaffected, damaged and heavily damaged stands in the Upper Palatinate compared to yield table values of Wiedemann (1948) for moderate thinning, in schematic representation (Pretzsch 1989).

1.3. Beech Stands

For beech, similar investigations failed to prove that a definite correlation exists between leaf loss and increment loss (Fig. 5). There is one exemplary representation of the development of the standardized basal area increment of beech trees in the forest district Kehlheim/Bavaria with and without foliage loss. Increment curves for single stems were related to average increment within the reference period (from 1933 to 1987). Figure 5 shows the following development: first, basal area increments have increased since the 1960s and 1970s, showing a development which is atypical of their age and emphasized by late accretion thinning. Second, this development can be considered a trend which is independent of diagnosed leaf loss. Third, there is no evident correlation between leaf loss and increment loss. Even leaf losses between 60 and 89% can lead to a basal area increment that exceeds that of trees without leaf loss. Compared to spruce, pine and fir, the species beech and oak behave in a manner totally different from the conventional model representation that foliage and increment are correlated.

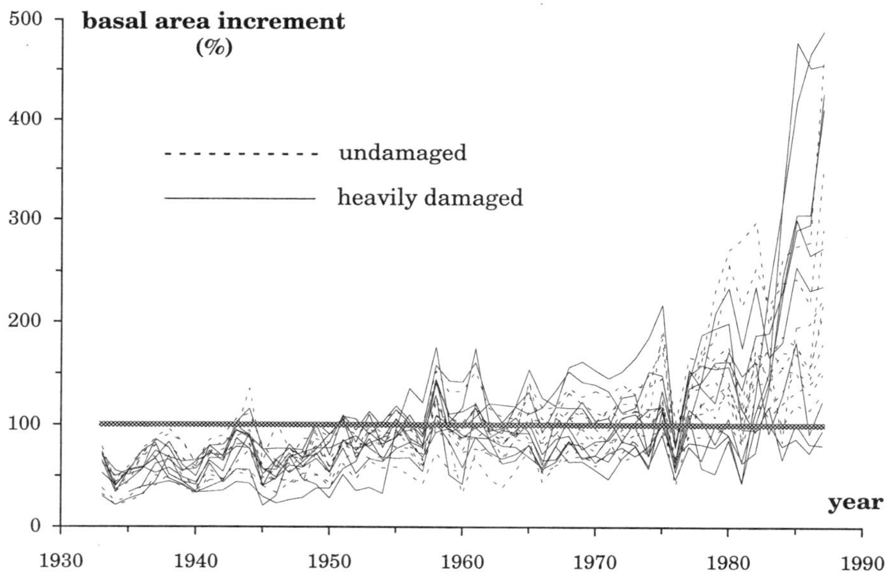

Fig. 5. Standardized development of basal area increment (reference period between 1933 and 1987) of beech suffering from moderate and high leaf loss in the Bavarian forest district Kelheim. Even with high foliage loss, there are no clear correlations between the phenotype and increment behaviour.

2. Widespread Observation of Increment Increase

2.1. Increment Reactions of Different Species on Long-Term Experimental Plots

Growth and yield investigations carried out in forest stands damaged by immissions have contributed towards the recognition that increment has developed along lines which appear to lie outside conventional models. Whenever increment loss had to be diagnosed and quantified, the following questions arose (Pretzsch 1987): what is the actual increment trend - the normal trend to be expected under undisturbed conditions? Are yield tables suitable for the correct assessment of increment depression? How is "normal" foliage density defined? To find answers to these questions, a series of healthy forest stands was observed and examined with regard to growth development. One fact, astonishing initially, proved a reflection of the general development: whereas considerable increment loss could be assessed in smaller forest regions heavily damaged by immissions, larger areas were shown to have developed increment increases for all main species. Since the 1960s and 1970s, it has become apparent that there is a highly positive deviation in increment and growth development from expected yield table values (Pretzsch 1992a; Röhle 1994; Schöpfer, Hradetzky and Kublin 1994).

2.2. Spruce Stands

The growth behaviour of spruce on experimental areas in the Bavarian Alps does not coincide with general developments, for the following reason: even in stands with moderate or more severe damage, increment is significantly above expected yield table values. In Garmisch-Partenkirchen (Fig. 6) increment growth is still consistently on the increase - despite the advanced age of the experimental area (149 years). Merely in the early 1970s was increment recorded to be retarded due to bad climatic conditions (severe droughts, particularly in 1974 and 1976). Increment development was nevertheless considerably above the level of yield tables by Assmann/Franz (1963) and von Guttenberg (1915). This kind of increment behaviour of alpine spruce (increment increase despite considerable needle loss) is basically different from diagnosed increment reactions (caused by forest damage) of spruce in other parts of Germany.

Comparisons of the volume increment recorded on 26 experimental units of the long-term plots Denklingen, Eglharting, Ottobeuren and Sachsenried/South Bavaria with the yield table shown in Fig. 7 indicate a development that is in many details, typical of more or less healthy spruce stands in Southern Bavaria (Röhle 1994). From the establishment of the experimental plots in 1880 up to the 1950s volume increment (influenced by climatic fluctuations) generally corresponded to yield table values. Since the 1960s volume increment has risen sharply, reaching levels of between 200 and 240% of the yield table, with the Assmann/Franz yield table (1963)

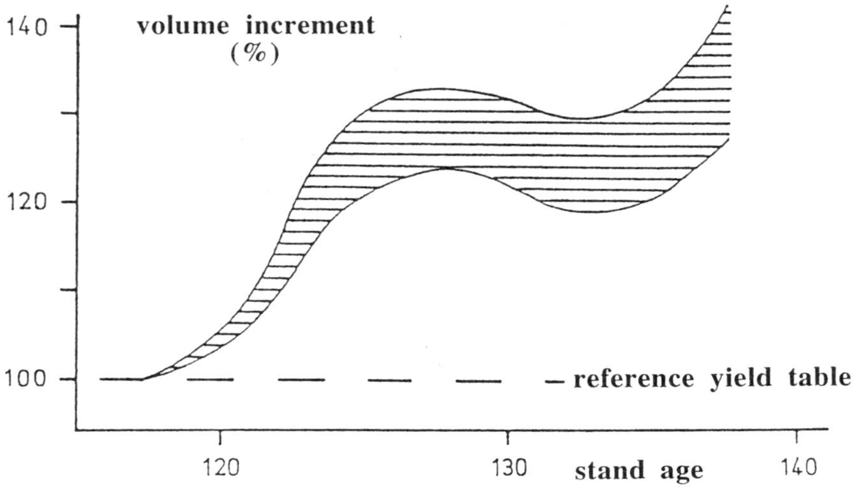

Fig. 6. Stand volume increment in percent of the standardized yield table increment for a damaged spruce stand in the forest district Garmisch-Partenkirchen in the Bavarian Alps (Röhle 1987). Indicated is the volume increment (100% level). The upper border line of the hatched sector indicates increment loss calculated according to the table developed by von Guttenberg (1915), the lower line denotes loss according to the table by Assmann/Franz (1963).

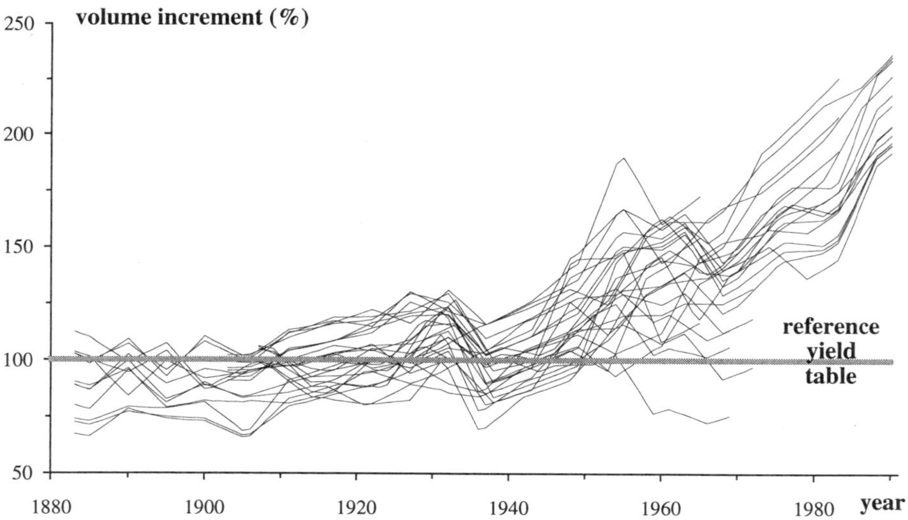

Fig. 7. Development of volume increment on 26 sample plots of the permanent experimental areas Denklingen, Eglharting, Ottobeuren and Sachsenried, compared to the yield table by Assmann/Franz (1963) over the calendar year (Röhle 1994).

used as reference. For over 30 years there has thus been a continually widening gap between actual growth behaviour and expected yield table values. So far no reversal in this development is discernible, i.e. there is no tendency towards a return to the increment level as set down in yield tables. Accordingly, absolute volume increments between 15 and 20 m^3 ha^{-1} y^{-1} are being recorded for stands between 120 and 130 years of age.

Growth conditions have improved considerably in the past decades as can be verified by comparing previous and successive stands. In the experimental areas network in Southern Bavaria there are a number of long-term plots where the second stand generation can now be observed on the same location. For example, the experimental area Sachsenried 03 was followed by the experimental area Sachsenried 602 on the same location (Fig. 8). Aged 29 years, the successive stand (with a top height of 14.5 m corresponding to yield class 39 by Assmann/Franz 1963) is obviously much taller than the previous stand with a top height of only 13.8 m at age 33. The superiority of successive stands is even more impressive with respect to volume development. Volume on experimental plot Sachsenried 602 reached solid timber amounts of 339 m^3 at age 29, while the previous generation on the experimental plot Sachsenried 03 achieved only 50% of this volume on the same location, i.e. 180 m^3 at age 33. These results correspond well to those gained by Kenk, Spiecker and Diener (1991). According to the latter investigations, also conducted on long-term trial plots in Baden-Württemberg, in particular on poorer sites, considerably better growth had been observed on successive stands than in preceding stands.

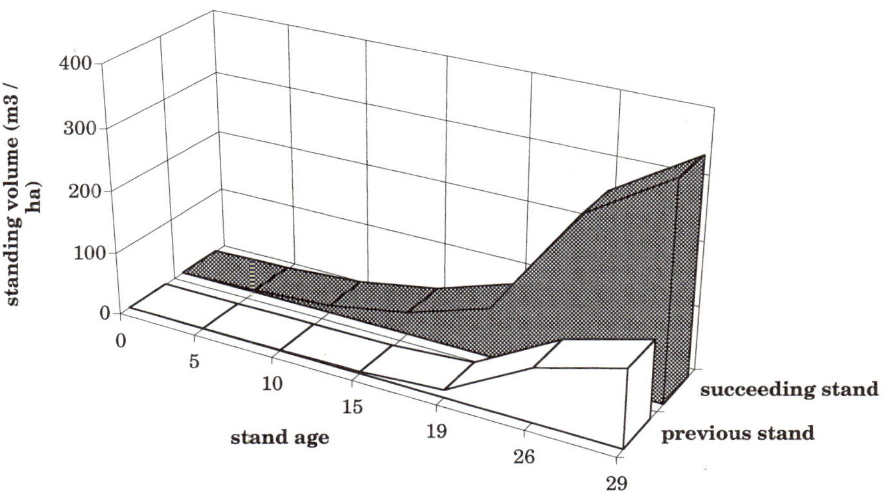

Fig. 8. Volume development of the previous and the successive stand of the long-term spruce experimental area Sachsenried 03/602 in Southern Bavaria (Röhle 1994).

2.3. Pine Stands

In pine stands, too, a significant growth increase was observed in the course of past decades (especially in originally poorer locations in the Upper Palatinate, Franconia and Brandenburg). One of the external signs is that long-shooted peaks between 3 and 5 m in height are seen to project from the round, vaulted, domed crowns of older pine stands (Pretzsch 1985) - a characteristic typical of dominant trees in middle-aged and older stands (Fig. 9). Figure 10 shows exemplary height growth developments on experimental areas in the forest district Schnaittenbach/Upper Palatinate, which accounts for a crown form rather atypical of the age of the trees presented in Fig. 9. The top height developments of the experimental pine stands which exhibit these symptoms are compared to height curves according to Franz (1983). Since the early 1960s, a significant yield class improvement (between 1 and 2 relative classes) can be found as regards height increment in all examined stands, independent of specific stand age. For the past two or three decades height increment development in the examined stands was no longer in line with trends described in the yield tables.

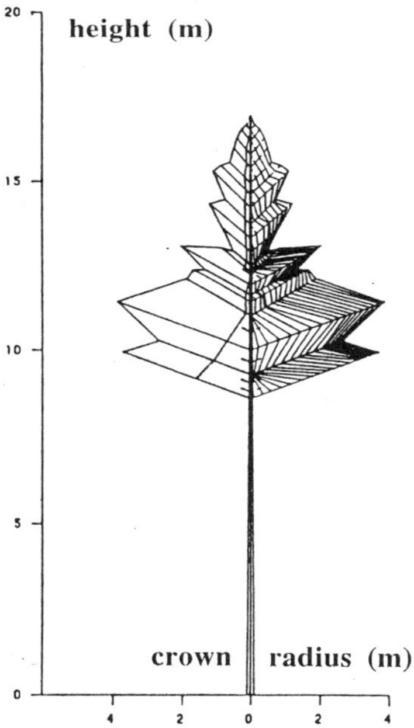

Fig. 9. Crown structure of a pine of the forest district Burglengenfeld/ Upper Palatinate (Pretzsch 1985).

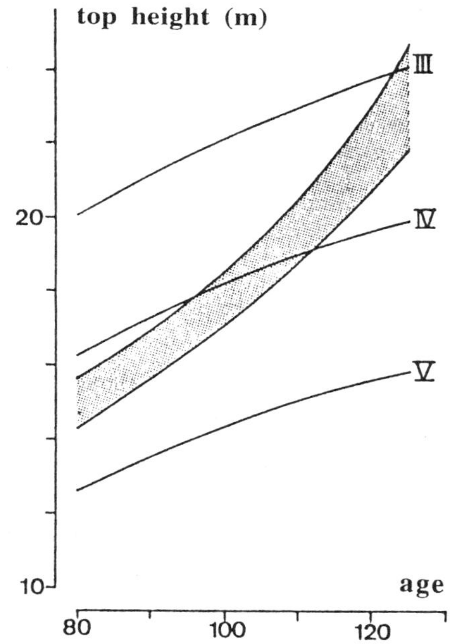

Fig. 10. Top height development on pine experimental plots near Schnaittenbach/Upper Palatinate compared to top height developments according to Franz (1983) for moderate thinning (Pretzsch 1987).

This development is also reflected in basal area increment curves (Fig. 11), which represent basal area increment on experimental pine stands in the Upper Palatinate, compared to the table by Wiedemann (1948) for moderate thinning from below (Pretzsch 1992a). Up to the 1960s, the curves of the annual basal area increment of these pines corresponded to expected yield table values. Since then, a significant improvement in increment level was observed in a great number of the pine stands under investigation. For two or three decades, basal area increment no longer coincided with expectation values of yield tables and reached levels of as much as 200 and 250% of yield table values according to Wiedemann, with a final level that even exceeds expected values for the period of increment culmination in the early years of the stand. These experimental plots demonstrate an increment increase that is absolutely contrary to the normal age trend, independent of stand age.

The causes of this significant growth increase in past decades may be assumed to be the following: increased nitrogen input, increase in carbon-dioxide concentration in the air, improved stand treatment (leading to yet more increment) and - on several locations - the cessation of litter raking (Hofmann et al. 1990; Klopries and Beckmann 1989; Thomasius 1991; Ulrich and Puhe 1992). Not always can a single dominant disruptive factor be found with an influence on increment behaviour as evident as in the case of pine stands in the East German lowlands as examined by Hofmann et al. (1990). The latter and other investigations reveal that both positive and negative increment reactions can follow upon one another as a result of nitrogen input. Nitrogen input, small doses of which can have an increment-promoting effect may, upon passing a saturation phase, initiate the destabilization and devastation of entire stands.

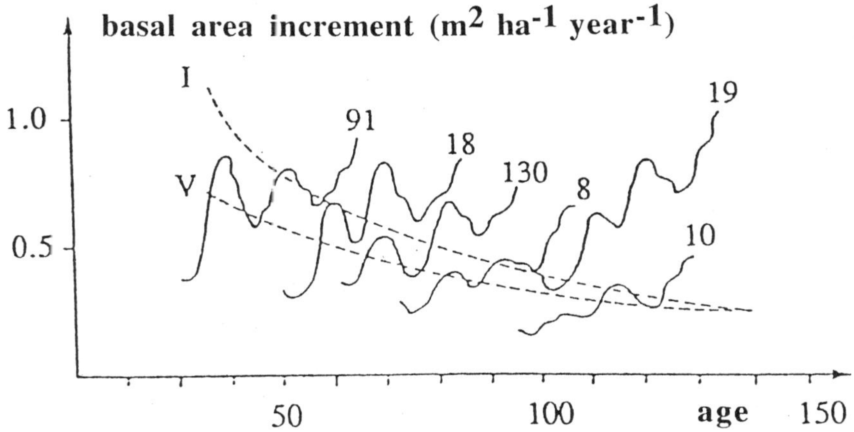

Fig. 11. Increase in annual basal area increment on indicator plots in pine stands of the Upper Palatinate compared to the yield table by Wiedemann (1948) for moderate thinning, yield classes I. and V. (Pretzsch 1992a).

2.4. Beech Stands

Growth trends on the long-term experimental plot Fabrikschleichach 015 in the forest district Eltmann in the Steigerwald show how far the actual growth behaviour of beeches may deviate from what one would expect. The experimental plot covers the treatment variants light, moderate and heavy thinning from below (A, B, and C degree). Figure 12 demonstrates the development of the periodic annual stand volume increment since the area was first assessed in 1870. This is compared with the expected increment values for yield classes I and III according to the yield table by Schober (1967) for moderate thinning. All three thinning variants show age devel-

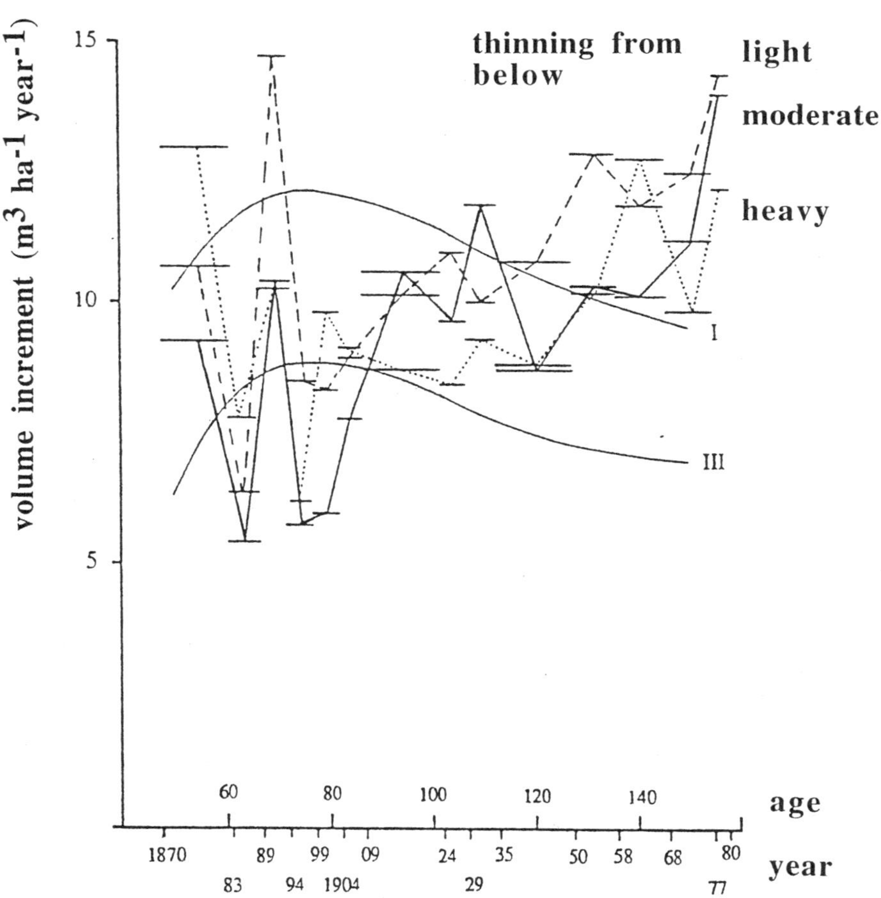

Fig. 12. Volume increment development on the beech thinning experimental area Fabrikschleichach 015 in the forest district Eltmann/Steigerwald. Comparatively, increment curves for the yield classes I. and III. according to Schober (1967) for moderate thinning are indicated (Pretzsch 1992a).

opments contrary to the typical age trend given in the table: in the first third of the observation period of over 120 years, increment trends covered the entire yield class range. In the following period, where, according to the yield table, increments were expected to decline, volume increments on all three areas rose considerably and have continued to do so to this day.

Following an initial peak at age 60 to 80, there was another one between ages 140 to 160 with values ranging from 12 to 14.5 m^3 ha^{-1} y^{-1} (under A, B and C degree conditions). Here, increment increase cannot be attributed to the fact that thinnings caused growth acceleration as it was also evident on the A-degree plot. The actual rise in increment amounts to 140 to 160% of expected yield table values. Similar trends (atypical of respective stand age) and deviations from table values were also observed on other experimental plots (Franz, Röhle and Meyer 1993; Foerster 1993).

2.5. Fir Stands

Similar growth behaviour was diagnosed for fir. As for spruce, pine and beech, fir shows increment decrease and decline in regions heavily damaged by immissions (Kenk 1984; Schöpfer and Hradetzky 1986); on the other hand, in other regions with pure and mixed fir stands there are positive deviations from the expected growth behaviour. Experimental area Wolfratshausen 97 in the forest district Starnberg is one of the rare pure stands of fir under long-term observation (Pretzsch 1992b). Based on its crown vitality, this stand is categorized as "moderately damaged". Nevertheless, in the past two to three decades the 140-year-old stand has achieved a considerable increment increase that is not typical for its age. As far as height increment is concerned, this A-degree area rates between yield classes II and III according to the fir yield table by Hausser (1956) for moderate thinning. As for standing volume development (Fig. 13, left) and volume increment (Fig. 13, right), this area has by far exceeded yield table values in the past two or three decades. Similar to spruce, pine and beech, increment increased in a manner that was atypical of the respective age during the 1960s and 1970s and has kept this up to the present day.

2.6. Oak Stands

For the species oak, severely damaged in some regions, there is every reason to assume that increment also increased in other regions. Such growth behaviour will be illustrated with the help of basal area developments on six plots of the oak thinning experiment Waldleiningen 88 in Rhineland-Palatinate (Fig. 14). Two of the six plots represent light thinnings, two moderate thinning, and two heavy thinning. The lightly thinned plots 2 and 5 obviously approached the maximum basal area typical of the respective location at age 80 to 90. Since then, the basal area has varied between 36 and 42 m^2. This high growth level accounts for the fact that even with moderate

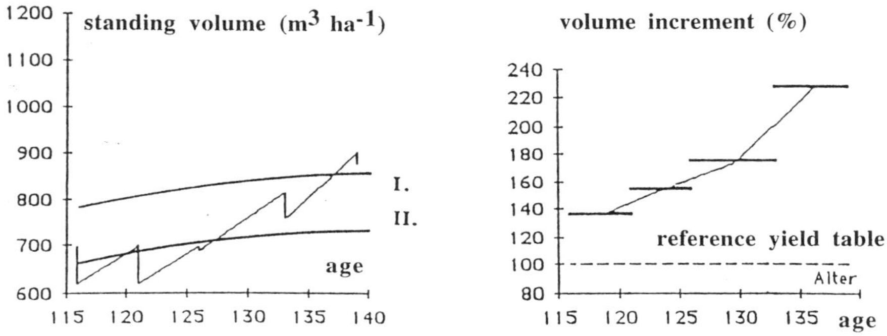

Fig. 13. Development of standing volume (left) and percent volume increment (right) on the fir experimental plot Wolfratshausen 97/1 compared to the yield table by Hausser (1956) for moderate thinning (Pretzsch 1992b). Comparatively, the volume increment for the yield classes I. and II. is indicated (left); volume increment is shown in relation to that of the yield table (right).

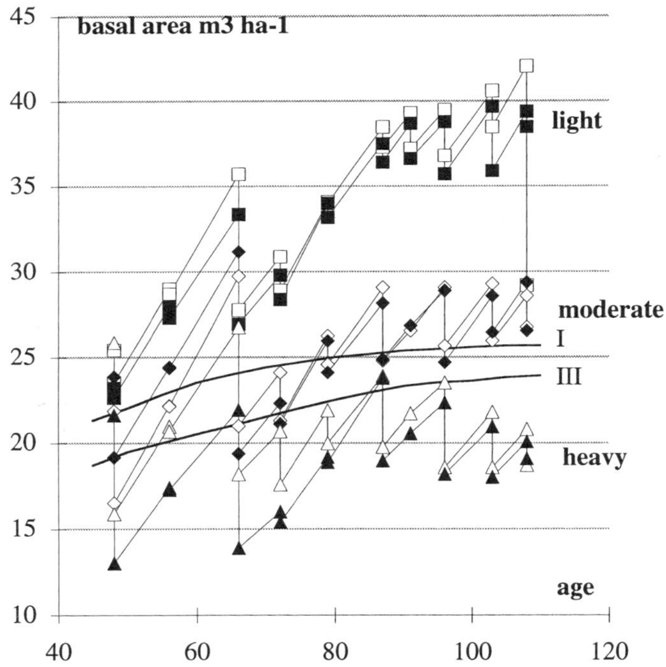

Fig. 14. Development of basal area on six plots of the oak thinning experiment Waldleiningen 88/Palatinate (light thinning, moderate thinning and heavy thinning) compared to the oak yield table by Jüttner (1955) for moderate thinning and yield classes I. and III. (Pretzsch and Utschig 1995).

or heavy thinning and a decrease of 30 and 50% respectively, in the referential basal area of the A-degree, basal area values at Waldleiningen temporarily surpass those of yield tables by Jüttner for moderate thinning. Both basal area and volume development are, except for the heavy thinning, considerably higher than given in the yield table according to Jüttner (1955) for moderate thinning and for yield classes I and III. Standing volume on experimental area 88 has increased to 600 m^3 ha^{-1} (A-degree) and development curves show no tendency to level off. As a result, volume at 108 years of age is the same as for veneer oak stands 300 years old and over. The expected level of standing volume is thus surpassed by 60%. Correspondingly, the level of total volume production is high and amounts to between 694.0 and 824.0 m^3 ha^{-1} on these experimental plots. High basal area and standing volume cast doubt on the suitability of conventional thinning concepts for oak. Neither the conventional numbers of selected trees nor conventional basal area concepts will be capable of exhausting the potential performance of the species oak at the described location. Preuhsler and Stögbauer (1990), Utschig et al. (1993) and Pretzsch and Utschig (1995) reported similar hypertrophic increment processes in oak stands. Again, this development was ascertained to have been sustained over the past two or three decades.

3. Site Class Development on Long-Term Experimental Plots in Germany

For a representative view of site class development of the main tree species in Germany (and to give a presentation of this development in relation to site conditions), Kahn (1994) compiled data on top height development of a considerable number of long-term experimental plots (some observed since 1860) of various German research stations and several university institutes. Figure 15 shows top height development of 47 plots in spruce stands, 89 plots in beech stands, 81 plots in pine and 37 plots in oak stands. This extensive data material for assessing height development is comparable to the outstanding Wiedemann evaluations made in the 1930s and 1940s. The experimental plots cover a wide range of locations: north-south extension between Schleswig-Holstein and Switzerland; east-west dimension between the Bavarian Forest and the Rhineland/Palatinate; height classes ranging from "better than I" to "worse than IV". Figure 15 provides information on the top height development of the world's oldest experimental area network.

The following basic statements about the site class development in Germany can be deduced from the four graphs (Fig. 15): first, a considerable number of the experimental plots is above the yield class I of respective yield tables (as far as height growth is concerned). The older the stands, the more numerous the positive deviations from the yield tables. Second, positive deviations prevail at all age levels, whereas there is practically no decrease in height development in relation to yield

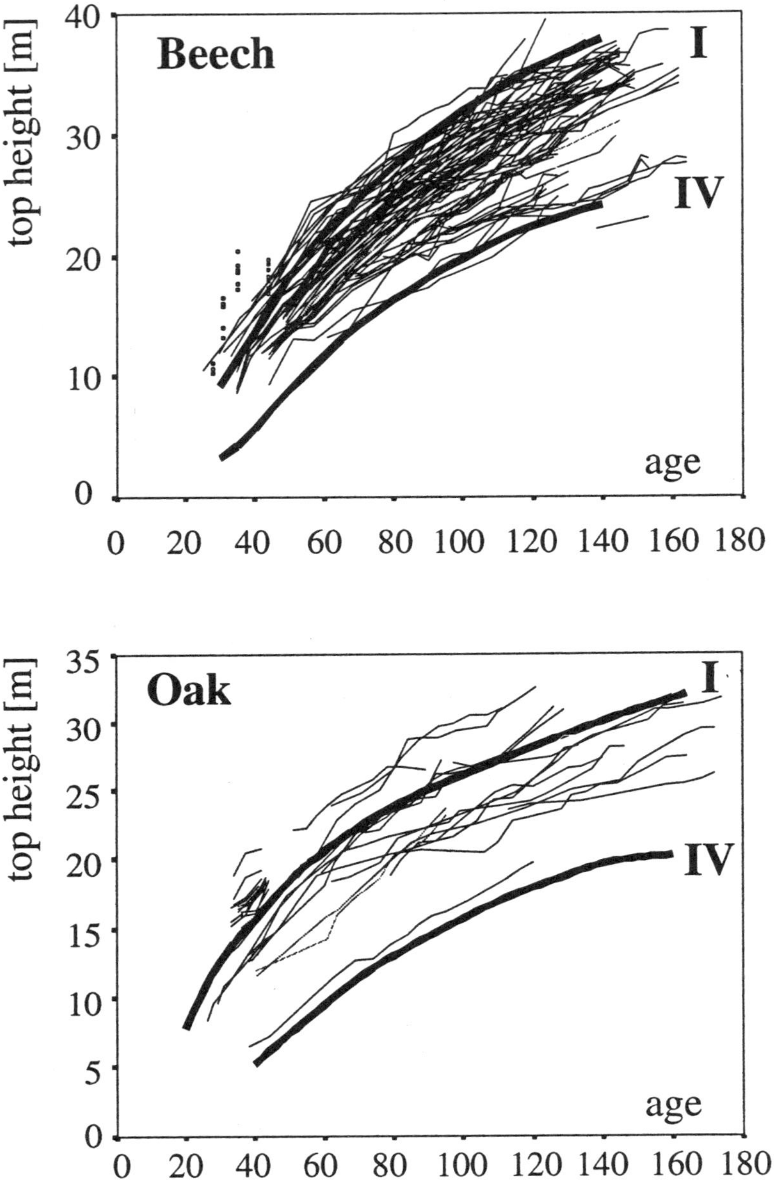

Fig. 15. Top height growth development on the long-term experimental plots for spruce, beech, pine and oak in Germany and Switzerland (according to Kahn 1994). Comparatively, the height curves for spruce, beech, pine and oak according to Wiedemann (1936/42), Schober (1967), Wiedemann (1943) and Jüttner (1955), each for moderate thinning, are indicated.

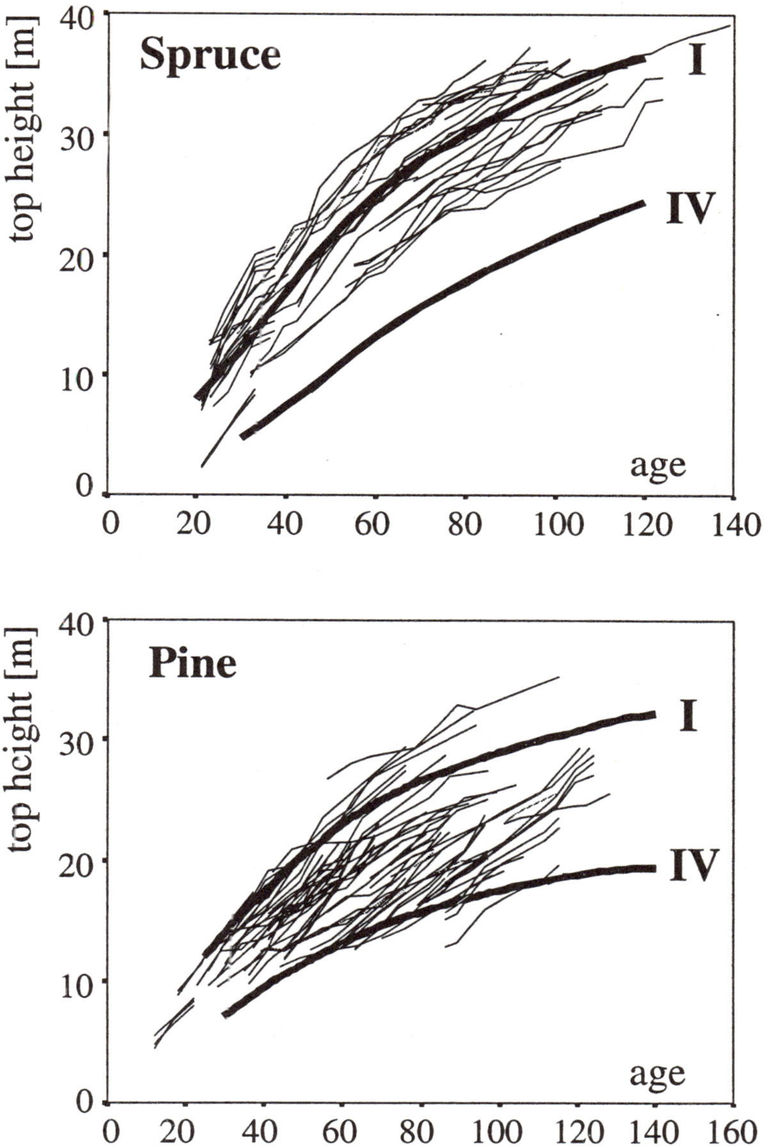

Fig. 15. Continued

tables. Third, young crops of oak and beech by far surpass expectation values of the yield table. Fourth, positive trend deviations and superiority to yield tables are most remarkable for pines, which mainly grow on poorer locations.

Changes in height growth level as well as described deviations from yield tables are by no means of marginal importance but represent a fundamental change in growth conditions, as compared to those assumed in the yield tables. Such changes in growth conditions must be taken into account whenever yield tables are used in forest management. Moreover, these changes can have important consequences for the care of pure and mixed stands. A change in height increment and height growth development has a decisive impact on the competitive behaviour between tree species and calls for the transition to adequate thinning regimes in the control and regulation of mixed stands.

4. Growth Development of the Main Tree Species According to the Results of Repeated Inventories

Reliable information is available on yield class and increment developments of the main tree species in Bavaria in the course of the past 20 years. Bavaria is the most densely wooded part of Germany (over 2.5 million hectares of woodland). It is the only federal state in Germany where two successive forest inventories with comparable inventory sampling methods were carried out. The first extensive inventory was made in 1971, and another in the years 1986-1990 (fixed day 10/1/1987) when Bavarian woodlands were assessed in the course of the Federal Forest Inventory; 17 vegetation periods have passed between these two inventories, a period long enough to permit reliable conclusions to be drawn on yield classes and increment development (Foerster, Böswald and Kennel 1993; Foerster and Böswald 1995).

4.1. Development of the Height Yield Classes

Figure 16 illustrates yield class results over age classes for the first inventory in 1971 (solid line) and the Federal Forest Inventory in 1987 (broken line). Yield class data of the tree species spruce, pine, beech and oak is based on yield tables by Wiedemann (1936/42), Wiedemann (1943/48), Wiedemann(1932) and Jüttner (1955), each for moderate thinning. A close look at the yield class data of the two inventories in dependence on age class, and at yield class development between the first and second inventory reveals that similar trends exist for the four main tree species (spruce, pine, beech and oak). First, all species show a remarkable decrease in yield class with advancing age. This development may perhaps be attributed to better growth conditions of young crops growing at the present time (due to more careful forest treatment methods and a more favourable nutrient situation, e.g. nitrogen in-

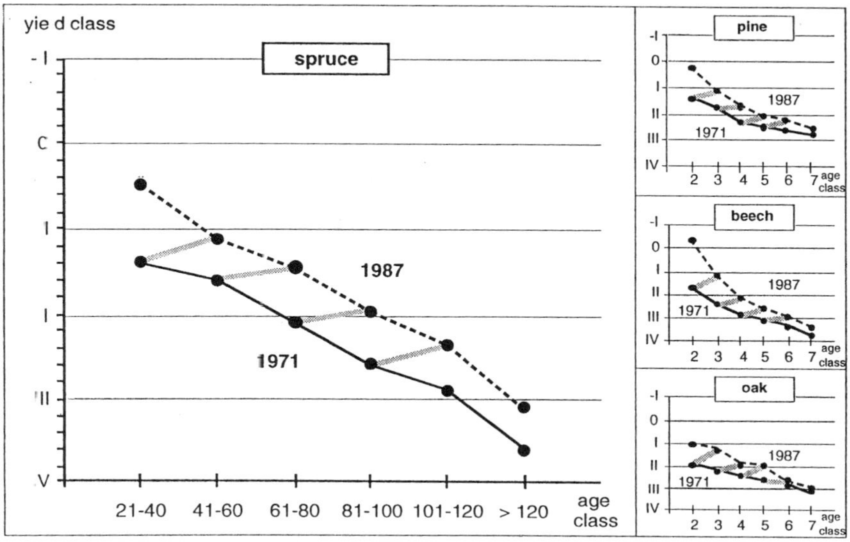

Fig. 16. Yield class development of Bavarian spruce, pine, beech and oak stands for the period covering the Bavarian inventory 1970/71 (solid line) to the Federal Forest Inventory 1986-1990 (dotted line) according to Foerster and Böswald (1995). The yield class changes of the different age classes of spruce, pine, beech and oak refer to the yield tables by Wiedemann (1936/42), Wiedemann (1943/48), Wiedemann (1932) and Jüttner (1955), each for moderate thinning.

puts) compared to the situation of old stands. Second, yield classes improved considerably in the course of the observation period of 17 years. As expected, this improvement is more clearly manifest in young crops than in old ones. The fact that this yield class increase took place within the short time of 17 vegetation periods gives an idea of the extent of the divergence between expected and actual growth development of our main tree species. This improvement indicates that stand development is not in agreement with developments to be expected from yield tables.

It can be deduced from the figures that there is a higher yield class increase when considering the change within individual age classes rather than the shift between age classes. The yield class of spruce in the first age class (age 21 to 40) was improved by almost one relative site class (from I.5 to 0.5) between 1971 and 1987. A comparison of identical stands and site class shifts in the period between the first and second inventory shows a slightly smaller increase in yield class increase (grey arrows). This implies that there may be a combination of the two following effects: first, young crops are generally established on more favourable sites. Second, modern thinning regimes, nitrogen inputs, less litter raking etc. all contribute towards improving site conditions of existing stands.

4.2. Standing Volume and Volume Increment

Based on the volume situation recorded in the first inventory in 1971, the second inventory in 1987 and yield data during this period (including a roughly calculated loss of 15%), calculations can be made for the increment in the period between 1971 and 1987 and offset against respective growth level of yield tables. Table 1 provides the following information on volume and increment development in the observation period: In 1971 the standing volume per hectare of the main tree species lay between 232 and 344 m^3 and had risen to between 263 and 415 m^3 by 1987. The average periodic increment for spruce, pine, beech and oak of 12.6, 8.5, 6.4 and 5.7 m^3 ha^{-1} and annum respectively were by no means absorbed by felling quantities. If real increment is compared with yield table increment, a real increase is apparent of between 10 and 40% above expectation values of the yield tables. If forest practice adheres to the thinning regimes and quantities of yield tables the logical consequence will be an even greater volume accumulation. Overstocking is responsible for a decrease in stability and a decline in increment (Assmann 1961). The deviations established in Table 1 between expected and real increment behaviour yet again underline the necessity to elaborate new growth models which are sensitive to local site conditions, i.e. growth models are required that are more easily adaptable to respective growth conditions than is the case with yield tables.

Table 1. Standing volume and increment development of the species groups spruce, pine, beech and oak according to the results of the Bavarian inventory 1970/71 and the Federal Forest Inventory between 1986 and 1990 (Foerster and Böswald 1995). The percental increments for spruce, pine, beech and oak refer to the yield tables by Wiedemann (1936/42), Wiedemann (1943/48), Wiedemann (1932) and Jüttner (1955), each for moderate thinning.

Growth and yield data 1971 - 1987 for Bavaria	spruce/ fir	pine/ larch	beech/other broadleaves	oak*
standing volume 1971, (m^3 ha^{-1})	344	240	232	233
standing volume 1987, (m^3 ha^{-1})	415	275	271	263
volume rise 1971 - 1987, (m^3 ha^{-1})	+ 72	+ 35	+ 39	+ 30
regular yield 1971 - 1987, (m^3 ha^{-1})	123	95	61	58
other losses, (m^3 ha^{-1})	18	14	9	9
total volume growth 1971 - 1987, (m^3 ha^{-1})	214	144	109	97
mean annual growth 1971 - 1987, (m^3 ha^{-1} y^{-1})	12.6	8.5	6.4	5.7
yield table prognosis 1971 - 1987, (m^3 ha^{-1} y^{-1})	9.6	5.9	5.7	4.5
real growth versus yield table growth, (%)	+ 31	+ 43	+ 12	+ 27

*While the data for spruce, pine and beech refer to the Bavarian State Forest, growth and yield information for oak reflects the situation in the entire Bavarian forest region. Reference to the whole Bavarian forest area provides more solid estimation values for oak.

5. Conclusions

5.1. Conclusions for Forest Practice

With the comparison of damage classes, yield tables, previous and successive stands, a number of different methods for the diagnosis of increment trends was used, whose advantages and deficiencies were discussed by the German Association of Forest Research Stations (DVFFA 1988). Information on discrepancies between yield tables and reality are of the greatest importance to practical forest management planning, which continues to rely largely on yield tables. Hence the comparison with yield tables was used as a reference for all observed tree species. The various reference methods led to the following well-correlated results: forest growth, interpreted rather statically in the past, is a reflection, to a much greater extent than hitherto supposed, of dynamically changing environmental conditions. Environmental conditions such as air pollution, frost and draught cause growth reductions which are superimposed by effects of rising carbon dioxide concentrations and nitrogen inputs. Both act as growth boosters, so that a decline in growth below the level of yield tables is merely observed at higher elevations in the German Middle Mountains in stands subjected to severe air pollution. Increment depression in these locations versus increment increase elsewhere: up to the present time the main point of interest has been the actual development and very little has been done other than to collate symptoms and possible underlying causes. However, there is good reason to suppose that this is not merely a period where growth trends are temporarily thrown out of gear, at the end of which conditions will return to normal and conventional prognostic models can once again be applied. Global changes in productive conditions seem to be becoming the norm whereas growth conditions that were hitherto assumed to be static will become the exception (Thomasius 1991; Ulrich and Puhe 1992). However, despite ambiguous production conditions and uncertain overall conditions, planning, realization and control in forest practice must continue. Economic decisions have to be taken even if there is a lack of detailed knowledge about ecological frame conditions. In damaged stands struck by increment depressions and mortality, economic measures must be aligned primarily to maintain these stands, e.g. by fertilizing and thinning operations to stabilize the stand, advance planting and pest control. If there is further decline and devastation of the stand, new plantations will have to be established as promptly as possible. Numerous studies deal with forest stabilization and the reafforestation in damaged regions. The consequences which should result from the diagnosed increment increase have not been taken into proper consideration. As regards increased increment rates the following silvicultural alternatives are recommended, some of which have already been put into practice: intensification of thinnings, prolongation of rotation periods especially under poor site conditions, transition to more intensive cultivation and to higher numbers of selected trees, and the establishment of mixed stand structures that take into account the changes in growth rhythms and the competitiveness of the respective tree species.

5.2. Conclusions for Future Growth and Yield Models

The present situation can be described as follows: it has been recognized that growth developments in our forests in many respects deviate from yield tables and model approaches. The point has not yet been reached where our deficient working basis can be replaced by a better and more adequate information system. However, a new model prototype has successfully been developed with whose help the kind of information and prognostic system can be outlined that will be required in the future under the changing growth conditions mentioned above. The first, essential frame condition for the establishment of a new model generation consists in the use of prognostic models that are capable of responding to changing climate and site conditions. Growth conditions can no longer be assumed to remain static. Second, the formation and intensive care of mixed stands has become an important factor in forest practice. Up to the present, we have had no suitable model approaches for mixed species stands (Franz 1987). Third, subsequent to the transition to modern thinning programmes that are mainly quality- and stability-oriented, the need for information has undergone a basic change, i.e. from mean and summation values for stands to detailed information on the growth of single trees under varying thinning conditions.

In response to these trends, the position-dependent stem simulator SILVA 1 was developed in recent years (Pretzsch 1992c). It is capable of modelling the growth of single stems in pure and mixed forest stands of all age classes. The current static management model SILVA 1 will be combined with water and nutrient supply models and a radiation model. In the present version of the model, increment prognosis for the single tree assumes a potential increment related to respective site conditions. Potential increment is reduced to real increment by competition indices (determined geometrically) that describe the competitive situation of the single tree. In the future model, SILVA 2, single tree development will be controlled in a more causal way: the increment potential can be estimated in dependence on the water and nutrient supply, as competitive conditions are basically the result of nutrient and water supply as well as of the light regime. Increment reductions are also specific to site and climate conditions. SILVA 2 will be a planning and research instrument, calibrated mainly on the basis of empirical data material obtained in the course of long-term observations of increment developments. These new quasi-causal approaches have, of course, some weak points (e.g. in the determination of the interactions between climatic, nutrient and treatment effects). Nevertheless, they will be the only reasonable alternative to conventional yield tables in the near future. It is only with the help of new site- and climate-sensitive models that quantitative estimations of increment reactions will be possible, even under conditions of global changes and their impact on site conditions and growth trends.

References

Assmann E (1961) Waldertragskunde, BLV Verlagsgesellschaft, München-Bonn-Wien, p 490

Assmann E, Franz F (1963) Vorläufige Fichten-Ertragstafel für Bayern. Institut für Ertragskunde der Forstlichen Forschungsanstalt München, p 104

Athari S, Kramer H (1983) Zur Problematik von umweltbedingten Zuwachsverlusten in Fichtenbeständen. Der Forst- und Holzwirt, 38. Jg., H. 8: 204-206

DVFFA (1988) Empfehlungen zur ertragskundlichen Aufnahme- und Auswertungsmethodik für den Themenkomplex "Waldschäden und Zuwachs", AFJZ, 159. Jg., H. 7: 115-116

Foerster W (1993) Der Buchen-Durchforstungsversuch Mittelsinn 025. Seit 122 Jahren ertragskundlich beobachtet, AFZ, 48. Jg., H. 6: 268-270

Foerster W, Böswald K, Kennel E (1993) Vergleich der Inventurergebnisse von 1971 und 1987. Überraschend hoher Zuwachs in Bayern, AFZ, H. 48: 1178-1180

Foerster W, Böswald K (1995) Vorratsstruktur, Zuwachs und Bonität in den bayerischen Waldbeständen nach den Ergebnissen der Bundeswaldinventur 1986 - 1990 in Verbindung mit einer Abschätzung der gespeicherten Kohlenstoffvorräte und der Möglichkeit zu deren Erhöhung. Lehrstuhl für Waldwachstumskunde, Univ. München, unveröff. Abschlußber. gem. LMS v. 2.12.1993, Nr. V2-FV 100 A-2470, p 74

Franz F (1983) Zur Behandlung und Wuchsleistung der Kiefer, Forstw. Cbl., 102. Jg.: 18-36

Franz F (1987) Zum Aufbau eines neuzeitlichen Informationssystems für die Forstwirtschaft. Forstarchiv, 58. Jg.: 131-137

Franz F, Pretzsch H (1988) Zuwachsverhalten und Gesundheitszustand der Waldbestände im Bereich des Braunkohlekraftwerkes Schwandorf. Forstl. Forschungsber. München, Nr. 92, p 169

Franz, F, Röhle H, Meyer F (1993) Wachstumsgang und Ertragsleistung der Buche. 120 jährige Beobachtung des Durchforstungsversuches Fabrikschleichach 15, AFZ, 48. Jg., H. 6: 262-267

von Guttenberg A (1915) Wachstum und Ertrag der Fichte im Hochgebirge, Wien, Leipzig, p 153

Hofmann G, Heinsdorf D, Krauss HH (1990) Zunehmende Stickstoffeinträge in Kiefernbeständen als Schadfaktor, Forstwirtschaft, 40. Jg., H. 2: 40-44

Hausser K (1956) Tannen-Ertragstafel. In: Schober R (ed) (1975) Ertragstafeln wichtiger Baumarten. J. D. Sauerländer's Verlag, p 154

Jüttner O (1955) Eichen-Ertragstafel. In: Schober R (ed.) (1975) Ertragstafeln wichtiger Baumarten. J. D. Sauerländer's Verlag, p 154

Kahn M (1994) Modellierung der Höhenentwicklung ausgewählter Baumarten in Abhängigkeit vom Standort. Forstl. Forschungsber. München, Nr. 141, p 221

Kenk G (1984) Zum Problemkreis Walderkrankung und Wachstumsforschung. Der Holz-und Forstwirt, 39. Jg., H. 18: 435-437

Kenk G, Spiecker H, Diener G (1991) Referenzdaten zum Waldwachstum. Kernforschungszentrum Karlsruhe, KfK-PEF 82

Klopries B, Beckmann G (1989) Der Anstieg der Kohgledioxidkonzentration in der Troposphärenluft - ein Kardinalproblem der Menschheit. Forst und Holz, 44. Jg., H. 8: 191-199

Pretzsch H (1985) Wachstumsmerkmale süddeutscher Kiefernbestände in den letzten 25 Jahren. Forstl. Forschungsber., München, Nr. 85, p 183

Pretzsch H (1987) Zur Frage des "Normalwachstums" der Kiefer in der Oberpfalz. Forst- und Holzwirt, 42. Jg., H. 11: 286-293

Pretzsch H (1989) Untersuchungen an kronengeschädigten Kiefern (Pinus sylvestris L.) in Nordost-Bayern. Forstarchiv, 60. Jg., H. 2: 62-69

Pretzsch H (1992a) Zunehmende Unstimmigkeit zwischen erwartetem und wirklichem Wachstum unserer Waldbestände. Konsequenzen für zukünkfige ertragskundliche Informationssysteme. Forstw. Cbl., 111. Jg.: 366-382

Pretzsch H (1992b) Leistung und Struktur des Tannen-Fichten-Mischbestandsversuchs Wolfratshausen 97. Waldwachstumskundliche Ergebnisse nach 25 jähriger Beobachtung. In: Preuhsler T, Röhle H, Utschig H, Bachmann M (eds) Festschrift zum 65. Geburtstag von Herrn Prof.Dr.Dr.h.c. Friedrich Franz am 5. August 1992 gewidmet von seinen Schülern, Mitteilungen aus der Waldwachstumsforschung in München, H. 3: 121-137

Pretzsch H (1992c) Konzeption und Konstruktion von Wuchsmodellen für Rein- und Mischbestände. Forstliche Forschungsberichte, München, 115, p 358

Pretzsch H, Utschig H (1989) Das "Zuwachstrend-Verfahren" für die Abschätzung krankheitsbedingter Zuwachsverluste auf den Fichten- und Kiefern-Weiserflächen in den bayerischen Schadgebieten. Forstarchiv, 60. Jg., H. 5: 188-193

Pretzsch H, Utschig H (1995) Die Eichen-Versuchsflächen im Forstamt Elmstein-Nord, AFZ, 50. Jg., H. 2: 67-70

Preuhsler T, Stögbauer K (1990) Strukturmerkmale des Furniereichen-Bestandes "Eichhall" im Bayerischen Forstamt Rohrbrunn. Forst und Holz, Jg. 45, H. 11: 283-289

Röhle H (1987) Entwicklung von Vitalität, Zuwachs und Biomassenstruktur der Fichte in verschiedenen bayerischen Untersuchungsgebieten unter dem Einfluß der neuartigen Walderkrankungen. Forstl. Forschungsber., München, Nr. 83, p 122

Röhle H (1994) Zum Wachstum der Fichte auf Hochleistungsstandorten in Südbayern. Ertragskundliche Auswertung langfristig beobachteter Versuchsreihen unter besonderer Berücksichtigung von Trendänderungen im Wuchsverhalten. Habilitationsschrift Forstwiss. Fak. Univ. München, p 249

Schober R (1967) Buchen-Ertragstafel. In: Schober R (ed) (1975) Ertragstafeln wichtiger Baumarten. J. D. Sauerländer's Verlag, p 154

Schöpfer W, Hradetzky J (1984) Der Indizienbeweis: Luftverschmutzung maßgebliche Ursach der Walderkrankung. Forstw. Cbl., 103. Jg., H. 4: 231-248

Schöpfer W, Hradetzky J (1986) Zuwachsrückgang in erkrankten Fichten- und Tannenbeständen - Auswertungsmethoden und Ergebnisse. Forstw. Cbl., 105. Jg.: 446-470

Schöpfer W, Hradetzky J, Kublin E (1994) Wachstumsänderungen der Fichte in Baden-Württemberg. Forst und Holz, 49. Jg., H. 21: 633-644

Thomasius H (1991) Mögliche Auswirkungen einer Klimaänderung auf die Wälder in Mitteleuropa. Forstw. Cbl., 110. Jg., H. 5: 305-330

Ulrich B, Puhe J (1992) Auswirkungen der zukünftigen Klimaveränderungen auf mitteleuropäische Waldökosysteme und deren Rückkopplung auf den Treibhauseffekt. In: Studienbericht für die Enquete-Kommission "Schutz der Erdatmosphäre" des Deutschen Bundestages, Göttingen: Forschungszentrum Waldökosysteme, p 145

Utschig H (1989) Waldwachstumskundliche Untersuchungen im Zusammenhang mit Waldschäden. Auswertung der Zuwachstrendanalyseflächen des Lehrstuhles für Waldwachstumskunde für die Fichte (Picea abies (L.) karst.) in Bayern. Forstl. Forschungsber. München, Nr. 97, p 198

Utschig H, Bachmann M, Dursky J (1993) Wirkung verschiedener Durchforstung auf Struktur und Wachstum. Traubeneichen-Durchforstungsversuch Rohrbrunn 620, AFZ, 48. Jg., H. 6: 271-274

Wiedemann E (1932) Buchen-Ertragstafel. In: Die Rotbuche 1931, Hannover
Wiedemann E (1936/42) Fichten-Ertragstafel. In: Schober R (ed) (1975) Ertragstafeln wichtiger Baumarten. J. D. Sauerländer's Verlag, p 154
Wiedemann E (1943/48) Kiefern-Ertragstafel. In: Schober R (ed) (1975) Ertragstafeln wichtiger Baumarten. J. D. Sauerländer's Verlag, p 154

Has Site Productivity Changed? A Case Study in the Eastern Swabian Alb, Germany

Hans Untheim
Institute of Forest Growth, University of Freiburg,
Bertoldstr. 17, D-79085 Freiburg, Germany

Abstract

Stem analyses of dominant spruce (*Picea abies* [L.] Karst.) and beech trees (*Fagus sylvatica* L.) on the site unit Elymus beech forest on moderately fresh loam from limestone in the growth district Northern part of the Eastern Swabian Alb were conducted to detect and quantify site-related growth changes. Height growth served as an indicator for site productivity and its changes. A statistical sampling design guaranteed the representation of growth of trees with different germination dates on the site unit within the growth district. During the past decades, height growth as well as radial growth of both tree species showed increasing trends. While several factors such as altered silvicultural management, site productivity changes, and, in the case of spruce, changes in genetics may be responsible for increased radial growth, increased height growth of beech is probably caused only by site productivity changes. This conclusion can be drawn because management influences on beech height growth are negligible for the trees analyzed and changes in genetics are not probable because beech has always been naturally regenerated.

1. Introduction

Environmental changes in the last few decades such as rising atmospheric CO_2-level, high deposition rates of air pollutants and climatic change are likely to have an effect on forest ecosystems. Research activities on forest decline in the 1980s were expected to reveal a decline in growth, but in many studies long-term increase of forest growth has been observed. Scientists and foresters have been reporting about this phenomenon on local, regional and national levels for several years. Spiecker et al. (1994) gave a brief overview of the subject concerning data types, methods and literature. Increased forest growth on a regional and national scale may be caused by changes in age distribution of forests, tree species composition, and

European Forest Institute Research Report No. 5
Growth Trends in European Forests
Edited by H. Spiecker, K. Mielikäinen, M. Köhl and J. P. Skovsgaard

distribution of forested area. On a local scale, site factors, silvicultural management and genetics may be responsible for changes in growth. It is possible that negative influences on forest growth such as air pollution or extreme weather conditions might be overcompensated by forces of increased growth that are unknown in detail and which vary according to region and site.

The objective of this project is to investigate growth trends in the forest landscape Eastern Swabian Alb, Baden-Württemberg, Germany. The two most common tree species of the region, spruce (*Picea abies* [L.] Karst.) and beech (*Fagus sylvatica* L.), were studied. Special emphasis is put on the detection and quantification of possible growth trends induced by changes in site productivity. Tree height growth is used as an indicator of site productivity. The following hypothesis is tested: young dominant trees show increased growth in comparison to older trees at the same age. It is assumed that all sample trees have been grown on the same site unit and have been dominant during their lifetime.

2. Material and Methods

2.1. Detecting and Quantifying Growth Trends

This study primarily aims to detect site productivity changes by analyzing tree growth. For this reason, influences of changes in silvicultural management and tree genetics on tree growth should be excluded as far as possible. Height growth of dominant trees is assumed to be influenced less by stand management than other growth parameters (Assmann 1970). It is therefore chosen as an indicator for site productivity. Additionally, radial growth is investigated. All sample trees are selected from one site unit: Elymus beech forest on moderately fresh loam from limestone. The tree species comparison that is included in the study will lead to further empirical conclusions about factors influencing possible growth trends.

The research approach of the study is to compare the growth of single trees at the same age but with different germination dates. It combines cross-sectional observations (trees with different ages) with time series of tree growth parameters to a longitudinal data analysis in a retrospective view. Possible changes of growth patterns are studied in an age-time plane. For this reason, linear regression functions are fitted to age-related growth parameters under the simplified assumption that possible growth trends show linear behavior. An important advantage of this approach is that no questionable reference such as e.g. a yield table is required (Spiecker et al. 1994).

Table 1. Description of stem analysis data within the six sampling groups

	Spruce			Beech		
Age group	1 (young)	2 (medium-aged)	3 (old)	1 (young)	2 (medium-aged)	3 (old)
Number of trees [n]	12	12	14	10	17	11
Germination date [year] *	1957 (1945-1965)	1922 (1914-1936)	1901 (1878-1913)	1931 (1918-1949)	1899 (1882-1912)	1862 (1840-1882)
Total age [year] *	38 (29-49)	73 (58-80)	93 (81-116)	63 (45-76)	95 (82-112)	132 (112-154)
Age at breast height [year] *	27 (18-34)	62 (47-69)	82 (70-103)	51 (37-65)	85 (72-98)	122 (105-139)
Height [m] *	18.3 (13.5-21.7)	30.1 (25.7-33.7)	33.8 (30.9-36)	21.5 (19-23.3)	28.5 (25-32)	33.3 (31.4-35.8)
Diameter [m^{-2}] *	26.2 (20.3-35)	43.6 (33.1-53.1)	50.9 (45.2-67.2)	25.8 (20.4-31.5)	40.8 (33.6-51.4)	53.4 (48.8-64.3)
CV$^{\psi}$ of height [%] * **	7 (6-11)	9 (7-11)	10 (5-18)	16 (4-22)	9 (3-16)	9 (3-16)
Double Std.error of mean height ($t_{1-0.05/2,n-1}\, S_{\bar{x}}$)$^{\varphi}$ [%] * **	5 (4-7)	6 (5-8)	6 (3-10)	11 (3-16)	5 (2-8)	6 (2-11)
CV$^{\psi}$ of 10 year ih [%] * ** τ	10 (8-12)	13 (6-19)	17 (8-30)	18 (9-28)	17 (14-23)	21 (13-36)
CV$^{\psi}$ of diameter [%] * **	14 (10-23)	16 (10-31)	20 (16-26)	23 (17-42)	19 (8-39)	17 (11-33)
CV$^{\psi}$ of 10 year ir [%] * ** τ	20 (11-29)	20 (11-32)	23 (15-33)	21 (10-30)	21 (10-34)	21 (11-32)

* mean and range in brackets

** related to the time series of observations versus breast height age

$^{\tau}$ ih: 10 year height increment, ir: 10 year radial increment

$^{\psi}$ CV: Coefficient of variation

$^{\varphi}$ $t_{1-0.05/2,n-1}$: t-value with p=5%, two sided df=n-1; $S_{\bar{x}}$ standard error of mean

2.2. Tree Data

In 1993, spruce and beech trees were sampled in young, medium-aged, and old pure even-aged stands (Table 1). Thus six strata were formed. The formation of these strata was intended to simplify the sampling procedure and to guarantee sampling over a wide age range. For purposes of statistical analysis, the strata were decomposed and individual tree data were considered. A sampling element consisted of a stem analysis tree and a surrounding temporary plot sample. Population parameter of interest is the mean height of dominant trees at certain tree ages within each of the six strata. The required sample size within a stratum depends on the expected precision of the estimated mean (*E%*). *E%* was fixed to be equal to or less than 5% on a 5%-significance level [Eq. (1)].

$$n = \frac{t^2 CV\%^2}{E\%^2} \tag{1}$$

n: sample size
t: t-value with p=5%, two sided df=n-1
CV%: coefficient of variation
E%: precision of the estimated mean.

The coefficient of variation (CV) of the age-height relationships was preliminarily estimated in a pilot study. The average site-related CV for both tree species and various ages on defined sites seemed to be about 10%. According to Eq. (1), a sample size of about 20 observations per stratum is required.

A network of sample stands has been assembled with the aid of a systematic grid which was superimposed on the region. The grid consisted of 20 dots in the forested region with a distance of approximately 3 km between dots. One sample tree of the required age on the investigated site unit was selected in the stand nearest to each dot. Management and site maps were used for a priori information. In the ideal case, six stands around each dot should have been found (2 species · 3 age classes). This was generally not the case. Therefore the number of observations is markedly below 60 per tree species (Table 1). Although actual sample sizes are below the ones estimated in the pilot study, the double standard error of mean height within each age group does not much exceed the expected precision of 5%.

The sampling of only one tree per stand-site combination is efficient in estimating population data, which is the main objective of the study. The total amount of sampling trees was restricted by economic factors and corresponded with the initially calculated sample size needs. Some characteristics of the stem analysis trees are described in Table 1.

The sample tree within each stand was selected by applying a clearly defined search algorithm. Sample trees had to be dominant and without injuries. The tree nearest to the dot was evaluated. Biasing effects on the results due to this sampling procedure are not likely because (1) growth differences need not be assumed between possible sample trees within a homogenous stand-site combination, (2) sampling elements at the border of the stand-site combination and near stand openings were excluded, and (3) the area of the stand-site units is usually very small. A larger number of sample plots with one observation each was preferred to fewer sample plots with more observations.

The location of the stump of each sample tree was the center of a temporary plot sample established by relascope with a basal area factor of 2 for beech and 4 for spruce plots. Height and diameter of all trees within each temporary plot were measured. The plots consisted of 10-20 trees. These observations were useful in verifying the results derived from the single stem analysis tree.

Stem analysis includes the measurement of annual terminal shoot lengths. With spruce, this could be done downwards along the stem until breast height. With beech, annual shoot growth could only be measured up to 3-4 m from the top. For both species cross-sectional discs were collected at stump height, breast height and in further distances of about 5 m along the stem. The section length was restricted by economical reasons. Annual shoot growth of spruce was verified by tree ring counting on cross-sectional discs. Height growth of beech was reconstructed using a combination of approaches described by Newberry (1991). It is based on linear interpolation between two subsequent discs correcting the shoot length out of which these discs were sampled by assuming that the discs were cut in the middle of the shoot. The approach was tested by applying it to the sampled spruce trees (Table 1). The error (true height minus estimate) was studied. It could be shown that the approach overestimates true heights in the three lower sections of the trees (0.097 m), and that underestimation occurs in the three upper sections (-0.14 m). This is consistent with the knowledge about the common shape of height growth curves. The bias is irrelevant for this particular study. No relation between germination date and error was found. Radial increment was measured on cross-sectional discs at breast height along eight radii supported by an image analysis system.

Number and size of living and dead branches, actual crown lengths and crown shapes, and the height-diameter ratio of trees in even-aged stands were used to prove whether older standing trees had had dominant social status throughout their lives (Abetz 1985). To test whether older trees had been suppressed in their youth, height increment between stump height and breast height as well as radial increment in breast height were checked. An individual examination of the height-diameter ratio versus tree age had to confirm the constancy of a dominant social status throughout lifetime.

2.3. Survey Sites

The growth district Northern part of the Eastern Swabian Alb was selected for investigation. This region is a part of southeastern Baden-Württemberg, which is a state in southern Germany. The selection considered macroclimatic homogeneity and importance as a forest landscape. Essential characteristics of the region are listed in Table 2. Mühlhäusser et al. (1983) give a detailed description of the system of forest site classification and mapping in the state of Baden-Württemberg. Spruce

Table 2. Description of the investigation region.

Characteristic	Description*
Growth area (name and reference number)	Swabian Alb (6)
Growth district (name and reference number)	Northern Part of Eastern Swabian Alb (6/01)
Regional forest community	continental-montane beech forest
Landform	Karstic depression, southeastern inclination, plateau divided by dry valleys, ancient surface
Elevation	460-778/650 m asL
Geology and parent material	Malm limestone, clay-with-flint, locally tertiary sediments
Precipitation (year)	750-1070 mm
Precipitation (May-Sept.)	440-520 mm
Mean air temperature (year)	6.5-7.2°C
Mean air temperature (May-Sept.)	13.2-14.2°C
Days with more than 10°C	140-150

* Data sources: Mühlhäusser et al. 1985; Schlenker et al. 1973

and beech are widespread tree species in the region. Spruce is not indigenous but is managed locally in the third rotation. Beech is the most important species of the regional forest community but has been replaced by spruce on many sites.

On the local level of the site classification system, site units are distinguished within each growth district. They are identified by a particular combination of physical and biotic site factors. Field delineation criteria are differences in topography, soil factors, microclimate, and vegetation. In this study, the widespread site unit Elymus beech forest on moderately fresh loam from limestone (LFL) was investigated (Table 3).

The author checked site unit delineation in the vicinity of each sample plot; thus it is ensured that all sample plots represent the investigated site unit. The influence of certain variations among the sites of the sample plots, which cannot be quantified, is negligible, because sampling was based on a statistical design. The sampling strategy was so designed as to avoid sampling of older trees that would have been grown on poorer sites. To ensure this, the upper age limit of each oldest age group was chosen well below the common rotation age on the investigated site unit.

Fertilization and liming has never taken place on LFL. Litter raking and forest pasturage have usually not played an important role because the study sites are remote from settlements (Koch et al. 1939). However site regeneration after former excessive land use cannot generally be excluded as a factor influencing growth. Afforestation sites were excluded from sampling.

Field methods used to delineate site units are not able to detect long-term or medium-term site changes and site unit definitions therefore allow for certain site

Table 3. Description of the investigated site unit Elymus beech forest on moderately fresh loam from limestone (LFL).

Characteristic	Description*
Local forest community	Elymus europaeus beech forest
Soil character	20-45 cm loamy-clayey solution rest over weathered Malm limestone with gaps, decalcification generally deeper than 20 cm
Topography	plane or hardly inclined
Moisture regime	moderately fresh (karst drainage, rooting in gaps)
Nutrition availability	good
Humus	mull humus
Soil type	Terra fusca (chromic Cambisol)
Ecological species group	Elymus europaeus-group

*Data sources: Moosmayer 1957

changes. Such changes might include increased nitrogen deposition, rising atmospheric CO_2-level, or other climatic changes. Recent nitrogen deposition rates are high (about 30 kg ha^{-1} a^{-1} in spruce stands) in the research area, with high ammonium concentrations (60-70%) (v.Wilpert and Hildebrand 1994). Buberl et al. (1994) report that the poor base saturation of soils on non-carbonatic parent material within the region is probably caused by long-term acidic deposition. LFL has a rather good base saturation due to its carbonatic parent material.

3. Results

Dominant trees on the investigated site unit show increased heights when compared at the same age (Fig. 1). It can be deduced from Fig. 1 that at an age of 30 years (age at breast height) the mean height of spruce increased from 16 to 20 m between 1920 and 1990 and the mean height of beech from 12 to 14 m between 1880 and 1990. A height growth trend is detected for spruce during the age of 10 to 60 years, for beech during the age of 30 to 80 years; results are significant on a 5% level at least (Table 4). The obviously increased height of trees that germinated later seems to have developed continuously during their lifetime. This can be shown by analyzing height increments.

Ten year height increments show increasing trends for the ages of 10 to 40 years (spruce) and 20 to 70 years (beech); results are significant on a 5% level (Table 4). The average of 10-year increments was chosen so that strong influences of annual variation on the results could be avoided. Height increment trends also explain the

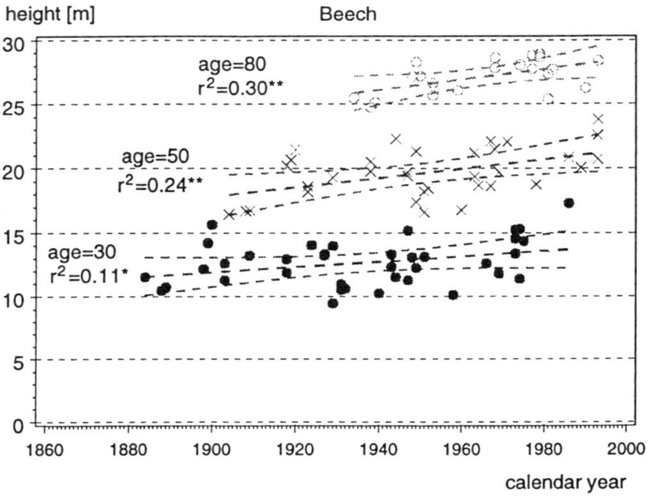

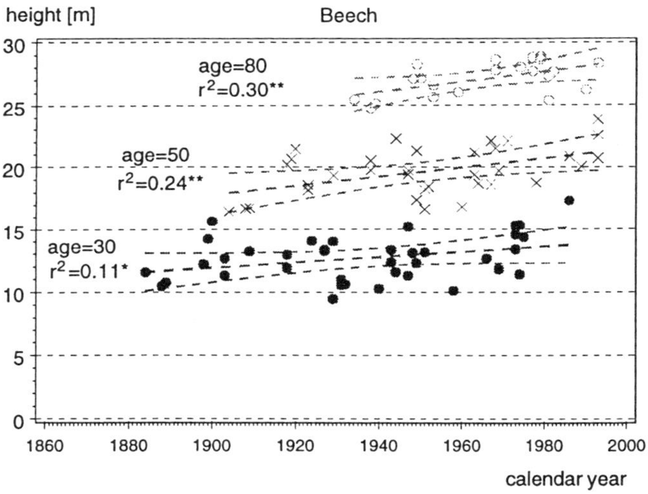

Fig. 1. Height growth trends of spruce and beech on LFL. Height of the sample trees at certain ages at breast height are plotted over the calendar year at which these ages were reached. Significance levels (* a<0.05, ** a<0.01, *** a<0.001) and 95% simultaneous confidence limits of linear regression models are shown.

age-dependent increase (statistically not significant) of slope coefficients of the height growth trend model. Analysis of the observed growth data generally does not indicate a deviation from linear trend behavior. Additional examination of 10-year radial increments also revealed increasing trends for most of the ages examined (Table 4). Exceptions are beech radial increments at the ages of 10 and 40 years.

Table 4. Height, mean 10 year height (ih) and radial increment (ir) regressed on the calendar year at the considered age. Increased heights of both species are linked with increasing height increments of spruce at the ages 10 to 40, and beech at the ages 20 to 70, each significant on a 5% level. Increasing radial increment trends are also observed for both tree species at several ages.

			Height				ih				ir			
Species	Age	n	a *	b *	r^2	p **	a *	b *	r^2	p **	a *	b *	r^2	p **
spruce	10	38	-40.7	0.0244	0.40	0.0001	-4.40	0.00256	0.38	0.0001	-36.2	0.0207	0.43	0.0001
	20	37	-88.7	0.0520	0.48	0.0001	-5.14	0.00294	0.36	0.0001	-27.5	0.0159	0.15	0.0094
	30	29	-98.0	0.0592	0.25	0.0029	-3.47	0.00206	0.21	0.0061	-37.9	0.0210	0.17	0.0120
	40	26	-131.3	0.0783	0.25	0.0046	-5.12	0.00286	0.38	0.0004	-71.5	0.0380	0.49	0.0001
	50	25	-163.5	0.0963	0.31	0.0019	-1.07	0.00071	0.02	0.2523	-67.1	0.0355	0.43	0.0002
	60	23	-186.9	0.1093	0.34	0.0012	-2.61	0.00148	0.19	0.0182	-53.3	0.0283	0.29	0.0039
beech	10	38	6.6	-0.0010	0.00	0.4176	0.59	-0.00012	0.00	0.4026	4.0	-0.0012	0.01	0.3457
	20	38	-8.8	0.0091	0.03	0.1413	-1.54	0.00100	0.12	0.0152	-16.7	0.0997	0.21	0.0023
	30	38	-27.5	0.0207	0.11	0.0214	-1.86	0.00116	0.24	0.0010	-11.7	0.0072	0.16	0.0075
	40	37	-36.7	0.0273	0.16	0.0076	-2.21	0.00133	0.27	0.0005	-4.2	0.0033	0.05	0.0944
	50	34	-50.5	0.0360	0.24	0.0018	-1.95	0.00118	0.26	0.0008	-11.8	0.0073	0.19	0.0041
	60	29	-53.5	0.0389	0.23	0.0047	-2.16	0.00127	0.21	0.0059	-17.4	0.0101	0.31	0.0008
	70	28	-79.5	0.0533	0.41	0.0002	-2.21	0.00127	0.21	0.0074	-28.6	0.0158	0.55	0.0001
	80	21	-53.8	0.0412	0.30	0.0048	-1.74	0.00102	0.10	0.0823	-25.3	0.0149	0.40	0.0006

* intercept and slope coefficient of a linear regression model (y=a+bx)

** probability for testing the null hypothesis that the slope coefficient of the linear regression model is equal to or less than zero

The analysis was conducted by applying SAS Proc Reg (SAS Institute Inc. 1987). The relationship between growth and germination date is examined only for those ages for which more than 20 observations exist across the 3 age classes. No statistical differences were found between the slope coefficients of regression models for the single growth parameters within a tree species. Single observations having a strong influence on the estimation of the regression models did not occur, the following criteria were applied to check this: Cook's D, DFFITS, leverage. The comparison of results from the two different tree species growing on the same site unit can be regarded as a validation of the regression models.

4. Discussion

A clear increase in height and radial growth has been observed for spruce and beech on the investigated site unit. Comparable results for beech height growth in a neighboring region on the same site unit are given by Abetz (1988). Some further studies on increasing growth trends of different species in southwestern Germany were published by Eichkorn (1986), Gerecke (1988), Kenk and Fischer (1988), Fischer and Rommel (1989), Moosmayer and Ripberger (1990), Kenk et al. (1991), Weise (1991), Schöpfer et al. (1994) and Spiecker (1995). There are a number of factors that may have influenced the positive growth trends observed. The main factors to be considered are changes in site, silvicultural management, and genetics. For most of these factors, long-term quantitative observations do not exist. In addition, interaction between factors may play an important role. Table 5 lists possible causes of growth trends which are relevant for this study. Some studies concerning each factor are cited. It is evident that the suite of factors is very complex.

It can be shown that several different site factors may be responsible for the observed growth trends. The occurrence of climatic change including rising atmospheric CO_2-level and increased deposition of atmospheric pollutants is widely accepted and discussed in literature (Houghton et al. 1990; Kenk and Fischer 1988; Schönwiese et al. 1993; Spiecker 1987). Long-term observations in the investigated region exist only for temperature and precipitation. Preliminary analysis of climate-growth relationships show that observed growth trends cannot be explained by these data. Several fertilization trials in the region (not on LFL) show only a moderate response of tree growth to nitrogen fertilization. Short-term influences of precipitation on the fertilization response were strong (Altherr et al. 1974; Spiecker 1991). It must nevertheless be assumed that long-term continuous nitrogen deposition may cause altered growth response in comparison to fertilization. Office records and detailed maps of the forestry administration made it possible to avoid sampling of limed, fertilized and afforestation sites. Existing historical growth data are not exact and reliable enough for an analysis in this study. It is furthermore impossible to refer such data to the single sampling locations.

Table 5. Evaluation of possible causes for observed height and radial increment growth trends in this study. It can be seen that a lot of subfactors of the main factor "site" possibly have an effect. Beech height growth trends seem to be caused merely by site factors.

Growth influencing factor		Spruce		Beech		
Main factor	Subfactor	Height growth	Radial growth	*Height growth*	Radial growth	Reference
Site	Climate including atmospheric chemistry (esp. CO_2-level)	effect possible	effect possible	effect possible	effect possible	Houghton et al. 1990; Schönwiese et al. 1993
	Deposition e.g. nitrogen	effect possible	effect possible	effect possible	effect possible	Kenk and Fischer 1988; Spiecker 1987; Houghton et al. 1990
	Regeneration after litter raking, pasturage and excessive use	effect possible	effect possible	effect possible	effect possible	Koch et al. 1939; Jänichen et al. 1951
	Number of generations on the same location	effect possible	effect possible	probably no effect	probably no effect	Moosmayer 1957
	Soil characteristics	effect possible	effect possible	effect possible	effect possible	v.Wilpert and Hildebrand 1994
	Liming and fertilization	excluded*	excluded*	excluded*	excluded*	office records and maps
	Afforestation sites	excluded*	excluded*	excluded*	excluded*	office records and maps
Silvicultural management	Regeneration method	effect possible	effect possible	probably no effect	probably no effect	Altherr 1966; Kenk 1990; Pollanschütz 1974
	Thinning regime	effect possible	effect possible	probably no effect	probably no effect	Assmann 1970; Dieterich 1924 and 1925; Wiedemann 1932 and 1937
Genetics	-	effect possible	effect possible	probably no effect	probably no effect	Koch et al. 1939

*excluded: influence of the factor excluded by sampling design

Site regeneration after changes of land use may have also taken place; yet it is impossible to determine to what extent this is the case for the single investigated sites; however, forest pasturage is recorded for the surroundings of the settlements until the beginning of the 19th century and litter raking until the end of it. Remote from the settlements, the forests were protected for hunting purposes; on the other hand, excessive forest use for iron foundries also occurred (Koch et al., 1939; Jänichen et al., 1951). Furthermore, the number of spruce generations on the same location may change site characteristics. Spruce is not an indigenous species. Moosmayer (1957), however, observed no reduction or even a slight increase of growth with the second or further spruce generation on LFL. With beech, the number of generations most likely did not cause site changes during the last decades, since this species has existed in the region for millennia. Several of the facts described above can, singly or in combination with each other, lead to changes in soil characteristics such as improved nutrition availability, which may cause increasing growth trends.

Influences of silvicultural management constitute another main factor which must be examined. Regeneration methods with spruce surely changed from seedings to plantings and to wider spacing of plants. A change of regeneration methods has not occurred for beech. This species has always been naturally regenerated without changes of the silvicultural system since the middle of the last century. An investigation of individual height growth between stump height and breast height for both species showed no trend related to germination date. An indication of wider spacing of spruce plantings is, however, the increased annual average radial increment already evident at the age of 10 (Table 4). Thinning regimes in the region have changed for both species, as in most forests of Germany. Broad-leaved forests have been managed as high forests since the middle of the last century. Formerly they were often structured as standards with coppice. Thinning developed from moderate low thinning to crown thinning with higher intensity. This must be at least partly responsible for the observed radial growth trends of both species. Height growth is less strongly influenced by silvicultural management practices, but exceptions were described in literature. Thus intensified thinning may result in reduced height growth of dominant beech trees (Assmann 1970, p. 46; Dieterich 1925; Wiedemann 1932). From these studies the conclusion can be drawn that an identical thinning intensity for all beech sample trees of different germination dates would possibly have led to an even stronger height growth trend than was observed here. Contrasting findings exist for spruce. Wider spacing and intensified thinning can result in better height growth (Altherr 1966; Dieterich 1924; Kenk 1990; Pollanschütz 1974; Wiedemann 1937). This kind of spruce reaction seems to be especially true on poorer sites. The investigated site unit, however, provides an intermediate water-storing capacity and good nutrition availability. For this reason, the possible differences among thinning regimes probably did not have an important influence on the height growth of spruce sample trees. Kenk et al. (1991) and Moosmayer and Ripberger (1990) found fewer evident height growth trends on better sites. It is assumed that growth trends on poorer site units tend to be stronger.

The last main factor to be discussed is genetics. Improved spruce provenances of different and usually unknown origins have been selected and planted for several decades. A certain contribution of genetics to the observed growth trends of spruce cannot be excluded, whereas genetic changes for beech are less probable within the relatively short observation period. This can be concluded from the unchanged natural regeneration methods described above.

Height growth is only one indicator for site productivity. The relationship height growth - total volume growth, also termed yield level (Assmann 1970), could give substantial additional information about site productivity changes together with the observed height growth trends. Moosmayer (1957) investigated yield level on permanent plots in the Eastern Swabian Alb. Based on case studies of some permanent plot data, Moosmayer and Ripberger (1990) supposed over 30 years later that known regional specific relationships may still be valid.

5. Conclusions

The following conclusions can be drawn from this study: (1) height growth trends in beech are most probably due to site productivity changes because other factors influencing growth play a negligible role; (2) part of the observed height growth trends in spruce on the same site unit as beech are probably also caused by site changes, but part of them may be caused by changes in silvicultural management and genetics; (3) radial increment growth trends of both tree species may be caused by site changes as well as by altered silvicultural management and are therefore less useful for detecting site productivity changes.

Further investigations have to be conducted on other site units to give insight into possible differences between site types and information on the driving site factors of observed growth trends. Yield level and its changes must also be investigated. This is only possible in an explorative way because long-term permanent plots would have to be examined and these were not established with statistical methods. Therefore, only some indications about development of yield level in the last decades can be expected.

It had been assumed for a long time that site-growth relationships were constant. Growth rates are obviously higher now. Factors influencing growth trends remain unknown and are surely complex. Besides site productivity changes, changes of stand management and genetic influences have to be considered. Observed growth trends require adaptive forest management in the future, and management decision tools have to be constructed which take these facts into account. The importance of growth as an indicator for environmental changes is obvious. However, a prognosis of future forest growth hardly seems possible, because there is a lack of information about past and future development of many growth influencing factors. Long-term consequences of increased forest growth on the stability and vitality of forests are also unknown. The fertilization trials with positive growth response described above, for example, revealed a higher susceptibility to environmental stresses (Spiecker 1987, 1991).

Acknowledgements

The author is grateful to the Forestry Administration of Baden-Württemberg, Germany, to several forest owners and to the European Forest Institute, Joensuu, Finland, which provided resources for this project. He would like to thank Prof. Dr. H. Spiecker, Prof. Dr. H.-U. Moosmayer, C. T. Scott, an anonymous referee and several other colleagues for their helpful comments and suggestions.

References

Abetz P (1985) Ein Vorschlag zur Durchführung von Wachstumsanalysen im Rahmen der Ursachenforschung von Waldschäden in Südwestdeutschland. Allgemeine Forst- und Jagdzeitung 156 (9/10): 177-187.

Abetz P (1988) Untersuchungen zum Wachstum von Buchen auf der Schwäbischen Alb. Allgemeine Forst- und Jagdzeitung 159 (11/12): 215-223.

Altherr E (1966) Die Bedeutung des Pflanzverband für die Leistung der Fichtenbestände. Allgemeine Forstzeitschrift 11/12: 191-200.

Altherr E, Evers FH, Linnert H. (1974) Erste Ergebnisse von Fichten-Düngungsversuchen auf Feuersteinlehmstandorten des Wuchsbezirks "Nördliches Härdtsfeld" (Ostalb). Allgemeine Forst- und Jagdzeitung 145 (6): 101-111.

Assmann E (1970) The principles of forest yield study. Studies in the Organic Production, Structure, Increment and Yield of Forest Stands. Oxford: Pergamon Press. 506 p.

Buberl HG, v.Wilpert K und Hildebrand EE (1994) Verteilung bodenchemischer Stabilitätskriterien - BZE-Ergebnisse aus Baden-Württemberg. Allgemeine Forstzeitschrift 4: 199-203.

Dieterich V (1924) Aus den Aufnahmeergebnissen von Durchforstungsversuchen in Fichtenbeständen. Forstl. Wochenschrift Silva 12: 26-30.

Dieterich V (1925) Aus den Ergebnissen von Durchforstungsversuchen in Buchenbeständen. Allgemeine Forst- und Jagdzeitung 101: 16-35.

Eichkorn T (1986) Wachstumsanalysen an Fichten in Südwestdeutschland. AFJZ 157 (7): 125-139.

Fischer H, Rommel WD (1989) Jahrringbreiten und Höhentrieblängen von Buchen mit unterschiedlicher Belaubungsdichte in Baden-Württemberg. AFZ 11: 264-265.

Gerecke E (1988) Zum Wachstumsgang von Buchen in der Nordpfalz. Mitteilungen der Forstlichen Versuchsanstalt Rheinland-Pfalz 88/8: 126 p.

Houghton JT, Jenkins GJ, Ephraums JJ (1990) Climate Change. The IPCC Scientific Assessment. Cambridge: University Press. 365 p.

Jänichen H, Müller S, Schlenker G, Sebald O (1951) Die Waldstandorte des nördlichen Härdtsfeldes (Nordostalb). Mitteilungen des Vereins für Forstl. Standortskunde 1: 1-35.

Kenk G (1990) Fichtenbestände aus Weitverbänden - Entwicklungen und Folgerungen. Forstwissenschaftliches Centralblatt 109: 86-100.

Kenk G, Fischer H (1988) Evidence from nitrogen fertilisation in the forests of Germany. Environmental Pollution 54: 199-218.

Kenk G, Spiecker H, Diener G (1991) Referenzdaten zum Waldwachstum. Kernforschungszentrum Karlsruhe, Projekt Europäisches Forschungszentrum für Maßnahmen zur Luftreinhaltung 82: 1-59.

Koch H, Schairer E, Gaisberg Ev (1939) Die Buche der Ostalb - Eine Standortsuntersuchung. Mitteilungen der Württembergischen Forstlichen Versuchsanstalt. 234 p.

Moosmayer HU (1957) Zur ertragskundlichen Auswertung der Standortsgliederung im Ostteil der Schwäbischen Alb. Mitteilungen des Vereins für Forstl. Standortskunde 7: 3-41.

Moosmayer HU, Ripberger M (1990) Überprüfung waldwachstumskundlicher Auswertungen von Standortsgliederungen anhand neuerer Versuchsflächendaten. Mitteilungen des Vereins für Forstl. Standortskunde 35: 9-14.

Mühlhäusser G, Hübner W, Stummer G (1983) Die Forstliche Standortskarte 1:10000 nach dem baden-württembergischen Verfahren. Mitteilungen des Vereins für Forstl. Standortskunde 30: 3-13.

Mühlhäusser G, Müller S, Stummer G (1985) Forstliche Wuchsgebiete und Wuchsbezirke in Baden-Württemberg. In: ARBEITSKREIS STANDORTSKARTIERUNG IN DER AG FORSTEINRICHTUNG: Forstliche Wuchsgebiete und Wuchsbezirke in der Bundesrepublik Deutschland. Münster-Hiltrup: Landwirtschaftsverlag Gmbh. 172 p.

Newberry JD (1991) A note on Carmeans estimate of height from stem analysis data. Forest Science 37 (1): 368-369.

Pollanschütz J (1974) Erste ertragskundliche und wirtschaftliche Ergebnisse des Fichten-Pflanzweiteversuches "Hauersteig". Forstliche Bundesversuchsanstalt Wien, 100 Jahre Forstliche Bundesversuchsanstalt: 100-171.

SAS Institute Inc. (1987) SAS/STAT™ Guide for Personal Computers, Version 6 Edition. Cary, NC, 1028 p.

Schlenker G, Müller S, Dieterich H, Hauff R, Jahn R, Mühlhäusser G, Schönnamsgruber H (1973) Erläuterungen zur Karte der Regionalen Gliederung von Baden-Württemberg. I. Teil (Wuchsgebiete Neckarland und Schwäbische Alb). Mitteilungen des Vereins für Forstl. Standortskunde 23: 3-66.

Schönwiese CD, Rapp J, Fuchs T, Denhard M (1993) Klimatrend-Atlas Europa 1891 - 1990. Berichte des Zentrums für Umweltforschung 20: 218 p.

Schöpfer W, Hradetzky J, Kublin E (1994) Wachstumsänderungen der Fichte in Baden-Württemberg. Forst und Holz 49 (21): 633-644.

Spiecker H (1987) Düngung, Niederschlag und der jährliche Volumenzuwachs einiger Fichtenbestände Südwestdeutschlands. Ergebnisse einer neuen Methode der Zuwachsermittlung. Allgemeine Forst- und Jagdzeitung 158 (4): 70-76.

Spiecker H (1991) Liming, nitrogen and phosphorus fertilization and the annual volume increment of Norway spruce stands on long-term permanent plots in Southwestern Germany. Fertilizer Research 27: 87-93.

Spiecker H (1995) Growth dynamics in a changing environment - long-term observations. Plant and Soil 168-169: 555-561.

Spiecker H, Mielikäinen K, Köhl M, Untheim H (1994) Growth trends of European forests - Has site productivity changed? European Forest Institute Working Paper 4: 1-68. Joensuu, Finland.

Weise U (1991) Ertragsniveau und Zuwachsgang der Weißtanne - Auf Versuchsflächen im badischen Schwarzwald. AFZ 4: 192-195.

Wiedemann E (1932) Die Rotbuche 1931. Mitteilungen der Preußischen Forstlichen Versuchsanstalt: 1-191.

Wiedemann E (1937) Die Fichte 1936. Zweiter Teil. Die Durchforstung und Lichtung.-Die neuen Ertragstafeln für verschiedene Durchforstung.-Die standörtlichen Untersuchungen.-Zusammenfassung. Mitteilungen aus Forstwirtschaft und Forstwissenschaft 8: 103-248.

Wilpert Kv, Hildebrand EE (1994) Stoffeintrag und Waldernährung in Fichtenbeständen Baden-Württembergs. Forst und Holz 21: 629-632.

Forest Decline and Growth Trends in Central Europe – a Review

Hubert Sterba
Institute of Forest Growth and Yield, Universität für Bodenkultur Wien
Peter Jordan-Strasse 82, A-1190 Vienna, Austria

Abstract

Reports of forest decline in the 1980s and recent reports of increasing increments seem to contradict one another. An investigation of the literature referring to Central Europe suggests that regional differences result from differences in views and survey methods rather than from site or climatic conditions.

Forest decline was first recognized in terms of needle loss, and this was defined as an indicator for decreasing vitality. Most of the research undertaken to find relationships between increment and crown transparency classes was methodically unable to detect anything other than increment decreases as the employed reference was usually a "best-growing" or healthy tree. Thus, trees with less needles than these could only show less or at most the same increment. The discussion focused only on the question at what degree of needle loss increment was affected.

Only few investigations at that time dealt with stand or plot increments, some of them observing positive deviations from the expectations taken from yield tables. Later observations especially of permanent plots where a second generation of stands was already being observed presented a tremendous increase in growth. These could well be regarded as singular events under singular site conditions. However, in Austria and Bavaria, regional and national forest inventories of sophisticated design have proved that increments increased by at least 25% between 1980 and 1990.

From this point of view, the observation that at the beginning of the 1980s needle losses suddenly seemed to exceed those expected might have to be interpreted as a sign of increasing differentiation between trees of different vigour together with competition effects due to increased per hectare increment.

Nevertheless, not enough is yet known about how or if site potential has changed to be able to make long-term forecasts of forest growth.

European Forest Institute Research Report No. 5
Growth Trends in European Forests
Edited by H. Spiecker, K. Mielikäinen, M. Köhl and J. P. Skovsgaard

1. Introduction

By the end of the 1970s, foresters increasingly began to observe a decrease in the health of trees and stands and a loss of tree vigour, defined in terms of needle or leaf losses or yellowing. In several cases, trees and stands seemed to die unexpectedly, and very soon air pollution was considered to be the main cause for what was interpreted as forest dieback or at least a large-scale forest decline. While many research programs were being established in Europe on a national scale to monitor this forest decline and to investigate the causes of decreasing tree vigour and tree health, by the end of the 1980s there were already some reports of increased growth (increments) deriving from permanent research plot data and from forest inventories. Recent discussion of changes in site productivity has been concentrating on increasing rather than decreasing site potential. The objective of this review is to find reasons for these seemingly contradictory research results.

2. Forest Dieback (*Waldsterben*) and Growth

In the past, many researchers merely assumed that air pollution caused forest dieback (Waldsterben) and recorded needle or leaf loss, needle or leaf yellowing, or crown transparency as an indicator of its impact; most did not or could not measure pollution or deposition of pollutants associated with the leaf, needle or transparency data. Only a few authors presented a detailed discussion of the fact that air pollution might affect growth without being manifested in needle loss and that needle loss might be caused by processes other than air pollution (Pollanschütz 1985; Sterba 1988; Eckmüllner 1988a).

Certain authors pointed out that what was really being observed was that some trees have less needles than others, and that it was not the syndrome of needle or leaf **loss** that was being recorded. They therefore proposed abandoning the concept "needle loss" and thinking in terms of tree vitality (e.g. Bert and Becker 1990) or, even better, in terms of crown aspect (e.g. Pollanschütz 1985; Becker et al. 1990) or crown transparency (Schneider et al. 1988) as an expression of tree health – independent of the cause for reduced vitality or poorer health. It was also recognized that a double reference system was needed, one that incorporated a measure for the crown aspect of a healthy tree and an increment.

Since both crown aspect and increment, and their relationship, depend on many factors, for example site, age, tree size and stand density, much effort has been put into defining standard measures. While some experts who monitor trees use standardized pictures to compare trees in the field directly or via aerial photographs (Hartmann 1984; Pollanschütz 1985; Bosshard 1986), others use the number of needle age classes (e.g. Lick and Krapfenbauer 1986), or total needles counted on a twig in a standardized position in the crown (Knabe 1981) or the needle mass related to a stem or crown dimension (Kramer and Dong 1985; Eckmüllner 1988a).

As it seemed clear that on the average trees will show superior growth when they have more needles or leaves (all other conditions beeing equal), a double index was often used as an indicator of increment; first of all the increment of the given tree at a reference growth period in the past when the tree was thought to be unaffected by any symptoms of dieback or reduced vitality (König 1979; Röhle 1987a,b; Eckmüllner 1988a, and many others); and secondly, the growth of a reference tree of comparable size, but without or with only negligible needle losses. Others, for example Abetz (1984) and Gerecke (1988), artificially constructed the reference tree from the height/age relationship of future crop trees taken from yield tables and from assumptions about their height/diameter ratio development (i.e. as an open-grown tree totally surrounded by other non-interfering trees, the "gezähmte Solitär" of Abetz).

Whatever kind of reference was defined, the standard method of seeking relationships between forest dieback and growth became the derivation of the correlation of an index that described the crown aspect of a tree relative to a "healthy" counterpart with another index, usually a double index similar to that described above (e.g. Athari and Kramer 1985; Röhle 1985; Schöpfer 1986; Becker 1987; Sterba and Eckmüllner 1988). Clearly, however, by assuming that trees in a healthy forest would all be similar to the 100% reference tree, this method could by definition only recognize "negative" aspects (i.e. values <100%), and this fed the concept of forest dieback.

These investigations were mostly performed only at the tree level, seeking answers to questions such as:

- From what date do crown condition classes differ in relation to tree growth (Röhle 1985; Becker 1987; Schöpfer and Hradetzky 1987; Bert and Becker 1990)?
- From what level of needle loss does growth reduction occur (Eckmüllner 1988b; Spelsberg 1988; Beyschlag et al. 1994)?

Few studies were conducted that actually compared the increment of "healthy stands" with those of stands demonstrating the symptoms of forest dieback (e.g. Athari and Kramer 1985). Most of these comparisons were made by presuming model stands consisting of appropriate proportions of trees of different crown transparency classes (based on actual forest counts), contrasting these with a theoretical "healthy" stand consisting of a mix of needle-loss classes or a range of crown aspects as the 100% reference (Schöpfer and Hradetzky 1987; Eckmüllner 1988b). However, even though these reference stands allowed the existence of a range of tree health categories, they were still always superior to an actual stand, and so, although the difference between the healthy reference stand and the actual stand was reduced, one could still only record a lowered increment for the actual stand.

The only point of discussion in the above context was the degree of needle loss or proportion of trees with poor crown aspects in a stand that led to marked growth reductions (Athari and Kramer 1985; Kenk 1987; Eckmüllner 1988b).

It is clear that the phenomenon of forest dieback (later also termed forest decline or new forest disease) became not only the explanation for the increased susceptibility of stands to insect and fungal damage, new risks and unexpected mortalities, but was also predestined to be the basis for forecasts of decreased growth and reduced yields.

3. Hidden Indications of Increasing Increments in Forest Decline Studies

Although, as a result of the methodological approach sometimes influenced by expectations, the usual studies did not show growth increases, several of these studies do contain hints that – compared with expectations from age trends, not from crown transparency classes – some evidence for increasing as well as decreasing increments had been noted.

For example Sterba and Eckmüllner (1988) found a relationship between the ratio observed increment for Norway spruce (*Picea abies*) to increment expected from the age trend of the same trees on the one hand, and the needle index (i.e. the needle mass per square centimetre cross-sectional area at the base of the crown) on the other (Fig.1). Eckmüllner (1988b) noted that the maximum needle index was 150 g cm^{-2}, and that it was rather independent of site and age. Such trees had been used as the 100% reference but, as can be seen from Fig.1, 150 g cm^{-2} corresponds to an increment ratio of 110%, i.e. some 10% higher than that expected from the age trend.

Increment increases running counter to the age trend for Norway spruce were also reported by Kenk et al. (1991). In the Black Forest they observed in two permanent thinning experiments that, in one case, the current volume increment at the age of 196 years was higher than it was when the experiment was started at an age of 76 years and that, in the other, the increment increased nearly linearly from 8 $m^3 ha^{-1} a^{-1}$ in 1945 at an age of 73 years to 14 $m^3 ha^{-1} y^{-1}$ in 1988 when it was 116 years old, without any signs of culmination.

When Röhle (1986) showed that there was a distinct differentiation in basal area growth of his crown transparency classes as compared with expectations (i.e. trees with poorer crown transparency showed less growth), he also gave a diagram from which can be deducted that the first 3 classes (out of 5) had distinctly higher basal area increments after the reference period (1961-1970). Between 1981 and 1983, they showed on average 125% of their previous increment.

In his work on the "vitality increment and biomass structure of spruce in several Bavarian areas", Röhle (1987b) gives the following equation for stand increments, derived using regression analysis:

$$\frac{\text{increment}_{1980\text{-}83}}{\text{increment}_{1961\text{-}70}} \bullet 100 = 170 - 1.127\text{NL} - 0.000638\,\text{NL}^2 - 0.515\text{age} + 0.0000112\,\text{EL}^2$$

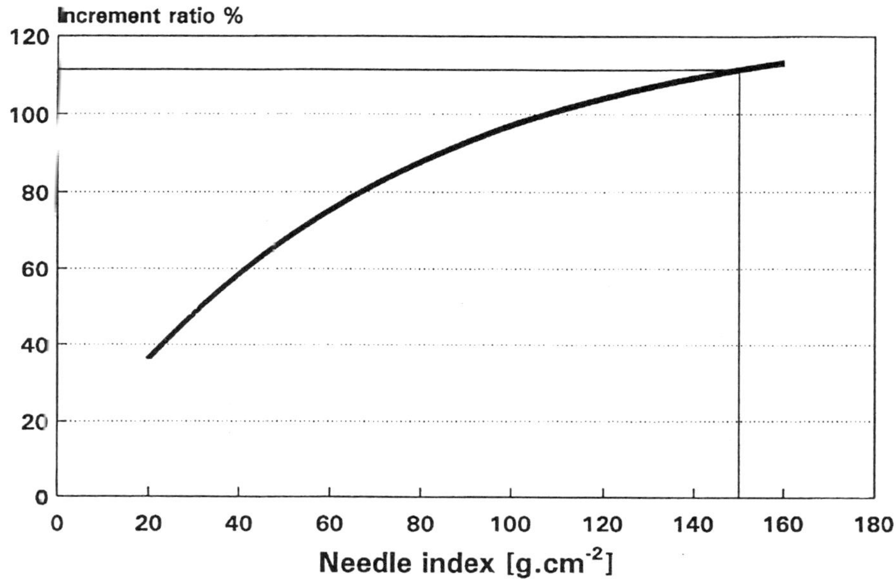

Fig. 1. Average relationship between needle index (gram needles per square centimetre of cross sectional area at the base of the crown) and increment ratio, i.e. the observed basal area increment in percent of the basal area increment expected from the age trend of past growth of the same tree (Eckmüllner 1988). At maximum needle index, found to be 150 g cm^{-2}, this index is about 110%.

NL being the estimated needle loss in percent, *age* the age of the stand, and *EL* the elevation (altitude) of the stand. To show the effect of needle loss on growth, he later used this equation to obtain the "reference increment" by putting *NL*=0. He only reported the ratio between the index calculated from the equation above for a given needle loss and that for *NL=0*. If only the first index, that is the ratio between the stand increment for 1980-1983 and the reference increment for 1961-1970 for *NL*=0 is calculated, it is greater than 100% for nearly all ages and elevations (Fig. 2). Thus, for stands without needle losses, the increment between 1980 and 1983 is distinctly higher than during the decade before. The above equation can also be used to estimate the needle loss at which the increment in 1981-1983 is the same as that for 1971-1980 (Fig. 3). It is obvious that a stand could have rather high needle losses, and a very poor transparency class distribution, without revealing increment losses in terms of the increment of the reference period.

A similar result must underlie the statement of Franz and Pretzsch (1988), when they write for Scots pine:

"the increment of the stands which seem damaged today had increments in the 50s and 60s which were about 50-100% higher than the yield tables would have expected. Therefore the yield tables are of no help to evaluate growth reductions,

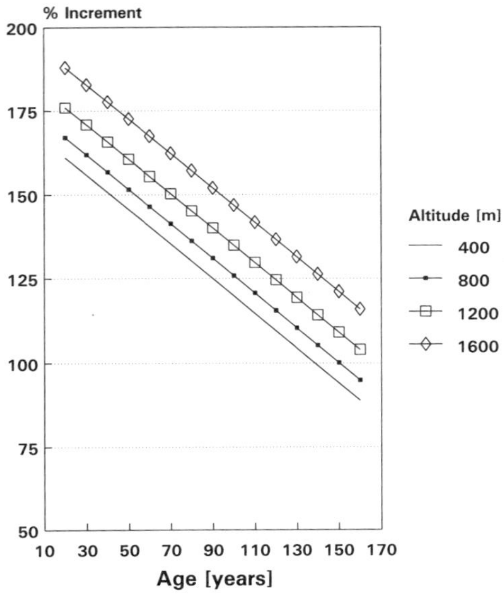

Fig. 2. Basal area increment 1981-1983 in per cent of reference increment 1961-1981 for trees with no needle losses depending on altitude and age, calculated from the equations given by Röhle (1987b).

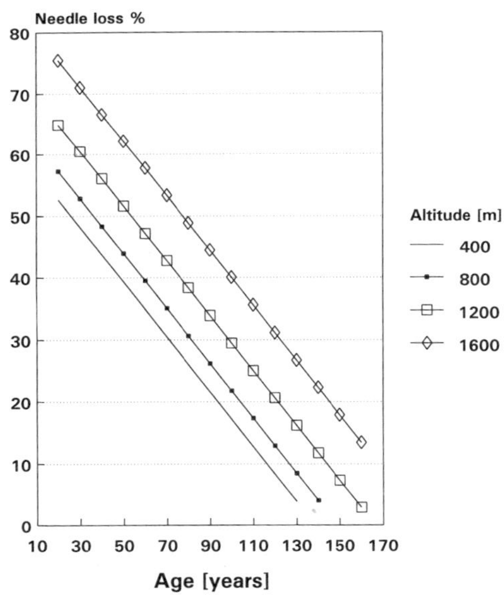

Fig. 3. Needle loss in percent, where the increment for 1981-1983 equals the reference increment depending on altitude and age, calculated from the equations given by Röhle (1987b).

because a Scots pine stand with 20-30% increment reductions still would have increments higher than the yield table expectations".

Here again, two different comparisons are considered: one with "healthy" trees in the same increment period, and past increments of the same stands.

For the European silver fir (*Abies alba*), Becker (1987) in the Vosges and Bert and Becker (1990) in the French Jura used a "growth index" that had previously been introduced by Abetz (1984). It relates the tree-ring width in a given tree in a given year to the average tree-ring width of trees of the same age rather than of trees in the same year, thereby examining whether trees of the same age have the same or a different increment at different dates. This index already differs noticeably from the results of needle-loss data since about 1935. It is also obvious that only trees with more than 20% needle loss exhibit decreasing "growth indices" since about 1940; all other crown transparency classes show increasing increments.

4. Yield Tables as Stand Growth References

The site index from yield tables is thought to predict increment, as it results from standard treatment that is near an "optimum treatment" (in the sense that it results in the maximum increment possible at a given site (Assmann 1961)). Thus, the tables are supposed to indicate site quality or the growth potential of a site. With only few exceptions (Athari and Kramer 1985), forest-dieback studies indicated more stand growth than expected from the yield tables.

4.1. Norway Spruce (*Picea abies*)

For repeatedly observed stands in Bavaria, Utschig (1989) reports that, while two had approximately the same volume increment as predicted by the yield table, three others exhibited 30-100% more growth. Also for Bavarian stands, Franz et al. (1986) found that, since 1957, the volume increment per hectare had remained constant or had increased, while according to the yield tables it should have decreased. Similar observations were made by Spiecker (1987) in Black Forest spruce stands. In permanent plots, Kenk (1987) found that, for up to 35% needle losses, the increment per hectare did not decrease; in most of the stands studied, increments between 40% and 85% greater than the yield table predictions were measured. The steadily increasing volume increments in old spruce stands observed by Kenk et al. (1991) have already been mentioned. In the Austrian part of the Bohemian Massif, Sterba (1992) and Hasenauer (1994) report from stem analyses that the height increments of spruce were about twice or even three times those predicted by the yield tables of Guttenberg (1896) that had been developed from stem analyses for the very same site. Spelsberg too (1995) found that the site index of spruce in Nordrhein Westfalen had increased, especially in the older stands, while the height increment from stem analyses were for the most part greater than predicted by the appropriate yield table (Wiedemann 1936, 1942).

4.2. European Silver Fir (*Abies alba*)

Probably because many forest scientists consider that the yield tables available for the silver fir do not represent its growth exactly enough, as pure even-aged fir stands are rare, and age determination by stem analysis of dominant firs which were severely suppressed in their youth is difficult (Gerecke 1988), no reports of growth comparisons with yield table projections for fir were found. Only Spiecker (1991) reports that mixed age, mixed stands of spruce and fir exhibit large differences in growth; he explains this with climatic events. No general trend, either increasing or decreasing, could be deduced for these stands from his data.

4.3. Scots Pine (*Pinus sylvestris*)

Already in the late 1960s, Schmidt (1969) had noted that, as of 1955, based on stem analyses, the height growth of Scots pine had increased rapidly compared to the site-index curves that he himself had developed from analyses of the very same stems at a previous time. Pretzsch (1985) also found for Scots pine stands in southern Germany that their height growth was greater than that derived from Wiedemann's yield table (1948). The volume increment had increased dramatically since the 1960s, independent of age and site index. From 1975 to 1980, while the volume increment had decreased, it never again reached the low values predicted by the yield table. Similar results presented by Franz and Pretzsch (1988) have already been discussed above. When Foerster (1990) evaluated the results of old fertilizer experiments with Scots pine, he found that the control plots all displayed about 50% more volume increment than predicted by the yield tables. The volume increment in these plots had increased counter to the age trend, and the site index determined by age and dominant height had generally improved. Sterba (1992) and Hasenauer (1994) also found that the height increment of dominant pines vastly exceeded the predictions of Guttenberg's (1896) yield table derived from stem analyses at the very same site.

In order to compare the results of Hasenauer (1994) and Foerster (1990), the "relative" site classes of Wiedemann (1948) are here converted into absolute site indices (mean annual volume increment at age 100 years) as they are given in the Scots pine yield table of Marschall and Sterba (1975) using the site index curves of Guttenberg (1896). Linear regression analysis was used to evaluate the site index improvement of Scots pine (Foerster 1990, Fig. 182, p.253). The comparison of the reevaluated results is given in Fig. 4 (Hasenauer 1994, broken line; Foerster 1990, full line). It shows the improvement percentage in the site index with age. Though in both studies the site index improvement is highest in young stands of low initial site

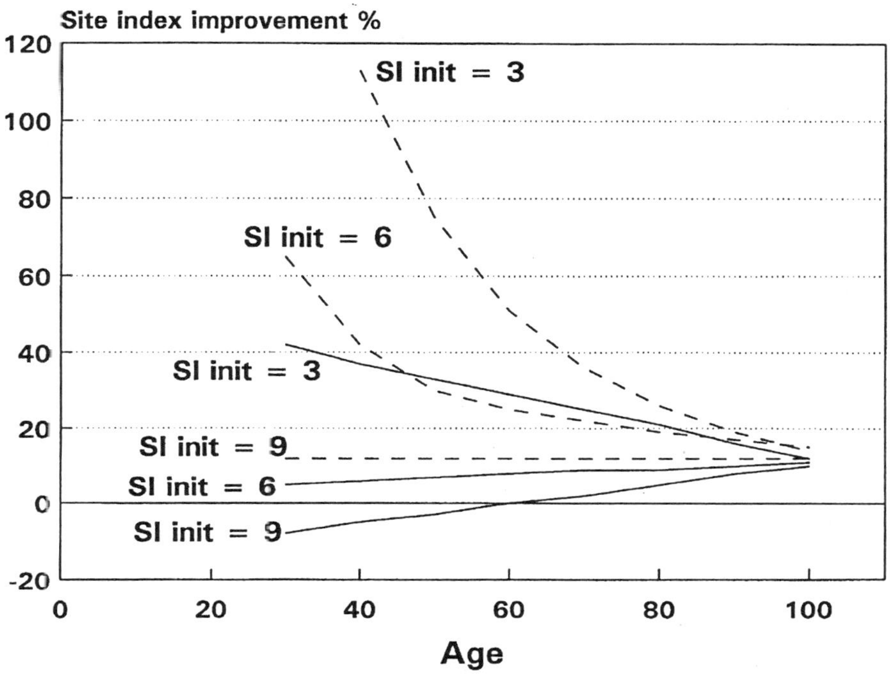

Fig. 4. Relative site index improvement (% of initial site index) depending on age and initial site index (SI init). Full lines are calculated from Foerster (1990), dashed lines are taken from Hasenauer (1994). Site index is defined as mean annual increment at age 100 in m^3a^{-1} ha^{-1}.

class, the magnitude of the effect is much greater for the Hasenauer study than for that of Foerster.

4.4. Beech (*Fagus sylvatica*)

For stands in the Schwäbische Alb, Abetz (1988) found that on all sites the site index for the approximately 50-year-old stands was distinctly higher than that for the old (100-year) stands on the same site (14-40% site index difference). However, the site index change within a given stand was not particularly large. On sites with good water supply, he showed only small site index improvements, while on dry sites the site index diminished.

An example of very large increment deviations from the yield table predictions is given by Pretzsch (1992). In 1929, an old thinning experiment started at age 100 years to show increased stand growth as compared with yield table expectations as well as with the stand's own past development. The then 160-year-old stand exhibited a current volume increment of 14 m^3 ha^{-1} y^{-1}, as compared with the 9 $m^3 ha^{-1} y^{-1}$ average between 1870 and 1940.

5. Indications of Increasing Growth from Other Data

While increment deviations from the yield table expectations could be caused not only by site quality improvement but also by poorly developed yield tables, a number of studies on tree ring series for trees of different ages show similar results. Becker (1987), Becker et al. (1990), Bert and Becker (1990), Becker et al. (1994), and, most recently, Neumann and Schadauer (1995) use a similar principle for their methodological approach. They compare the increments of trees of the same age in different calendar years. With few exceptions, the general discovery has been that trees of the same age had smaller tree ring widths earlier in the century. For example, Neumann and Schadauer (1995) demonstrated that spruces in Austria in 1977 of about 120 years of age had a tree ring width of about 1.8 mm when they were 30 years old in 1887. However, 80-year-old spruces in 1977 had a 2.5 mm average tree ring width in 1927 when they were 30 years old. Thus, 30-year-old trees in 1944 had an approximately 40% greater tree ring width than trees of the same age in 1887. They recognize this increase of tree ring width for the same cambial age with time to be rather regular for over 100 years, and find it for nearly all strata chosen on the basis of site characteristics. Similar results are reported for fir in the Vosges (Becker 1987) and Jura (Bert and Becker 1990), and for beech at the Lorraine side of the Vosges (Picard 1995).

A last indication of the tendency for better growth of trees are the national and/or regional forest inventories. In a 1100-ha district of the Bavarian State forests, some 500 plots for a permanent inventory were established in 1984 (Fuchs and Kennel 1994). The audit in 1994 showed that the volume increment observed (approx. 11 $m^3 ha^{-1} a^{-1}$) was 80% higher than predicted by the appropriate yield tables. For the whole of Austria, Schieler and Schadauer (1993) found that the total volume increment had increased by about 30% compared with the last inventory taken in 1971-1980; this in turn had already shown a higher increment than that for the previous inventory period (1961-1970). While the inventories in 1961-1970 and 1971-1980 were inventories with temporary plots only, the final audit in 1990 was already based on permanent plots and was thus more reliable than earlier ones.

6. Discussion

As already indicated, most of the investigations on increment and forest dieback carried out in the past 30 years were methodologically unable to detect an increment *increase*. Several of these studies clearly describe their derivation of the "reference growth", and can thus be used to demonstrate that increasing increments certainly existed in some of the research areas. From others, it seems that there was actually an increment depression during the late 1960s and early 1970s; partly attributable to climatic events, partly explainable by hypothesizing air pollutants of some kind as the underlying cause. A number of cases were found with increments higher than expected from age trends and/or yield tables. An explanation why both types of result — forest dieback and increasing increments — could be correct has been suggested by Spelsberg (1987), Gerecke (1988) and Spelsberg (1988). They describe the "forest dieback phenomena" as an effect of differentiation of vitality classes. This coincides with the findings of Löchelt and Franke (1993) and Löchelt (1995). They noted significant genetic differences between crown transparency classes in spruce as well as fir, thus indicating that there may be genetic factors that influence the sensitivity of trees to whatever the external cause for this differentiation may be. The idea that increasing growth alone could, because of greater competition stress, lead to differentiation and higher mortality rates was the result of a simulation study by Sterba (1995).

Yet other investigations that concentrated on the effect of increasing growth mostly differed concerning the interpretation of the results. The questions raised were:

- What are the causes for the large differences between expected growth and actual growth?
- Is there a long-term trend towards increasing growth or only a short-term one, but with large variations?

Only tentative answers to these questions can be given. At the beginning of the 1960s, Eidmann (1961) already believed yield table projections to be of no value because weather conditions might overrule the "growth laws" underlying yield table construction. Even Assmann (1961) admitted that, owing to climatic variations, increments might deviate seriously from yield table predictions.

Among the investigations cited in this report, several authors conclude that nearly all, or at least many of the observed increment variations could be explained by climatic variations or singular climatic events (König 1979; Becker et al. 1990; Spiecker 1986, 1987, 1991; Abetz 1988; Kahle et al. 1993). Some state, however, that there are factors additional to those of climate that affect increment. A number of authors put forward air pollution (Eckstein and Bauch 1987) and carbon dioxide (Becker et al. 1994). Sterba (1993) found significant though weak correlations of the height increment of dominant spruces and Scots pines with climatic variables. While most of the authors are not able or willing to decide whether their data and methodological approaches point to specific long-term trends or not, Becker et al. (1994) present time series of increment, precipitation and temperature that seem to correlate for at least 100 years. Citing climatic studies, Hager (1994) found similar — but not yet statistically significant — trends for increasing temperature and decreasing precipitation for central Europe. Nonetheless it is not clear why increasing temperature in combination with decreasing precipitation should generally lead to higher increments.

Nitrogen deposition has also been suggested as an explanation for symptoms of forest decline (see e.g. Nihlgaard 1985; Oren and Schulze 1989; Skeffington and Wilson 1988) as well as an explanation for increasing increments (Kenk and Fischer 1988; Dralle and Larsen 1995). Decreasing response to nitrogen fertilization as a result of excess nitrogen deposition and thus nutritional disharmony seems to be evident in northern Europe (Dralle and Larsen 1995). In respect to central Europe, however, Stefan (1995) still finds that more than half of Austria's forests suffer from nitrogen deficiency, and Kenk and Fischer (1988) have evidence of a still continuing fertilization effect of nitrogen deposition on increment in parts of France and Germany.

The steady increase in CO_2 concentration in the atmosphere over the past 100 years has also been confirmed for central Europe, at least since 1972 (Fricke and Wallasch 1994). According to Rosenberg et al. (1990), increment should proportionally increase in C_3 plants, to which group our tree species belong.

The findings of Becker (1987) , Becker et al. (1990), Bert and Becker (1990), Becker et al. (1994), and Neumann and Schadauer (1995), namely that the radial growth of trees of the same age was much less 20 to 100 years ago, correlate well with those who find a long-term site quality improvement, indicated by better site indices of younger stands as compared with older ones (Foerster 1990; Spelsberg 1995; Abetz 1988). While increasing diameter increments could well also be the effect of thinning, there seems to be no reason to assume that thinning frequency and/or intensity have increased continuously over the last 100 years. Since the 1960s, earlier and heavier thinnings have often and strongly been recommended in Germany (e.g. by Abetz 1975) as well as in Austria (e.g. Pollanschütz 1971). While on the one hand, this is proof that between the late 19th century and the middle of the 20th century, during which period increasing diameter increments have also been measured, there must in fact have been little thinning (otherwise such recommendations would not have been necessary!); on the other hand, at least for Austria, a comparison of the last three observation periods (1961-70, 1971-80 and 1981-90) for the Austrian forest inventory has shown that the volume recommended for removal by thinning increased from $20 \times 10^6 m^3$ in 1970, to $29 \times 10^6 m^3$ in 1980, and to $65 \times 10^6 m^3$ in 1990 (Forstliche Bundesversuchsanstalt 1985, 1993). However, even if views concerning the required thinning intensity have changed, these figures cannot be interpreted as signs of increasing thinning intensity actually performed.

While the improvement of diameter increment could well be associated with a long-term improvement of silvicultural management, this should not seriously affect the height increment of dominant trees. Only in very young stands has accelerated height growth through initial wide spacing been reported (Pollanschütz 1974), and even this effect is only temporary; later, the dominant height of the different initial spacing treatments approach each other.

The fact that younger stands have better site indices is well known to every forest surveyor (Mantel 1959, p. 151) and has led to the "dynamic approach of site index determination" in young stands. One reason for higher site indices of younger stands may be that stands on poor sites are usually required to grow longer and are thus, on average, older than those on better sites. This effect could also have contributed to the very clear relationship between time and radial increment of trees of the same age found by Becker (1987), Becker et al. (1990), Bert and Becker (1990), Becker et al. (1994). In contradistinction, at least for some sites, Neumann and Schadauer (1995) and Abetz (1988) have found a similar effect *within* the same site; and in a few areas on well documented permanent plots that had already borne two generations of stands, the second generation has exhibited distinctly larger site indices than the first (Kenk et al. 1991).

7. Conclusions

From the results of a great number of growth and yield studies, the question of whether there is a more or less continuous trend of increasing increments since the late 19th century or only a rather short period of higher increments between 1980 and 1990 is open to discussion. Even studies which emphasized forest decline symptoms can be shown to have indicated greater increments as compared with previous reference periods. Whether there is a long-term trend or whether we are merely seeing an episode of greater increments, later to be followed by decreasing increments, will depend both on the site and the causes. Increasing CO_2 concentration will probably lead to a more steadily growing increment. The same would be true for increasing temperatures if not accompanied by decreasing precipitation, i.e. droughts. Nitrogen deposition would, on many sites, sooner or later result in nitrogen saturation and nutrient imbalances. Whether there is any interaction between these effects in relation to increment and susceptibility of trees to pests cannot yet be answered. I believe that, on present evidence, it is not possible at this time to make a serious long-term forecast of future forest growth.

References

Abetz P (1975) Eine Entscheidungshilfe für die Durchforstung von Fichtenbeständen, Allg. Forstz. 33/34: 666-668.

Abetz P (1984) Die physiologische Belastbarkeit der Waldbäume. Zum Erkennen aus waldwachstumskundlicher Sicht, Allg. Forstztg 95(11):322-325.

Abetz P (1988) Untersuchungen zum Wachstum von Buchen auf der Schwäbischen Alb, Allg. Forst- Jagdztg 159:215-223.

Assmann E (1961) Waldertragskunde, BLV, München, 490 pp.

Athari A, Kramer H (1985) Ergebnisse von Wachstums- und Strukturanalysen in erkrankten Fichtenbeständen, In: Inventur und Wachstum in erkrankten Fichtenbeständen Kramer H, Athari S, Akca A, Dong PH (ed), Schr. Forstl. Fak. Univ. Göttingen 82:6-96.

Becker M (1987) Bilan de santé actuel et retrospectif du sapin (Abies alba Mill.) dans les Vosges. Etude écologique et dendrochronologique, Ann. Sci. 44(4):379-402.

Becker M, Bert GD, Bouchon J, Picard JF, Ulrich E (1994) Tendences à long terme observées dans la croissance de divers feullus et resineux du nord-est de la France depuis le milieu du XIXe siecle, Rev. For. Fr. XLVI (4):335-341.

Becker M, Bräker OU, Kenk G, Schneider O, Schweingruber FH (1990) Aspect des houppiers et croissance des arbres au cours des dernières decennies dans les régions frontalières d'Allemagne, de France et de Suisse, Rev. For. Fr. XLII (3):284-300.

Bert GD, Becker HM (1990) Vitalité actuelle et passée du sapin (Abies alba Mill.) dans le Jura, Etude dendrochronologique, Ann. Sci. 47:395-412.

Beyschlag W, Ryel RJ, Dietsch C (1994) Shedding of older needle age classes does not necessarily reduce photosynthetic primary production of Norway spruce, Trees 9:51-59.

Bosshard W (ed) (1986) Kronenbilder mit Nadel- und Blattverlustprozenten, Eidg. Anst. forstl. Versuchswes. Birmensdorf, 98 pp.

Dralle K, Larsen JB (1995) Growth response to different types of NPK-fertilizer in Norway spruce plantations in Western Denmark, Plant and Soil 168-169:501-504.

Eckmüllner O (1988a) Zuwachsuntersuchungen an Fichte im Zusammenhang mit neuartigen Waldschäden, Diss. Univ. Bodenkultur Wien, Vienna, 129 pp.

Eckmüllner O (1988b) Neuartige Waldschäden. Benadelung und Zuwachs der Fichte, Kooperationsabkommen Forstwirtschaft-Platte-Papier, Vienna.

Eckstein D, Bauch J (1987) Dendroklimatologische Untersuchungen an Fichten aus einem Waldschadensgebiet im Forstamt Farchau/Ratzburg, GSF-Bericht 10/87:156-164.

Eidmann FE (1961) Langperiodische Klimaänderungen und ihr Einfluß auf ertragskundliche Tatbestände, Allg. Forst- Jagdztg. 132:137-156.

Foerster W (1990) Zusammenfassende ertragskundliche Auswertung der Kiefern-Düngungsversuchsflächen in Bayern: ein Beitrag zur Beschreibung des Kiefernwachstums in Süddeutschland, Forstl. Forschungsber. München 105:328 pp.

Forstliche Bundesversuchsanstalt (1985) Österreichische Forstinventur 1971-80. Inventurgespräch, Mitt. Forstl. Bundesversuchsanstalt, 154/II: 100 pp.

Forstliche Bundesversuchsanstalt (1993) Österreichische Forstinventur 1986-1990, Diskette, Forstliche Bundesversuchsanstalt, Wien.

Franz F, Pretzsch H (1988) Zuwachsverhalten und Gesundheitszustand der Waldbestände im Bereich des Braunkohlekraftwerkes Schwandorf, Forstl. Forschungsber. München 92:156 pp.

Franz F, Preuhsler T, Röhle H (1986) Vitalitätsmerkmale und Zuwachsreaktionen erkrankter Bergwaldbestände im bayerischen Alpenraum, Allg. Forstz. 39.

Fricke W, Wallasch M (1994) Atmospheric CO_2 records from sites in the UBA air sampling network, In: Trends '93. A compendium of data on global change, Boden TA, Kaiser DP, Sepanski RJ, Stoss FW (eds) CDIAC Oakridge, TN, pp.135-147.

Fuchs A, Kennel E (1994) Erste Wiederholungsaufnahme einer permanenten Betriebsinventur im bayerischen Staatswald, Forstwiss. Centralbl. 113:224-235.

Gerecke KL (1988) Herleitung und Anwendung von "Referenzbäumen" zur Beschreibung des Wachstumsganges vorherrschender Tannen, Diss. Univ. Freiburg/Breisgau.

Guttenberg Av (1896) Die Aufstellung von Holzmassen- und Geldertragstafeln von Stammanalysen, Österr. Vierteljahrsschr. Forstwes. Wien, 36:203-237; 319-345.

Hager H (1994) Mögliche Einwirkungen von Klimaveränderungen auf forstliche Ökosysteme, In: Klimaänderung in Österreich Geburek Th, Müller F, Schultze U (ed), FBVA Berichte 81, Forstliche Bundesversuchsanstalt, Wien, pp.7-18.

Hartmann G (1984) Waldschadenserfassung durch Infrarot-Farbluftbild in Niedersachsen, Forst- Holzwirt, 39:131-142.

Hasenauer H (1994) Ein Einzelbaumwachstumssimulator für ungleichaltrige Fichten-Kiefern- und Buchen-Fichtenmischbestände, Forstl. Schriftenr. Univ. Bodenkultur Wien, Vol. 8, Österr. Ges. Waldökosystemforsch. exp. Baumforsch., Univ. Bodenkultur Wien, Vienna, 152 pp.

Kahle HP, Hahn D, Spiecker H (1993) Witterungsverlauf und Wachstum von Fichten, Tannen und Buchen in Südwestdeutschland unter besonderer Berücksichtigung des Standorts und des Standraums, KfK-PEF 104:241-252.

Kenk G (1987) Der Vergleich von Wachstumsdaten aufeinander folgender Waldbestandsgenerationen, KfK-PEF 12:87-99.

Kenk G, Fischer H (1988) Evidence from nitrogen fertilization in forests of Germany, Environm. Pollut. 54:199-218.

Kenk G, Spiecker H, Diener G (1991) Referenzdaten zum Waldwachstum, KfK-PEF 82:59 pp.

Knabe W (1981) Immissionsökologische Waldzustandserfassung in Nordrhein-Westfalen, Allg. Forstz. 36:641-647.

König E (1979) Entwicklungstendenzen bei der Tannenerkrankung. Forst- Holzwirt, 16:361-366.

Kramer H, Dong PH (1985) Kronenanalyse für Zuwachsuntersuchungen in immissionsgeschädigten Nadelholzbeständen, Forst- Holzwirt, 40(5):115-118.

Lick E, Krapfenbauer A (1986) Terrestrische Waldzustandsinventuren und Probleme einer Objektivierung, Centralbl. gesamt. Forstwiss. 103: 227-241.

Löchelt S (1995) Bestimmung genetischer Merkmale von Fichten (Picea abies [L.] Karst.) mit unterschiedlich ausgeprägten Schadsymptomen auf baden-württembergischen Dauerversuchsflächen, Allg. Forst- Jagdztg. 165(2):21-27.

Löchelt S, Franke A (1993) Bestimmung der genetischen Konstitution von Waldbäumen mit unterschiedlich ausgeprägten Schadsymptomen auf baden-würrtembergischen Dauerbeobachtungsflächen zur Walderkrankung, KfK-PEF 108:115 pp.

Mantel W (1959) Forsteinrichtung, J.D. Sauerländer's Verlag, Frankfurt/Main, 262 pp.

Marschall J, Sterba H (1975) Die Erstellung der neuen Kiefernertragstafel-Litschau, Allg. Forstztg 86 (11).

Neumann M, Schadauer K (1995) Die Entwicklung des Zuwachses in Österreich an Hand von Bohrkernanalysen, Allg. Forst- Jagdztg. 166(12):230-234.

Nihlgaard B (1985) The Ammonium Hypothesis - An Additional Explanation to the Forest Dieback in Europe, Ambio 14 (1):1-8.

Oren R, Schulze ED (1989) 4-H Nutritional Disharmony and Forest Decline: A Conceptual Model, Ecol. Studies 77:425-443.

Picard JF (1995) Evolution de la croissance radiale du hêtre (Fagus silvatica L) dans les Vosges. Premiers résultats sur le versant lorrain, Ann. Sci. For. 52:11-21.

Pollanschütz J (1971) Durchforstung von Stangen- und Baumhölzern, Forstarchiv 42:257-259.

Pollanschütz J (1974) Erste ertragskundliche und wirtschaftliche Ergebnisse des Fichten-Pflanzweiteversuches "Hauersteig", In: 100 Jahre Forstliche Bundesversuchsanstalt Wien Egger J (ed), 99-172.

Pollanschütz J (1985) Zur Frage Nadelverlust und Kronenverlichtungen bei Fichte, Allg. Forstztg 96(5):110-114.

Pretzsch H (1985) Wachstumsmerkmale süddeutscher Kiefernbestände in den letzten 25 Jahren, Forstl. Forschungsber. München 65:173 pp.

Pretzsch H (1992) Zunehmende Unstimmigkeiten zwischen erwartetem und wirklichem Wachstum unserer Waldbestände, Forstwiss. Centralbl. 111:366-382.

Röhle H (1985) Ertragskundliche Aspekte der Walderkrankungen, Forstwiss. Centralbl. 104:225-242.

Röhle H (1986) Waldschaden und Zuwachsreaktion, dargestellt am Beispiel geschädigter Fichtenbestände im Nationalpark Bayerischer Wald, Forstwiss. Centralbl. 105:115-122.

Röhle H (1987a) Zur Reaktionskinetik der Fichte in verschiedenen Schadgebieten Bayerns, GSF-Bericht 10/87:144-155.

Röhle H (1987b) Entwicklung von Vitalität, Zuwachs und Biomassenstruktur der Fichte in verschiedenen bayerischen Untersuchungsgebieten unter dem Einfluß der neuartigen Walderkrankungen. Forstl. Forschungsber. München 83:112 pp.

Rosenberg NJ, Kimball BA, Martin P, Cooper CF (1990) From Climate and CO_2 Enrichment to Evapotranspiration, In: Climate Change and U.S. Water Resources, Rep. Am. Assoc. Adv. Sci., Panel on Climatic Variability, Climate Change and the Planning and Management of U.S. Water Resources Waggoner BA (ed), Wiley Series in Climate and Biosphere, Wiley, New York (496pp), pp.151-171.

Schieler K, Schadauer K (1993) Zuwachs und Nutzung nach der Österreichischen Forstinventur 1986/1990, Österr. Forstztg 104(4):22-23.

Schmidt A (1969) Der Verlauf des Höhenwachstums von Kiefer auf einigen Standorten der Oberpfalz, Forstwiss. Centralbl. 87(1):33-40.

Schneider O, Hartmann P, Schlaepfer R, Petter DA (1988) Relationship between tree ring width and crown transparency of spruce (Picea abies Karst.), Dendrochronologia 6:9-31.

Schöpfer W (1986) Zusammenhang zwischen Wuchsraum und Zuwachs in erkrankten Fichte- und Tannen-Beständen. Forst- Holzwirt, 41(12):315-319.

Schöpfer W, Hradetzky J (1987) Zuwachsrückgang in erkrankten Fichten- und Tannenbeständen: Auswertungsmethoden und Ergebnisse, Forstwiss. Centralbl. 105:446-470.

Skeffington RA, Wilson EJ (1988) Excess Nitrogen Deposition: Issues for Consideration, Environm. Poll. 54:159-184.

Spelsberg G (1987) Zum Problem der Beurteilung des Zuwachses in geschädigten Beständen, Allg. Forst- Jagdztg, 158: 205-211.

Spelsberg G (1988) Zur Frage der Wachstumsdifferenzierung bei vorherrschenden Fichten unter besonderer Berücksichtigung der Waldschäden, Forst- Holzwirt, 43: 538-542.

Spelsberg G (1995) Zum Höhenwachstum der Fichte in Nordrhein-Westfalen, Allg. Forst-Jagdztg 165(4):77-80.

Spiecker H (1986) Das Wachstum der Tannen und der Fichten auf Plenterwald-Versuchsflächen des Schwarzwaldes in der Zeit von 1950-1984, Allg. Forst- Jagdztg 157(8):152-164.

Spiecker H (1987) Düngung, Niederschlag und der jährliche Volumszuwachs einiger Fichtenbestände Südwestdeutschlands, Allg. Forst- Jagdztg 158(4):70-76.

Spiecker H (1991) Zur Dynamik des Wachstums von Tannen und Fichten auf Plenterwald-Versuchsflächen im Schwarzwald, Allg. Forstz. 21:1076-1080.

Stefan K (1995): Bioindikatornetz: Schwefel- und Stickstoffgehalte. In: Österr. Stat. Zentralamt u. Forstliche Bundesversuchsanstalt Wien (ed) Ökobilanz Wald Österreich:79-85.

Sterba H (1988) Waldschäden und Waldzuwachs, Modelle und Expertensysteme, In: FIW-Symposium 1988, Waldsterben in Österreich, 27-28 Oct. 1988 (Führer E, Neuhuber F ed), Univ. Bodenkultur Wien, Vienna, pp.117-126.

Sterba H (1992) Determining parameters of competition models under changing environmental conditions, In: Mathematical Modelling of Forest Ecosystems Franke J, Roeder A (ed), J.D.Sauerländer Verlag, Frankfurt/Main, pp.148-159

Sterba H (1993) The impact of the variation of potential height growth on the projections of a single growth model, In: Modelling Stand Response to Silvicultural Practices, Colloq. Forestry and Wildlife Resources: Burkhart HE, Gregoire TG, Smith GL (ed), Virginia Polytech. Inst. State Univ., Blackburg, VA, Publication FWS-1-93:158-171.

Sterba H (1995) Forest decline and increasing increments - A simulation study, Forestry 68(2).

Sterba H, Eckmüllner O (1988) Nadelverlust - Zuwachsrückgang: Doch eine Beziehung, Österr. Forstztg 99(10):52-53.

Utschig H (1989) Waldwachstumskundliche Untersuchungen im Zusammenhang mit Waldschäden. Auswertung der Zuwachstrendanalyseflächen des Lehrstuhls für Waldwachstumskunde für die Fichte (Picea abies (L.) Karst.) in Bayern, Forstl. Forschungsber. München 97:185 pp.

Wiedemann E (1936, 1942) Fichtenertragstafeln, reprinted in Schober, R. (1987) Ertragstafeln wichtiger Baumarten, J.D. Sauerländer's Verlag, Frankfurt/Main, 166 pp.

Wiedemann E (1948) Die Kiefer 1948, Verlag Schaper, Hannover, 337 pp.

Long-Term Growth Trends of Trees: Ten Years of Dendrochronological Studies in France

Vincent Badeau, Michel Becker, Didier Bert, Jean-Luc Dupouey, François Lebourgeois and Jean-François Picard
INRA, Forest Ecophysiology Unit. F-54280 Champenoux, France

Abstract

The study of long-term growth trends in French forests began 10 years ago at the Phytoecological Laboratory of the National Agronomic Research Institute (INRA). Very large surveys have been carried out in several regions and for many species, allowing the study of changes in radial growth of individual trees during the past 150 years. In all cases, a significant increasing growth trend appeared. It varied between +50% and +160% depending on species and location. Careful analysis of possible bias has been made. Beside these biases, possible causes of such trends are discussed. This chapter summarizes the main methods and results of our laboratory. For additional information, our readers may refer to the cited articles.

1. Introduction

The study of long-term radial growth trends in the French forests began 10 years ago within the framework of the national research programme DEFORPA (*Dépérissement des Forêts et Pollution Atmosphérique*). Consequent to observations of abnormal crown damages on conifers in the eastern France, this programme was launched in 1984 in order to understand forest decline in relation to natural and man-made factors (see a review of research findings since 1985 in Landmann & Bonneau 1995).

The first hypothesis was that a strong increase of the atmospheric pollution levels, especially of acidifying compounds, may have had disastrous consequences on some soil characteristics and on health and growth of trees (Becker 1985). However, while this explanation seemed sufficient in some cases of severe dieback associated with local and heavy pollutions, the role of air pollution was not convincingly established in the majority of the damaged regions studied. The possible role of natural conditions (climatic or edaphic variations) on forest dieback was assessed

European Forest Institute Research Report No. 5
Growth Trends in European Forests
Edited by H. Spiecker, K. Mielikäinen, M. Köhl and J. P. Skovsgaard

through large ecological and dendrochronological surveys (Bonneau 1987). The main objective was, on one hand, to describe and to analyse the ecological factors of several sites and, on the other hand, to reconstruct the growth of the trees on the same sites over the past decades. The combination of ecological studies (i.e. the analysis of physical and biotic growth conditions) and dendrochronological studies (i.e. the reconstruction of radial growth) provides a sound basis for the historical assessment of the vitality of the major forest species and the role of ecological factors.

Initially, these dendroecological studies had various goals:

- to analyse the functioning or malfunctioning of forest ecosystems;
- to explain observed crown damages;
- to understand the short-term radial growth dynamics according to climatic factors;
- to quantify the reaction of trees after fertilization.

The study of long-term radial growth trends was, as such, outside the scope of these studies. Nevertheless, a significant increasing growth trend during the last century was brought to light in various situations. The question of global change became a matter of increasing concern. Multilocal and multispecies approaches were needed in order to assess the impact of such changes on the functioning of ecosystems, especially forest ecosystems.

2. Initial Results: a Strongly Increasing Radial Growth Trend

Since the first dendroecological study made by Becker in 1987 on silver fir in the Vosges mountains, nine other surveys have been made in France by the Phytoecological Laboratory of the National Agronomical Research Institute (INRA). These studies concerned several forest species in different regions (Fig. 1 & Table 1): silver fir (*Abies alba* Mill.) in the Vosges and Jura mountains; Norway spruce (*Picea abies* [L.] Karst.) in the Vosges mountains; beech (*Fagus sylvatica* L.) in the Vosges mountains, the acidic Lorraine lowland and the calcareous Lorraine plateau; oaks (*Quercus robur* L. and *Quercus petraea* [Matt.] Liebl.) in the Lorraine lowland; mountain pine (*Pinus uncinata* Mill.) in the Pyrénées mountains, and Corsican pine (*Pinus nigra* Arnold ssp. *laricio* var. *corsicana*) in a region called Pays de la Loire in western France.

All these studies were based on large samples of trees (from 500 to 1500 trees) from stands chosen so as to provide a balanced sample of ecological diversity of the area (elevation, aspect, topography, soil, ground vegetation). Furthermore, the main criterion was to find trees for which the age was as varied as possible (between 20 and 200 years). This is an absolute condition to distinguish between the effects of aging on radial growth and possible trends due to concomitant changes in growth conditions according to calendar dates.

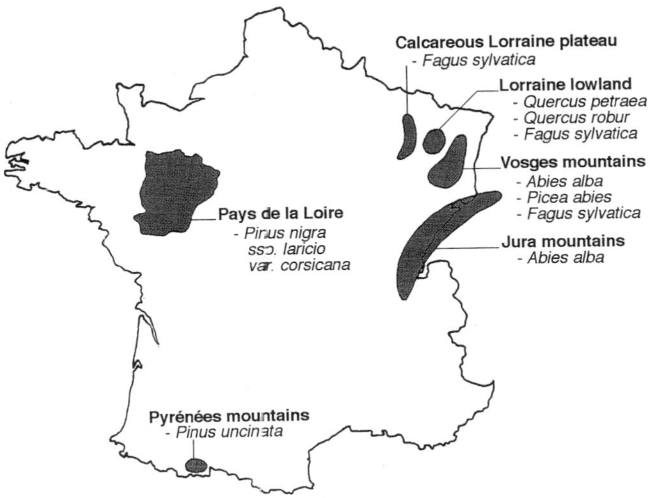

Fig. 1. Geographic location of regions and species studied.

The methods used to extract the part of variation due to ageing have been described comprehensively in various publications (Becker 1987, 1989; Becker *et al.* 1989, 1994, 1995a, b; Bert 1992; Dupouey et al. 1992; Badeau 1995; Lebourgeois 1995) and they will be concisely presented here (see Section 4.1). Annual tree rings are measured using a computerized device, then they are cross-dated carefully with the help of pointer years. Finally, ring widths (or ring areas) are standardized (i.e. converted into indices according to the cambial age of each tree ring) and averaged according to calendar dates in order to obtain a mean chronology excluding the cambial age effect.

Figure 2a to h displays the radial growth level expressed through radial growth indices for eight surveys according to calendar dates. Beside large interannual variations mainly controlled by climatic factors, these curves show a clear and positive long-term trend.

In the Vosges mountains, radial growth indices have increased by:

- +160% from 1850 to 1986 for silver fir (Becker 1987; Fig. 2a);
- +130% from 1865 to 1988 for Norway spruce (Becker et al. 1995b; Fig. 2b);
- +90% from 1850 to 1990 for beech (Picard 1995; Fig. 2c).

The results in the Vosges mountains are consistent with those obtained in the Lorraine lowland, on acid soils (Fig. 2d to f):

- +90% from 1850 to 1987 for sessile oak (Becker et al. 1994; Fig. 2e);
- +55% from 1850 to 1987 for pedunculate oak (Becker et al. 1994; Fig. 2f);
- +120% from 1860 to 1992 for beech (Becker, unpublished data; Fig. 2d).

Table 1. Overview of the data structure of the ten dendroecological studies.

Species	Geographic location	Silvicultural management	Number of sites	Number of trees	Number of rings	Age spectrum
Abies alba Mill.	Vosges mountains	high forest	277	1 475	120 000	50 to 180 years
Abies alba Mill.	Jura Mountains	high forest & selection forest	208	1 248	115 170	20 to 468 years
Picea abies (L.) Karst.	Vosges mountains	high forest	181	1 050	99 750	40 to 180 years
Fagus sylvatica L.	Vosges mountains	high forest	155	947	134 000	30 to 220 years
Fagus sylvatica L.	Lorraine lowland	high forest	66	258	57 300	50 to 180 years
Fagus sylvatica L.	Calcareous Lorraine plateau	high forest & coppice-with-standards	102	1025	93 971	11 to 203 years
Quercus robur L.	Lorraine lowland	high forest	115	505	80 800	10 to 332 years
Quercus petraea (Matt.) Liebl.	Lorraine lowland	high forest	121	529	91 000	11 to 239 years
Pinus nigra Arnold ssp. *laricio* var. *corsicana*	"Pays de la Loire"	high forest	183	1 808	49 379	15 to 70 years
Pinus uncinata Mill.	Pyrénées mountains	free growing trees	13	521	107 300	10 to 715 years

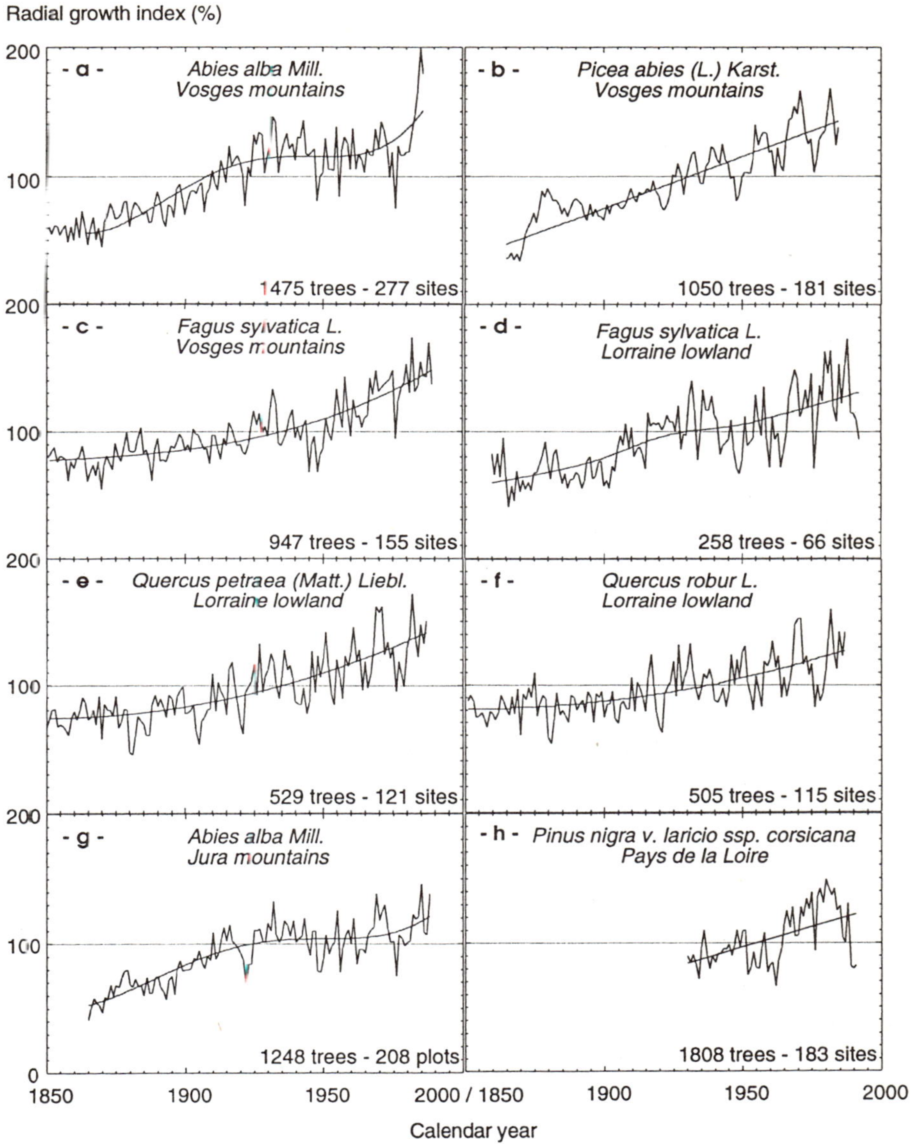

Fig. 2. Radial growth changes since the last century. For each species, radial growth indices are plotted against calendar years and smoothing functions are fitted to highlight the low frequency signal.

In the Jura mountains (Fig. 2g) silver fir showed an increasing trend of +130% (Bert 1993) and in the region Pays de la Loire in western France (Fig. 2h), the radial growth indices of Corsican pine have increased by +50% since 1940 onwards (Lebourgeois and Becker 1996).

3. Possible Causes of the Trends Observed

All these results show that, since the last century, an increasing trend in radial growth, independent of any biological effects related to the ageing processes, seems to be widespread. It has occurred for broadleaved and coniferous species, in plains or mountain regions, on rich or poor lands, etc.

In the context of a slowly and continuously changing environment, three hypotheses could explain these observed trends:

- climatic changes, i.e. temperature or precipitation increase;
- CO_2 fertilization effect;
- pollution effects like nitrogen fertilization through enhanced atmospheric nitrogen deposition.

However, other causes could explain these trends, especially a bias:

- during sampling;
- during the analysis of the data sets;
- finally, silvicultural changes (i.e. changes in stand structures due to thinning frequency or intensity) during the last century could also explain such trends.

The possible role of these non-environmental causes are studied specifically.

4. Attempts to Find Possible Biases

4.1. Methods Used to Identify and Separate the Long-Term Trends

Methods. Various factors may generate growth trends in tree rings: climatic factors, ageing, disturbances, stochastic errors, etc. Graybill (1982) and Cook (1987, 1990) summarized these causes as follow:

$$R_t = A_t + C_t + D_t + E_t$$

R_t is the observed tree ring width; A_t represents the growth trend associated with increasing age and date; C_t is the climatic signal which can vary at both low and high frequencies; and D_t is the tree disturbance signal, often divided into local and stand-wide disturbances. The separation of these different trends is crucial, but this problem has been considered as difficult (Briffa et al. 1988; Visser and Molenaar 1988 and 1990; Cook et al. 1990). In order to take into account the confounding effect between the age trend (i.e. A_t) and the date trend (i.e. D_t the region-wide disturbance associated with large-scale environmental changes), three methods have been tested.

The first method, known as at constant cambial age, is very simple and does not require any transformation of the initial data. It consists of considering simultaneously, in the whole sample, the tree rings developed at a given cambial age but at various calendar dates (Becker 1987; Briffa 1992; *the cambial age of a given tree ring is the age of the tree when this tree ring was built*). This simple method can obviously give interesting results but it is not sufficient. All the information contained in the tree rings is impossible to use and the figures cannot be compared among themselves because the average growth level of each cambial age is not constant, due to the age trend. Nevertheless, the comparison becomes possible if each ring width or ring area is converted against a reference in order to suppress this trend: the removal of A_t from tree ring series is known as standardization (Fritts 1976).

The second method is a calculation of *standardized* radial growth indices using a regional age curve (Mitchell 1967; Becker 1989). After alignment of all the available individual series according to their cambial ages, a regional age curve is built by averaging annual ring widths (or annual basal area increments). A mathematical model is adjusted to this curve. Then each initial ring width or ring area is converted into a growth index, which is the ratio expressed as a percentage of the initial value to the corresponding reference value given by the above model for the same cambial age. In relation to the first method, all the available tree rings can be taken into account. Yet, Cook et al. (1990) point out the main shortcoming of the regional age curve approach: it assumes that the shape of the age curve is independent of the time period. In order to overcome this limitation, a new approach was developed.

The third method is the analysis of variance (Dupouey et al. 1992; Badeau 1995; Badeau *et al.* 1995). Ring areas can be organized in a two-way table with cambial age in rows and calendar dates in columns. The ANOVA ideally requires an equal and sufficiently large number of observations for each combination of age and date. In order to obtain this balanced design, the age and date factors can be divided into a limited number of classes. Then, in each cell of the table, individual annual rings belonging to a given tree can be averaged in order to limit the dependence of the variables for each combination of age and date and in order to give each tree the same weight in the analysis. At last, the following simple linear model is used (Badeau et al. 1995):

$BAI_{tad} = A_a + D_d + Aa.D_d + E_{tad}$, where:

BAI_{tad} is the mean basal area increment of tree t at cambial age class a and belonging to date class d;
A_a is the effect of age class a;
D_d is the effect of date class d;
$A_a . D_d$ is the interaction between age class a and date class d;
E_{tad} is the residual.

This method is more demanding in terms of computer processing, but it produces a more reliable extraction of both age and date effects. Finally, although we used averaged BAIs from each core for each combination of age and date classes, some dependencies remain. Statistical models taking into account such dependencies are under development.

Results. Figures 3 to 5 present some results obtained for beech managed in high forest on the calcareous Lorraine plateau in northeastern France, using the three previous methods.

With the method at constant cambial age, ring areas of the same cambial age are plotted against calendar date (Fig. 3a, b). For both examples, ring areas have increased during the period studied. These two plots are only an example: for each available cambial age a figure can be built if there are enough calendar dates. For each study presented in the introduction, these plots were made as a matter of routine for cambial ages ranging for 10 to 100 years. In each case, a significant and positive trend was brought to the fore for each cambial age, using ring width or ring areas.

In the second method, a regional mean relation between radial growth and cambial age is built at first (Fig. 4a). The resulting mean curve is smooth because of the sample size, cambial age and calendar year diversity. Then ring areas of each dendrochronological series are standardized according to this curve and the resulting indices are averaged according to calendar dates in order to obtain a standardized chronology at the regional level (Fig. 4b). As in Fig. 2, this curve displays a high-frequency signal mainly controlled by interannual climatic variations (especially drought); growth depression of varying severity at the decade time scale (e.g. around 1948 or 1976); and finally, a positive and low-frequency trend in radial growth.

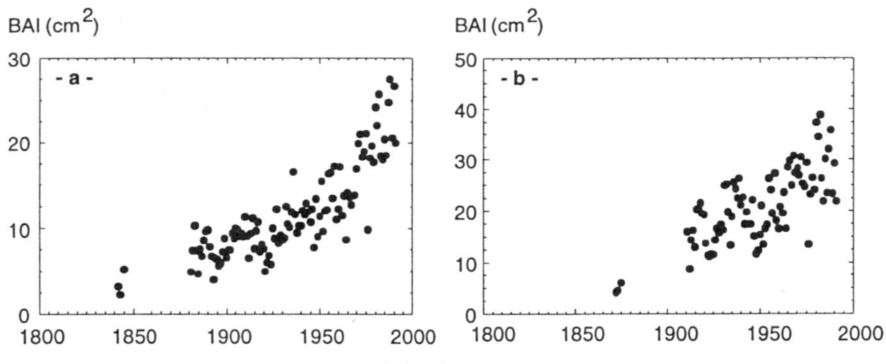

Fig. 3. Long-term trend in radial growth of beech on the calcareous Lorraine plateau at the age of 40 (*a*) and at the age of 70 (*b*).

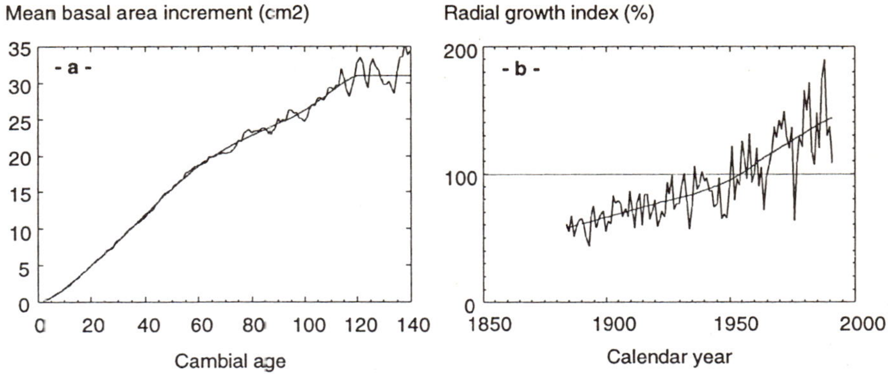

Fig. 4. Standardization from the regional mean relation between radial growth and cambial age for beech on the calcareous Lorraine plateau:

a age trend: regionally averaged basal area increments (BAIs) versus cambial age and fitted polynomial curve;
b date trend: radial growth indices versus calendar years and smoothing function.

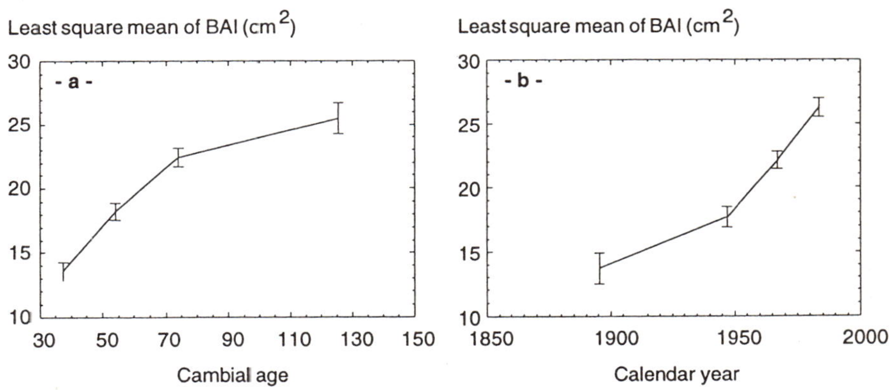

Fig. 5. Least square means of annual BAIs for cambial age effects (a) and calendar date effects (b) computed from an analysis of variance for beech on the calcareous Lorraine plateau. Confidence intervals at the 95% level are plotted.

In the third method, the whole sample of tree rings is divided into four age and date classes and the ANOVA is computed. Table 2 shows that the chosen model is highly significant. The cambial age and the calendar date integrate most of the total variance but, in the case of high forest, the part of variance explained by the inter-

Table 2. Analysis of variance of cambial age and calendar date effects on annual BAIs, computed with the Global Linear Model procedure of SAS (1988), for beeches managed in high forest.

Source of variation	Degrees of freedom	Mean Square	F Value	Pr > F
Model	15	11747.10	116.20	0.0001
Error	3141	101.09		
				$r^2 = 0.3569$

Source of variation	Degrees of freedom	Mean Square	F Value	Pr > F
cambial age effect	3	13587.46	134.40	0.0001
date effect	3	14145.72	139.93	0.0001
interaction	9	133.89	1.32	0.2184

action between age and date effects is not significant. Figure 5a and b shows the age effect means and the date effect means.

Without going into detail, the three methods proved to be effective in bringing to the fore long-term radial growth trends and gave similar results. These three methods have been applied in other studies with the same results, so it is now well established that the positive long-term trends observed are not due to an artefact introduced by the methods used for processing the data, even if each of them has advantages and disadvantages.

4.2. Are the Chosen Trees Representative?

Among the causes which may explain the long-term growth trends, another possible bias is that the productivity in their youth of the very old trees alive now may be not representative of the productivity of the whole forest in the last century. The oldest sampled trees may still be alive because they were not big enough to be harvested. Conversely, the trees of the same age but with a higher productivity may have been harvested already.

In order to check this hypothesis, Dupouey et al. (unpublished data) compared radial growth levels of oaks alive now and archaeological oak samples in the same region for different areas in Europe. In every region, a significant difference was found between living and ancient trees: at a given cambial age, living oaks displayed higher levels of radial growth. The origin and dominance status of ancient trees in such samples was unknown. However, these results are in accordance with trends observed in living tree growth.

Previous observations of increasing radial growth obtained from living trees are confirmed. This archaeological analysis shows that the possible bias during sampling alone cannot explain the observed long-term radial growth trends.

4.3. Changes in Silvicultural Practices

Silvicultural management could affect individual tree radial growth if, in the long term, the thinning intensity or frequency had increased, thus lowering the competition level between trees.

Two studies were carried out in order to test this possible effect on long-term radial growth:

- a comparison of radial growth between high and low density stands of beech (Badeau 1995);
- an analysis of free-growing trees at tree-line in high elevation mountains (Dupouey et al., unpublished data).

Comparison of High and Low Density Stands. In France, beech is managed in high forests. These stands are even-aged, with high competition pressures between trees during the whole of their life. In this case, silvicultural practices may have evolved since the last century. On the other hand, beech is managed in coppice-with-standards. In this type of stand, there are two different strata: the coppice, which is clearcut every 25-30 years, and a few tall trees, the standards, which grow at a very low density level. The standards are allowed to grow for several rotations of the coppice before being harvested. Here, silviculture maintains a constant sparsely populated area, with very low competition pressures between tall trees, and with a stable influence at the century time scale.

Figure 6 shows that a long-term growth trend is observed in both silvicultural systems, but it is stronger and more regular in high forests than in coppice-with-standards. The significant difference observed between the two silvicultural systems implies that the trends cannot be explained entirely by changes in environmental factors. However, the trend observed in the coppice-with-standards treatment implies that changes in silvicultural practices alone cannot explain the trends because an increase in the thinning intensity is unlikely in these stands. Management registers even suggest that the frequency of coppicing has been lowered since the 1960s, thus the stand density of coppice-with-standards has increased. The possible positive influence of nitrogen flushes on growth of the standards, following each removal of the coppice, has disappeared.

Study of Naturally Open Stands. In order to test the influence of silvicultural practices, another study has been carried out at tree line (between 2200 and 2400 m above sea level) in the Pyrénées mountains, with free-growing mountain pine (*Pinus uncinata* Mill.) at a very low competition level and without any human silvicultural practices.

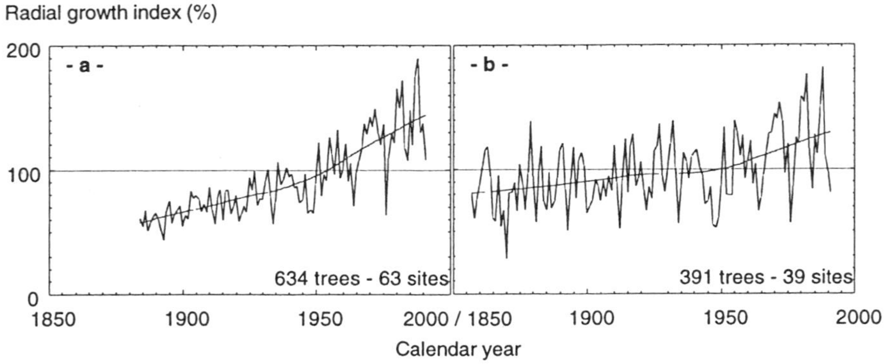

Fig. 6. Radial growth changes since the last century for beech managed in high forest (a) and coppice-with-standards (b) on the calcareous Lorraine plateau.

The increasing radial growth trend observed for the 521 sampled trees (Fig. 7) is still very steep (approximately +120%), so this result corroborates the possible role of environmental changes on radial growth. Nitrogen fertilization cannot explain this radial growth increase because atmospheric nitrogen deposition is low in this region.

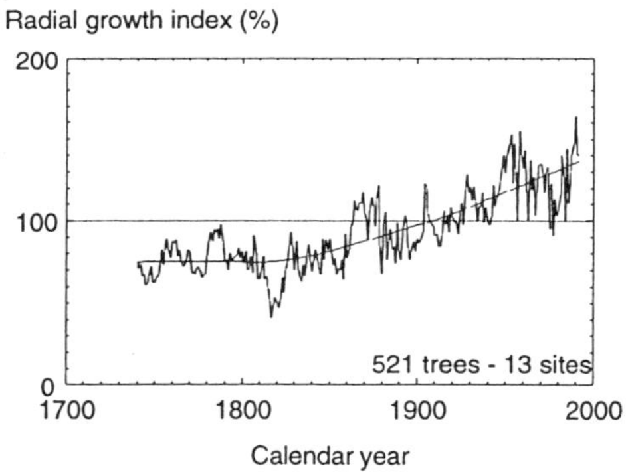

Fig. 7. Long-term growth trend of mountain pine (*Pinus uncinata* Mill.) at tree line in the Pyrénées mountains.

5. Conclusions

Ten studies, concerning seven forest species, have been carried out in France since the early 1980s. In all cases, a significant increasing long-term trend in radial growth appeared during the last 150 years. This trend varies between +50 and +160% for basal area increment indices, depending on the species and the location.

One can legitimately be surprised by the magnitude of the increases observed, and some uncertainties remain about the interpretation of the results. In order to test the reliability of the results, careful analyses of possible biases have been carried out. These analyses show that the results are not due to a bias introduced by the methods used for processing the data.

The chosen trees seem to be representative, according to a comparison between living and archaeological trees. Various attempts to identify a bias in the spatial structure of the sample and the ecological diversity of the corresponding site types have been made. In the Jura mountains for example, the older trees available are mainly located on poor sites, where radial growth is slow. However, for sessile oak, pedunculate oak and beech on the Lorraine plateau, this bias has been limited by a careful selection of trees within a limited ecological range. The stratification of the data in an attempt to isolate more satisfactory subsets did not result in fundamentally different results (Becker et al. 1995a).

Another possible bias, in closed stands, is that some of the currently dominant trees may have been suppressed previously. Such social changes in the hierarchical structure of stands are often considered as infrequent in managed forests, but their exact role on the observed long-term trends remains to be precisely quantified.

Silvicultural practices could explain part of the long-term trends observed, according to the results obtained for beech in high forest and coppice with standards. On the other hand, the results obtained for mountain pine in the Pyrénées mountains strongly suggest the role of environmental changes.

In order to gain a better understanding of the different causes of the long-term increasing growth changes currently observed in different parts of Europe, it may prove useful to extend these studies in different environments where the various factors could be separated. For example, northern boreal forests, which are free from silvicultural disturbances and where nitrogen pollution is very low, offer a good opportunity. It may also be necessary to study height growth trends in addition to radial growth. Dominant height growth is supposed to be lower depending on stand density. It could be an additional indicator of long-term environmental changes.

Whatever the causes, we strongly advocate the use of proper methods of standardization such as the regionally based age curve or the joint analysis of variance of date and age effects for the study of long-term growth trends, which allow reliable conclusions to be drawn.

References

Badeau V (1995) Etude dendroécologique du hêtre (*Fagus sylvatica* L.) sur les plateaux calcaires de Lorraine. Influence de la gestion sylvicole. PhD thesis, University of Nancy I, 224p + annexes.

Badeau V, Dupouey JL, Becker M, Picard JF (1995) Long-term growth trends of *Fagus sylvatica* L. in northeastern France. A comparison between high and low density stands. Acta Œcologica, 16 (5), 571-583.

Becker M (1985) Le dépérissement du sapin dans les Vosges. Quelques facteurs liés à la détérioration des cimes. Rev. For. Fr., 37 (4), 281-287.

Becker M (1987) Bilan de santé actuel et rétrospectif du sapin (*Abies alba* Mill.) dans les Vosges. Etude écologique et dendrochronologique. Ann. Sci. For., 44 (4), 379-402.

Becker M (1989) The role of climate on present and past vitality of silver fir in forest in the Vosges montains of northeastern France. Can. J. For. Res., 19, 1110-1117.

Becker M, Bert GD, Bouchon J, Dupouey JL, Picard, Ulrich E (1995a) Long-term changes in Forest Productivity: the Dendroecological Approach. In: Forest decline and atmospheric deposition effects in the French mountains, G Landmann, M Bonneau (ed), Springer Verlag, 143-153.

Becker M, Bert GD, Bouchon J, Picard JF, Ulrich E (1995b) Growth trends in various broadleaved and coniferous trees in Northeastern France from the mid 19th century. In: Functioning and dynamics of natural and perturbed ecosystems, Bellan-Santini D, Bonin G, Emig C (ed), Lavoisier Publishing, Paris, 35-48.

Becker M, Landmann G, Levy G (1989) Silver fir decline in the Vosges Mountains (France): role of climate and silviculture. Water, Air and Soil Pollution, 48, 77-86.

Becker M, Nieminen TM, Geremia F (1994) Short-term variations and long-term changes in oak productivity in northeastern France. The role of climate and atmospheric CO_2. Ann. Sci. For., 477-492.

Bert GD (1992) Influence du climat, des facteurs stationnels et de la pollution sur la croissance et l'état sanitaire du Sapin pectiné (*Abies alba* Mill.) dans le Jura. Etude phytoécologique et dendrochronologique. PhD thesis, Nancy I University, 200 p + annexes.

Bert GD (1993) Impact of ecological factors, climatic stresses, and pollution on growth and health of silver fir (*Abies alba* Mill.) in the Jura mountains: an ecological and dendrochronogical study. Acta Œcologica, 14 (2), 229-246.

Bonneau M (1987) Genése et contenu du programme DEFORPA. In: Les recherches en France sur le dépérissement des forêts. Programme DEFORPA, 1er rapport, ENGREF, Nancy, 6-10.

Briffa KR (1992) Increasing productivity of "natural growth" of conifers in Europe over the last century. In: Tree rings and Environment. Proceedings International dendrochronological symposium "Tree rings and environment", Ystad (Sweden), 3-9 September 1990, Lundqua report, 34, 64-71.

Briffa KR, Jones PD, Pilcher JR, Hughes MK (1988) Reconstructing summer temperature in Northern Fennoscandinavia back to A.D. 1700 using tree ring data from Scots Pine. Artic and Alpine Research, 20, 385-394.

Cook ER (1987) The decomposition of tree ring series for environmental studies. Tree ring bulletin, 47, 37-59.

Cook ER (1990) A conceptual linear aggregate model for tree rings. In: "Methods of dendrochronology", ER Cook, LA Kairiukstis (ed), Kluwer Academic Publishers, Dordrecht / Boston / London, 98-104.

Dupouey JL Denis JB, Becker M (1992) A new method of standardization for examining long term trends in tree ring chronologies. In: Tree rings and Environment. Proceedings International dendrochronological symposium "Tree rings and environment", Ystad (Sweden), 3-9 September 1990, Lundqua report, 34, 85-88.

Fritts (1976) Tree rings and climate. Academic press, vol. 1 (245p) & vol. 2 (567p).

Graybill DA (1982) Chronology development and analysis. In: Climate from tree rings, Hughes M.K. et al. (ed), Cambridge University Press.

Landmann G, Bonneau M (ed) (1995) Forest decline and atmospheric deposition effects in the French mountains. Springer Verlag, 461p.

Lebourgeois F, Becker M (1996) Dendroécologie du pin laricio de Corse dans l'ouest de la France. Evolution du potentiel de croissance au cours des dernières décennies. Ann. Sci. For., in press.

Lebourgeois (1995) Etude dendroécologique et écophysiologique du pin laricio de Corse (*Pinus nigra* Arnold ssp. *laricio* Poiret var. *corsicana*) en région Pays de la Loire. PhD thesis, University of Paris XI, 209p. + annexes.

Mitchell VL (1967) An investigation of certain aspects of tree growth rates in relation to climate in the central Canadian boreal forest. Univ. Wisconsin. Dept. Meteorol. Tech. Rep. 33. Task NR 387-022, ONR Contract 1202 (07), NSF GP-5572X, Madison.

SAS Institute Inc. (1988) SAS/STAT® User's Guide. Release 6.03 Edition, Cary, NC: SAS Institut Inc., 1028p.

Visser H, Molenaar J (1988) Kalman filter analysis in dendroclimatology. Biometrics, 44, 929-940.

Visser H, Molenaar J (1990) Estimating trends in tree ring data. Forest Science, 361, 87-100.

Growth Trends of Trees. Regional Study on Norway Spruce (*Picea abies,* [L.] Karst.) in the Swiss Jura

Olivier Schneider and Philippe Hartmann
DendrOrbe. Sur le Blétru 4, CH-1132 Lully, Switzerland

Abstract

The presented regional case study deals with tree growth as an indicator of tree vitality and site productivity. It is based on Norway spruce from the Swiss Jura. The aim is to identify any possible long-term trends.

Analysis revealed that the radial increment of single trees shows both positive trends over the past decades as well as growth levels higher than selected references which were made with data from the sample itself.

The study of growth in height leads to similar conclusions.

As possible explanations, a few hypotheses are briefly discussed: changes in climate (precipitation, temperature), fertilizing effects (CO_2, N), evolution of silvicultural practices, artefacts (data structure effect).

1. Introduction

In connection with forest decline and generally with forest ecosystem monitoring, the growth and growth trend of trees are indicators of vitality and site productivity.

Many recent research projects on different tree species and regions of Europe (e.g. Becker 1989; Becker et al. 1994; Bert 1992; Kenk 1988) and North America (e.g. Graumlich and Brubaker 1987; LaMarche et al. 1984; Parker 1987) show long-term variations of tree growth and site productivity that indicate an improvement in growth conditions.

In this chapter, recent growth measurements are compared with previous growth patterns of the sample itself and with dendroclimatic growth models. The results give insight into the development of site productivity. Additionally, the relationship between annual growth and climate, silviculture or atmospheric pollutants is discussed.

European Forest Institute Research Report No. 5
Growth Trends in European Forests
Edited by H. Spiecker, K. Mielikäinen, M. Köhl and J. P. Skovsgaard

2. Material and Methods

The basic material used in this research project is a combination of tree-ring data from stem analyses and from cross sections taken in the lower part of the crown.

The data material was collected in the western part of Switzerland (Fig. 1).

Fig.1. Situation of the study area in Switzerland. Vaud, Swiss Jura.

The situation and data type are as follows:

- region: Vaud, Swiss Jura
- tree-ring data from 1113 spruces on 200 stands with measurements on cross sections (2 or 4 radii; number of rings per cross section: 16-137; mean: 63; standard deviation: 20) mainly from the lower part of the crown and from 24 spruces with stem analysis, evaluated using Nagel and Athari's method (1982) (cross sections at 0.5 m / 1.3 m / every 5.1 m, 4 radii)
- altitude: 600 - 1300 m
- exposition: south-east and north-west
- topography: mountain, Jura
- geology: limestone sediments, moraines
- soil type: rendzina on limestone sediments, brown earth on moraines
- climate: annual mean temperature 8.5 °C at 600 m, 4.0 °C at 1300 m (-0.65 °C/ 100 m), annual precipitation 950 - 1500 mm
- climax vegetation: oak-type forest, beech-type forest, beech-silver fir-type forest
- stand: more or less uneven-aged, mixed, thinning from above, regeneration mostly natural
- site index: mainly 10-20 m (top height at age 50)
- site productivity: 3-10 $m^3\ ha^{-1}\ y^{-1}$

The investigations presented below are all based on Norway spruce (*Picea abies* [L.] Karst.).

The results are given at single-tree level. The dimension used is radial increment per year, measured on cross sections (2 or 4 radii). For practical reasons, the cross sections were taken at various heights, mainly in the lower part of the crown. The influence of this unconventional height on the results is discussed in Section 3.3.

The height growth of dominant trees is also examined. The observations are based on height measurements of every felled tree and on stem analysis of a small subsample.

3. Increment and Growth Analysis

3.1. Evolution of Radial Increment During the 20th Century

In this analysis based on 1113 dominant (90%) and codominant (10%) spruces, the reference is taken in the sample itself.

In Fig. 2, three graphs represent aging curves with tree-ring data from different periods: 1880-1950 / 1951-1993 (whole data-set and subsample of trees with needle loss ≤ 10%), and 1880-1930 / 1931-1960 / 1961-1993 (whole data-set). The pattern is similar if other periods are chosen.

For a given cambial age, the increase in radial increment between the successive periods is obvious (Fig. 2). This level difference is more pronounced for young cambial ages than for older ones.

The height growth curve of the dominant trees group is also higher after 1950 than before. This means that for the same tree age and site, the spruces are higher now than they were in the past.

After having indexed the aging effect with the idealised aging curve 1880-1950, the same data as in Fig. 2 are plotted as a function of the calendar year. The resulting curve displays a continuous and clear increase during the 20th century (see Fig. 3). The observed trend, tested by sign trend (Cox and Stuart 1955) and linear regression coefficient, is very highly significant. The test of the mean square successive difference (Neumann et al. 1941) also indicates a clear trend. This result is very similar (positive trend with comparable slope) to those obtained on silver fir (*Abies alba* Mill.) at 1.3 m, by Bert (1992) for the same region (Jura) and by Becker (1989) for the Vosges mountains in France.

If the indexation is done with the aging curve 1880-1993, the same pattern is observed but the trend's slope of the indexed curve is slightly more gradual.

It can be assumed that over time the subsample of trees with small needle loss (≤ 10% at felling time, 663 trees) shows more or less constant characteristics concerning foliage density. As expected in this case, the pattern is very similar too, but compared to the indexed curve in Fig. 3, the slope is steeper for the last two decades.

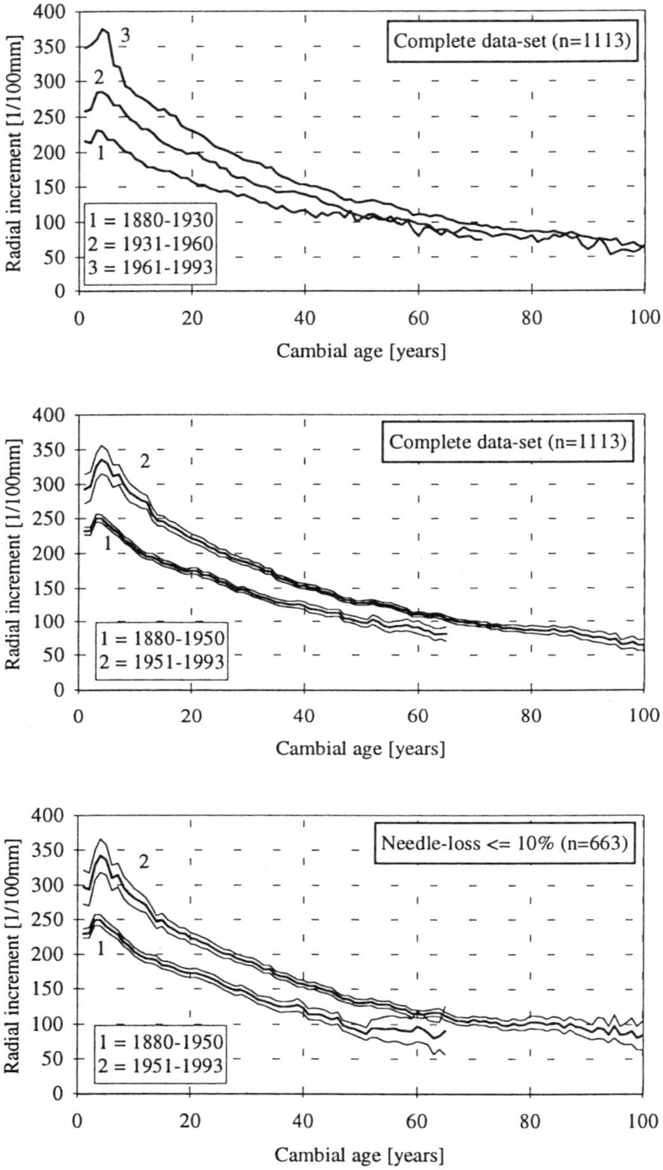

Fig. 2. Aging curves of 1113 dominant and codominant spruces, and of a subsample of 663 trees with small needle loss (≤ 10%). On both graphs with two periods (with confidence intervals of 95%), the lower curve is built with tree-ring data from the period before 1950. The upper curve is built with tree-ring data from the period after 1950. For a given cambial age, a tree ring produced before 1950 is smaller than a tree ring produced after 1950. The third graph is built on the same principle, but for three periods.

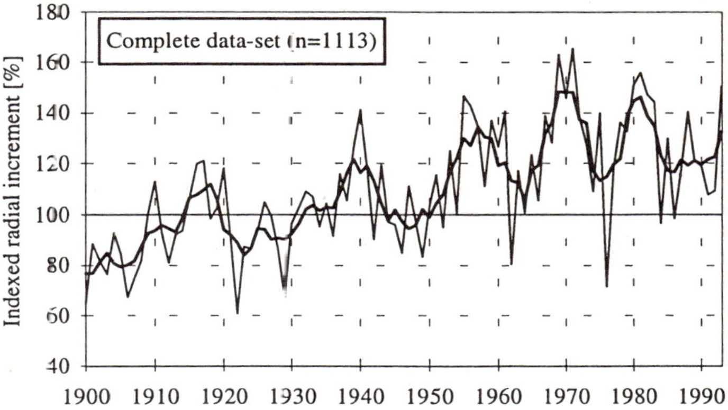

Fig. 3. Radial increment indices (reference: aging curve 1880-1950) of 1113 dominant and codominant spruces. Thin line: annual values. Thick line: moving average over five terms. The trees give their own reference (increment before 1950). The level difference between the aging curves (Fig. 2) is expressed here as a positive long-term trend.

For the interpretation, it is important to mention that 25% of the data-set are formed by trees which are more than 160 years old. The left part of the mean curve in Fig. 3 is mainly calculated from the oldest trees in the data-set, which tend to be found on poorer sites. Therefore, even if site productivity remained unchanged during the century, a positive trend would be theoretically observable. The study of the site index's influence on radial increment showed, however, that the differences between the low and high site indexes are too small to explain the whole observed increase (see Figs. 4 and 8).

Additionally, the subsample of the youngest trees issued from thinnings (age ≤ 120 years) has been investigated. Although the mentioned artefact can be excluded in this case, the result for this group (Fig. 5) confirms clearly the curve patterns of the whole data-set (Fig. 2).

Finally, another fact which has an opposite effect must be considered. The recent increment values are issued from trees with various vitalities, whereas the ones from the beginning of the century, which constitute the reference, theoretically come from healthy trees only (the weak ones disappeared progressively and therefore cannot be involved in the study). In our case, the only reference that includes all trees of this past period is the yield table. For our data-set, however, it is unsuitable for use as a reference (single trees).

It is difficult to quantify these two effects accurately. However, the mentioned difference of radial increment level between low and high site index on one hand, and the observed distribution of needle loss with its relation to radial increment (Hartmann et al. 1987) on the other, show (after calculation) that they have a comparable opposite influence on the trend (about 1/3 of the slope each, Fig. 3).

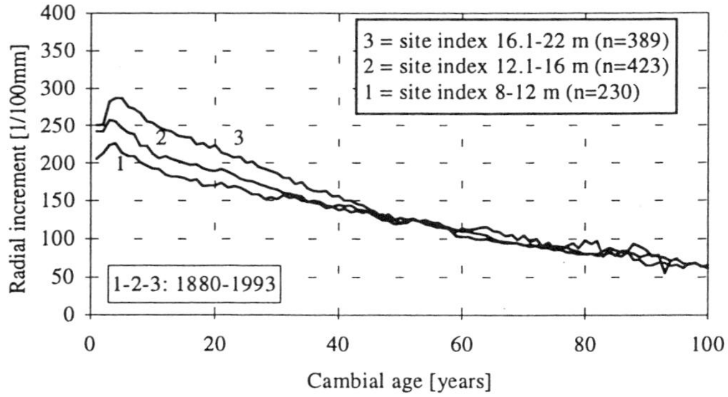

Fig. 4. Aging curves for three site index classes of 1042 dominant and codominant spruces. The differences in radial increment are relatively small among the three classes. The average tree ages of the three groups are: 166 years (1), 140 years (2) and 118 years (3).

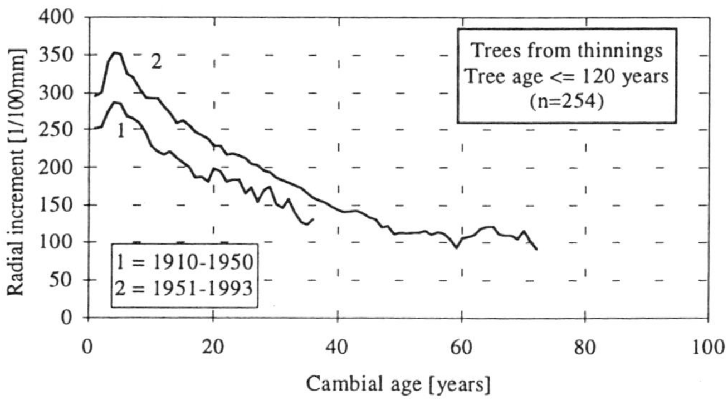

Fig. 5. Aging curves of 254 dominant and codominant spruces issued from thinnings. For these relatively young trees (≤ 120 years), where a site index / age artefact can be excluded, the curve patterns remain similar to those for the whole data-set (Fig. 2).

In a different way than in the Figs. 2 and 3, Fig. 6 shows again that for a given age of reference, growth is higher now than in the past. The trend is highly or very highly significant for each reference age, and especially strong for low cambial ages.

To interpret the observed positive trends, three hypotheses can be formulated:

- Climate (see dendroclimatic model below).
- Silviculture: changes in stand densities and evolution of silvicultural practices have had an influence on the competition situation.
- Effect of human activity on atmospheric composition. Consequence: higher atmospheric CO_2 level, higher N import (deposition), etc.

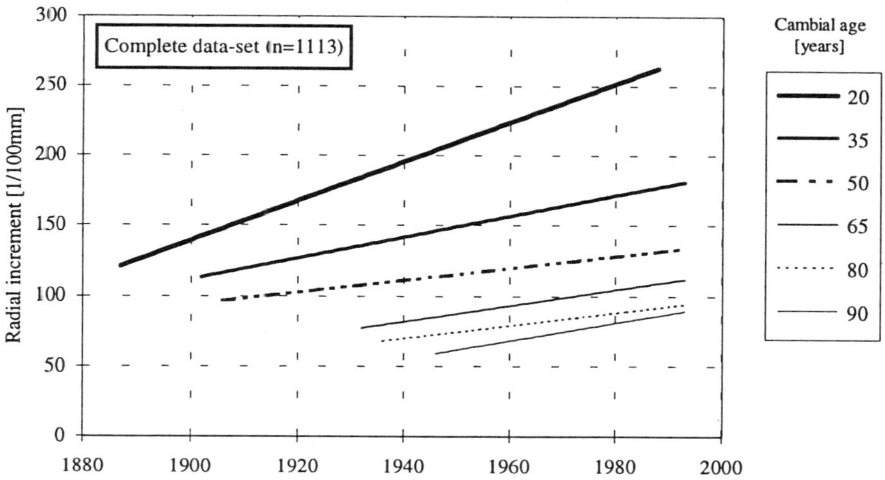

Fig. 6. Radial increment at constant age of reference over the calendar year, for 1113 dominant and codominant spruces. The observed values are idealised by linear regression. For a given cambial age, a tree ring produced recently is larger than a ring produced earlier.

To test whether the climatic factor might explain the positive long-term growth trend, several multiple linear regression models (MLR) were created. The dependent variable is the indexed annual radial increment (see Fig. 3, thin line). The independent variables are climatic parameters of the nearest meteorological stations: mean precipitation (P) of two stations established in the study area (Baulmes, 642 m, and Valeyres, 512 m), and temperature (T) of Neuchâtel (485 m), 40 km north-eastward. All of the calculated models display a clear positive trend in their residuals (significant). That means that the long-term radial increment trend cannot be explained by the influence of the considered climatic parameters, although several temperature values show a significant increase during the examined period (1904-1993), for example: + 0.75 °C for the annual mean temperature, + 0.67 °C for the mean temperature of March - October, + 0.89 °C for the mean temperature of November - February, + 1.42 °C for the mean temperature of October. The conclusion is that the climatic hypothesis can probably be rejected in this case.

With a comparable method, Bert (1992) obtained similar results for silver fir in the Jura. As he was unable to explain the observed positive trend with climatic parameters alone, he tested a model including the CO_2 level among the independent variables. The trend of the CO_2 level coincides well with the radial increment trend, thus allowing a direct influence of CO_2 level on the photosynthesis to be given as a plausible explanation, even if the causal relation has not been proved.

Silviculture may also have influenced the observed trend, because thinning methods have changed and stand densities have generally increased during the century. This may have an effect especially on the radial increment, which is strongly related to the competition conditions, but data to quantify it are not available for the analysed trees. However, height growth, which is theoretically less sensitive to silviculture, shows an evident increase in our case as well.

For these reasons, an effect of site productivity improvement factors like CO_2 or N increase could be a realistic explanation. According to Hostettler (1991), the increase of CO_2 concentration in the atmosphere has been about 25% since the beginning of industrialisation. According to several sources mentioned by Stadelmann (1992), the annual deposition of atmospheric N in the Swiss forest today reaches about 20 to 80 kg ha^{-1}.

Finally, to test whether the short- and middle-term increment variations are related to climate, the long-term trend of the indexed curve was removed by a linear regression model. With this detrended curve, a MLR model based on four precipitation (log values) and two temperature variables of the year y and y-1 was calculated (Table 1, Fig. 7). It also includes the y-1 value of the detrended radial increment curve (autocorrelation factor).

This model explains 62% of the detrended curve's variance and the residuals show no structure. It is stable and its independent variables are easy to interpret. In other words, unlike the long-term increment variations, the short- and middle-term fluctuations are well explained by climatic parameters. This result is confirmed by testing the model on a verification period (1970-1993) after calibration for 1905-1969.

Table 1. Dendroclimatic model for 1113 dominant and codominant spruces. Dependent variable: detrended radial increment indices 1905-1993. P: total precipitation [1/10 mm], mean of 2 stations; T: mean temperature [1/10 °C]; y: year of ring formation.

n	DF	F	p>F	r^2	Var. coeff.
89 radial increment indices (1905-1993)	7	18.73	0.0001	0.62	13.80

Parameter	DF	Estimate	Std. error of estimate	F (type III)	p>F
intercept	-	-256.7636	59.4110	-4.32	0.0001
radial increment $_{y-1}$	1	0.2784	0.0600	4.64	0.0001
log P_{y-1} (1st September - 30 November)	1	9.1095	3.2532	2.80	0.0064
log P_y (1st March - 15 May)	1	7.0011	2.9903	2.34	0.0217
log P_y (16 May - 15 July)	1	31.3399	3.7234	8.42	0.0001
log P_y (16 July - 15 September)	1	6.9317	3.1390	2.21	0.0301
T_{y-1} (16 June - 31 August)	1	-0.6647	0.1124	-5.91	0.0001
T_y (1st March - 15 May)	1	0.2705	0.1328	2.04	0.0449

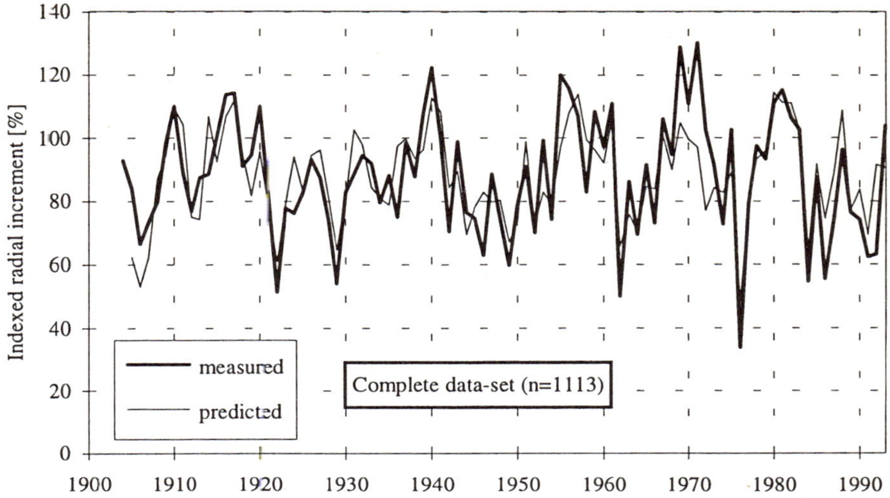

Fig. 7. Dendroclimatic model for 1113 dominant and codominant spruces: detrended radial increment indices (thick line) and predicted values according to the climatic model (thin line).

3.2. Influence of Site and Tree Parameters on the Growth Patterns

The last section showed that the increase of radial increment during the 20th century is very clear when the data-set is considered globally. It is interesting to investigate whether this evolution is identical when the data-set is divided into homogeneous subsamples built for different site and tree parameters.

The results presented in Section 3.2 are based on the same data-set as in Section 3.1. (1113 dominant and codominant spruces). The total number of analysed trees, however, is usually smaller because of some missing data among the site and tree factors, and because extreme values are not taken into account. The aim is to build groups that are as homogeneous as possible for the parameter under consideration.

Figs. 8 and 9 show that the trees growing on poor sites display an evolution of radial increment very similar to that of trees growing on more fertile sites. The observed growth pattern is nearly the same as for the global analysis discussed in the previous section. Investigations on other subsamples lead to the same conclusion: for small and quite homogeneous groups, the radial increment of single trees increases in any case throughout the century. The factors taken into account are: soil depth, soil type, altitude, thinning trees versus regeneration trees, cross-section age, tree age at cross-section height, tree age.

Concerning the last parameter, we wrote in Section 3.1 that the oldest trees often grow on poor sites and we mentioned that this point has an influence on the observed evolution of radial increment. For this reason, it was interesting to build

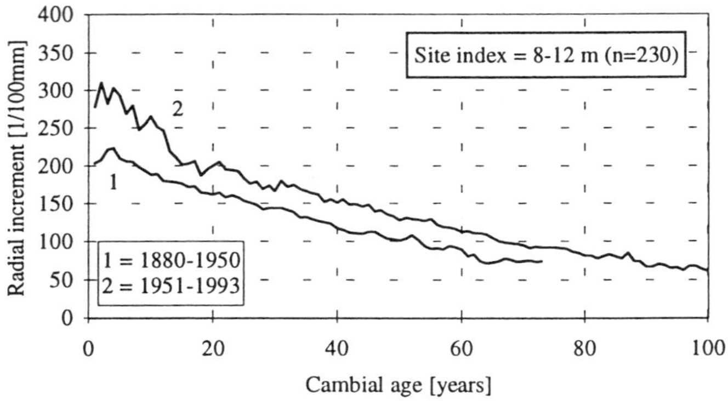

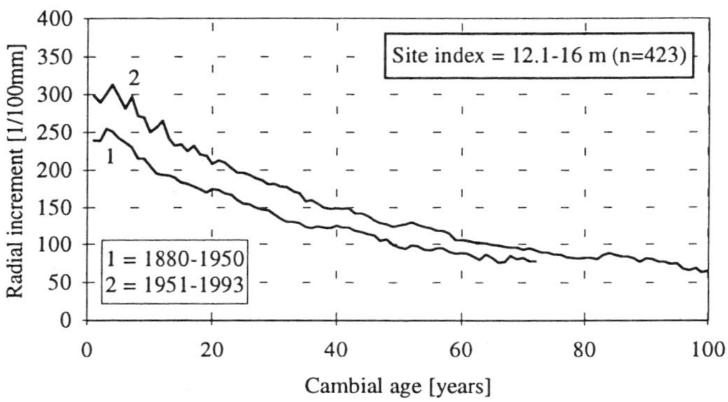

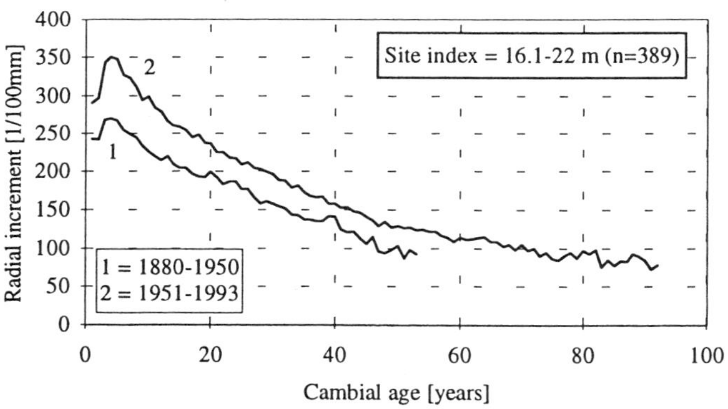

Fig. 8. Aging curves of 1042 dominant and codominant spruces stratified for site index. The three classes display the same growth pattern: for a given cambial age, a tree ring produced before 1950 is smaller than a tree ring produced after 1950.

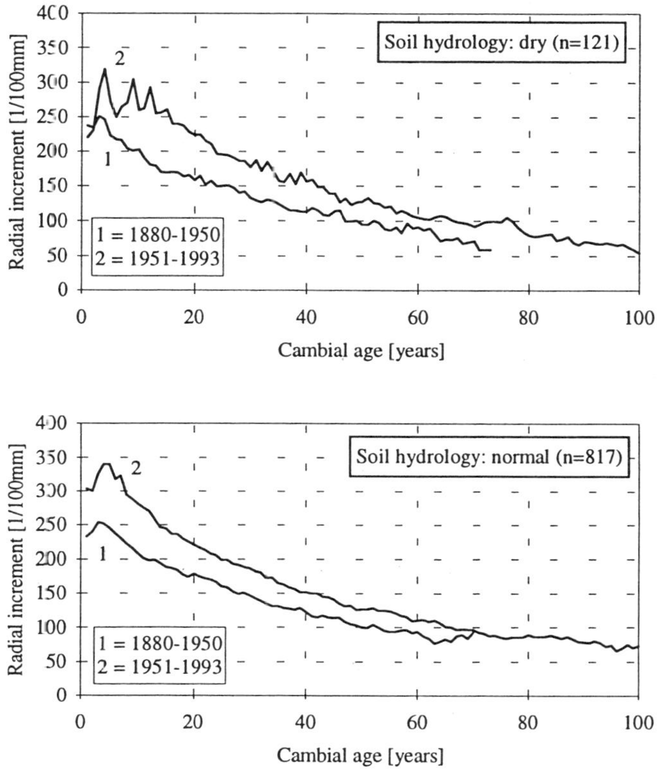

Fig. 9. Aging curves of 938 dominant and codominant spruces stratified for soil hydrology. Both classes display the same growth pattern: for a given cambial age, a tree ring produced before 1950 is smaller than a tree ring produced after 1950.

groups of trees having the same age. Six categories, including 978 trees, were created for following tree ages: 80-100, 101-120, 121-140, 141-160, 161-180, 181-200. As a result, we observe another level for every group, but the difference between the two aging curves always remains.

3.3. Growth Patterns for Various Heights Along the Stem

Cross sections or cores for radial increment measurements are traditionally taken at 1.3 m. In the present study, however, the cross sections come from different heights, which mostly coincide with the lower part of the crown. The mean height of the cross sections is 21.5 m (standard deviation: 4.3 m).

The variability of the radial increment along the stem is well known (e.g. Topcuoglu 1940, Athari 1980, Gerecke 1988). For this reason, we have to take this factor into account in order to examine whether or not it influences the interpretation of the results.

Before analysing this point, it is necessary to remember that our general purpose is to analyse site productivity changes, which are in fact expressed by the evolution of the total forest biomass production. In practice, only indicators of this evolution are usually measured, i.e. growth in volume, basal area, diameter or height. In this context, the radial increment at different heights along the stem constitutes an indicator which is as valuable as at breast height. Actually, the choice of 1.3 m as the conventional reference height is useful for comparisons, but considering the tree biology and the indicator value for biomass evolution, it is not the only possible reference.

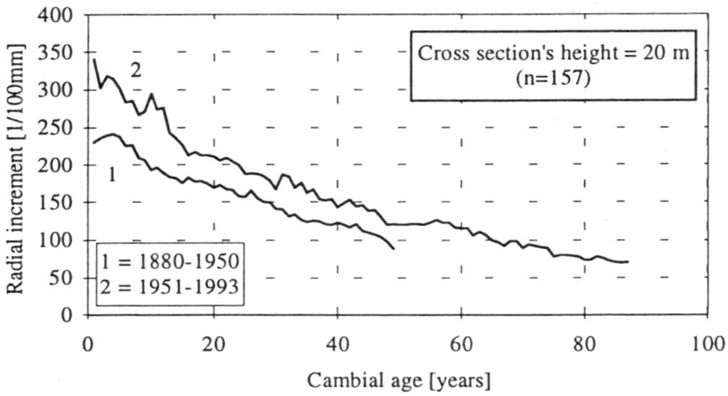

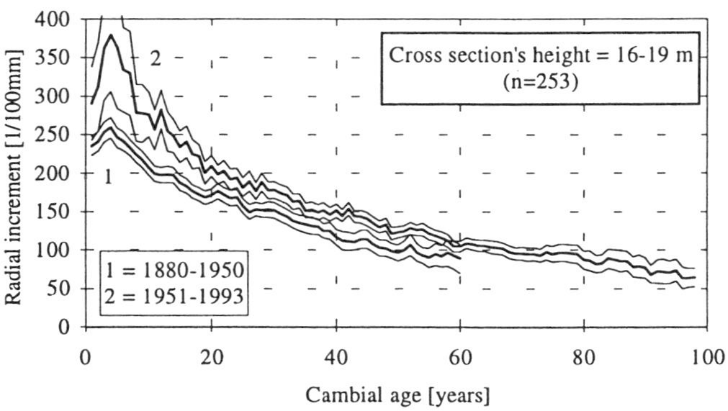

Fig. 10. Aging curves of 889 dominant and codominant spruces with confidence intervals of 95%. Cross section height 20 m, and classes 16-19 m, 20-22 m, 23-25 m. For a given cambial age, a tree ring produced before 1950 is smaller than a tree ring produced after 1950 in each group.

A detailed analysis of the data-set shows that the general level and the difference between two aging curves 1880-1950 and 1951-1993 are not identical at different heights along the stem. But the radial growth patterns are nevertheless very similar. Fig. 10 illustrates this point: for the three classes of cross section heights 16-19 m, 20-22 m, 23-25 m, as well as for a constant height (20 m in this example), a difference between the growth level 1880-1950 and 1951-1993 is always observed. Consequently, a positive trend in radial increment during the 20th century is present in each data subgroup.

To test the relationship between radial increment and cross section height, the increment at constant cambial age (20, 40, 60, 80 years) and at constant calendar year (1930, 1960, 1990) has been plotted as a function of the cross section height. As a result, level differences were observed and no crossing between the curves was found for the values investigated. Fig. 11 illustrates this for a cambial age of 20

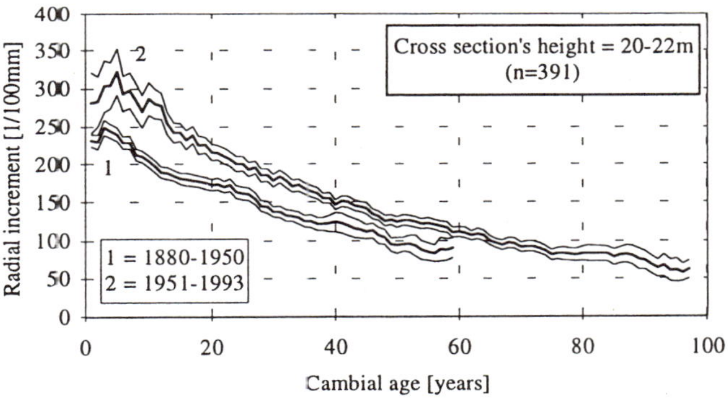

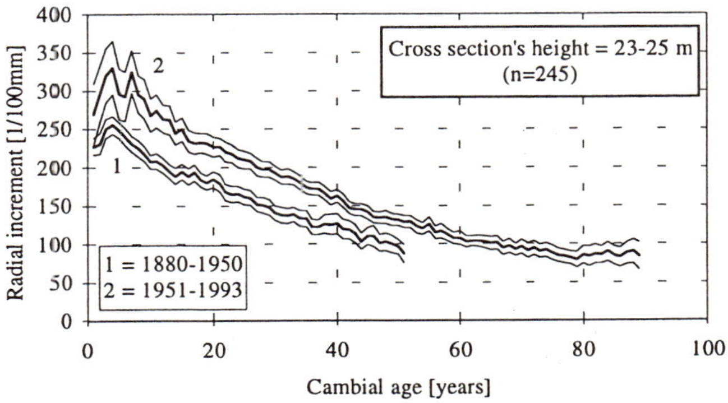

Fig. 10. Continued

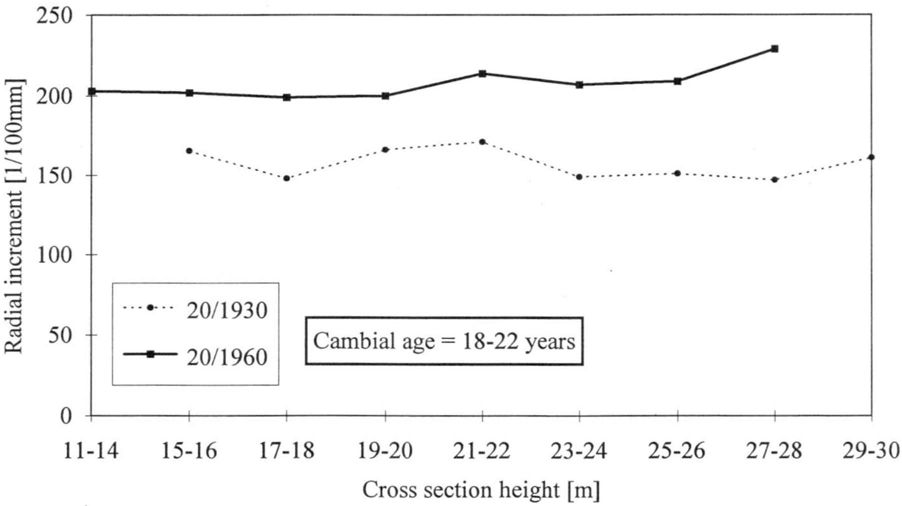

Fig. 11. Mean radial increment of 1113 dominant and codominant spruces along the stem (cross section height 14-30 m) for the calendar year 1930 and 1960 (± 2 years) and for the cambial age 20 years (± 2). Both curves show a different level and do not cross.

years (± 2) and for the calendar years 1930 and 1960 (± 2). As expected according to Fig. 10, the radial increment level is clearly higher in 1960 than in 1930 at every height along the stem.

Further, we have tested several multiple linear regression models which express the radial increment in a given calendar year (dependent variable) as a function of different parameters describing tree, stand and site. They explain more than 70% of the variance of the dependent variable, but the cross section height never has any significant influence in these models.

With regard to the purpose of this research project, to the graphs presented here (Figs. 10 and 11) and to the comparison with other studies (e.g. Becker 1989, Bert 1992), the growth patterns in higher parts of the stems, although unconventional, can be considered as valid indicators of vitality and site productivity.

4. Conclusions

For the case study presented here, no sign of a negative growth trend during the 20th century was found. On the contrary, a clear increase in the radial increment of single trees is observed. Many influencing factors (including silviculture) should be considered when attempting to explain this result, but it seems quite reasonable to admit a fertilizing effect of nitrogen or CO_2 in our case. This point underlines the fact that a positive trend may reveal a disturbance of the forest ecosystem, just as a negative trend would.

Of course, there is no evidence, and these results can be considered only as signs for a site productivity change, as they merely give indications about increment at single tree level. In this context, we can only specify that none of our research projects in the Swiss Jura and the Swiss Plateau (Hartmann et al. 1992, Schneider and Hartmann 1996) showed negative trends for the basal area increment at stand level (per ha).

References

Athari S (1980) Untersuchung über die Zuwachsentwicklung rauchgeschädigter Fichtenbestände. Dissertation Göttingen.

Becker M (1989) The role of climate on present and past vitality of silver fir forests in the Vosges mountains of northeastern France. Can. J. of Forest Research, 19, 9: 1110-1117.

Becker M, Nieminen TM, Gérémia F (1994) Short-term variations and long-term changes in oak productivity in northeastern France. The role of climate and atmospheric CO_2. Ann. Sci. Forest., 51, 5: 477-492.

Bert GD (1992) Influence du climat, des facteurs stationnels et de la pollution sur la croissance et l'état sanitaire du Sapin pectiné (Abies alba Mill.) dans le Jura - Etude phytoécologique et dendrochronologique. Thèse de doctorat de l'Université de Nancy I. 200 p.

Cox DR, Stuart A (1955) Quick design tests for trend in location and dispersion. Biometrica, 42: 80-95.

Gerecke KL (1988) Grundflächenzuwachs in verschiedenen Schafthöhen bei vorherrschenden Tannen. DVFFA, Sektion Ertragskunde, Tagungsber. 1988: 13/1-13/14.

Graumlich LJ, Brubaker LB (1987) Increasing net primary productivity in Washington (USA) forests during the last 100 years. In: Proceedings of the International Symposium on Ecological Aspects of Tree-Ring Analysis, 17-21 Aug. 1986, Marymount College, Tarrytown, NY. pp 59-69.

Hartmann P, Schneider O, Petter DA, Schlaepfer R (1987) Etude des relations, chez l'épicéa, entre la largeur du cerne et l'état sanitaire de la cime. J. for. suisse, 138, 11: 923-943.

Hartmann P, Schneider O, Petter DA, Zimmermann D (1992) L'accroissement en diamètre des arbres. Evolution des connaissances. Influence du climat, de la sylviculture et de la pollution. Bull. Soc. Vaud. Sc. Nat., 81: 131-151.

Hostettler M. (1991) Anthropogen verursachte globale Klimaveränderung und Schweizer Wald. J. for. suisse, 142, 11: 915-923.

Kenk G (1988) Zum Wachstum aufeinanderfolgender Fichtenbestände. DVFFA, Sektion Ertragskunde, Tagungsber. 1988: 9/1-9/6.

Lamarche VC, Graybill DA, Fritts HC, Rose MR (1984) Tree-ring evidence for growth enhancement in natural vegetation. Science, 225: 1019-1021.

Nagel J, Athari S (1982) Stammanalyse und ihre Durchführung. Allg. Forst- u. J.-Ztg., 153, 9/10: 179-182.

Neumann J von, Kent RH, Bellinson HB, Hart BI (1941) The mean square successive difference. Ann. Math. Statist., 12: 153-162.

Parker ML (1987) Recent abnormal increase in tree-ring widths: a possible effect of elevated atmospheric carbon dioxide. In: Proceedings of the International Symposium on Ecological Aspects of Tree-Ring Analysis, 17-21 Aug. 1986, Marymount College, Tarrytown, NY. pp 511-521.

Schneider O, Hartmann P (1996) Growth trends of trees and stands - Regional studies on Norway spruce in Switzerland. Report IUFRO XX World Congress, 6-12 August 1995, Tampere, Finland. In print.

Stadelmann FX (1992) Mögliche Wirkungen von Stickstoff aus landwirtschaftlicher Sicht. In: Forum für Wissen 1992, Waldschadenforschung in der Schweiz: Stand der Kenntnisse. pp. 113-116.

Topcuoglu A (1940) Die Verteilung des Zuwachses auf die Schaftlänge der Bäume. Tharandter forstl. Jahrb., 91. 485-554.

Growth Trends of Swiss Forests: Tree-Ring Data. Case Study Toppwald

O.U. Bräker
Swiss Federal Institute for Forest, Snow and Landscape Research,
Tree-Ring Research Group, Birmensdorf, Switzerland

Abstract

In cooperation with other colleagues, concerted research on recent growth trends based on data from tree analyses, permanent plots and from forest inventories was undertaken. This study is based on tree growth analysis of spruce and fir on the permanent plot Toppwald and the trees are compared to earlier sampled trees of the Swiss Sanasilva 1984 study. Climate models are used to detect known influences of weather patterns and climate change; these effects explain 9 to 37% of the growth variance. In the calculated, non-climatic residual data, a long-term time-dependent trend of 4 to 49% of the growth variance was found. This shows that growth of the two species has improved since the beginning of the century. In the most recent period the residuals show, for most cases, an increased growth since 1980. Analysis of the whole time series reveals neither an exceptional rise nor any growth levels which have not been seen or reached earlier in this century.

1. Introduction

This study on tree ring growth was undertaken in connection with research conducted by A. Zingg and M. Köhl on the permanent plot Toppwald. This plot has served our research institute as a long-term growth study area with standardised silvicultural treatment since 1905. It is, therefore, a most appropriate site to combine plot information with tree growth information. While management data for the area give information on growth for periods at 5- to 6-year intervals, changed in 1969 to intervals of 8 years, the single-tree data broaden the interpretation on a temporal scale. This gives more insight into possible competition effects and variability within and between trees.

Tree-ring growth was measured on discs taken from all felled trees of the most recent silvicultural treatment of 1993. In addition to the study by Kontic et al. 1990,

European Forest Institute Research Report No. 5
Growth Trends in European Forests
Edited by H. Spiecker, K. Mielikäinen, M. Köhl and J. P. Skovsgaard

this research effort documents tree-growth patterns on a different site in Switzerland. Data from this case study are compared with data from the Swiss Sanasilva Inventory of 1984, with representative coring material from sample plots in a 4 × 4 km grid over Switzerland (Bräker 1992). The combination of the data from a local case study site and the averaged data from a large area will show more reliably if there are any generalised trends or signals with a common source.

2. Data Description

The study site Toppwald is located at approx. 950 m.a.s.l. on a slight, west / northwest slope. The inventoried plot surface covers 3 ha and the main species are fir (*Abies alba* Mill.), spruce (*Picea abies* [L.] Karst.) and beech (*Fagus silvatica* L.). The management system maintains an uneven-aged, multi-storey structure of a mixed species population (selection forest). The site index for fir is 22 (estimated h_{dom} at age 50 in m). The felling follows a regulation regime to maintain a constant basal area growth pattern.

The disc samples for the tree-ring study were taken during the 1993 felling from the top end of each bole. There were 47 fir, 17 spruce and only 3 beech included in the sample. The selected trees cover a proportional part of the remaining stand in regard to species and dominant and co-dominant trees at different competitional stages. Since the felling height was around 0.2 m above ground, the ring pattern at the lower end of the boles was affected by root influences. As such disturbances should be avoided for growth studies, discs were chosen from the top end of the boles. This choice results in different disc heights on the stems which, in turn, affects the age interpretation of the data: the age of the discs does not directly reflect tree age. The distributions are shown in Fig. 1.

On each disc the radial increments of 4 perpendicular radii were measured, synchronised, dated and averaged. From these increment data (i_r), development of diameter (d), annual disc area increment (i_g), average disc area increment ($\bar{i}_g$) and disc area (g) were calculated. The corresponding formulas for the values are: $d(j) = 2*\text{sum of } i_r(j)$ for years $=1..j$, $g(j) = d(j)^2*\text{Phi}/4$, $(i_g) = g(j)-g(j-1)$, $(\bar{i}_g) = g(j)/j$. Since all relevant variables react similarly in the long-term trend, the presentation of curves is restricted to radial increment.

3. Methods of Growth Analysis

The radial growth pattern of a conifer in closed stands usually shows a distinct ageing signal. When average growth data of populations are calculated in a definite year, the tree ages are mixed. This mixed ageing effect disturbs further interpreta-

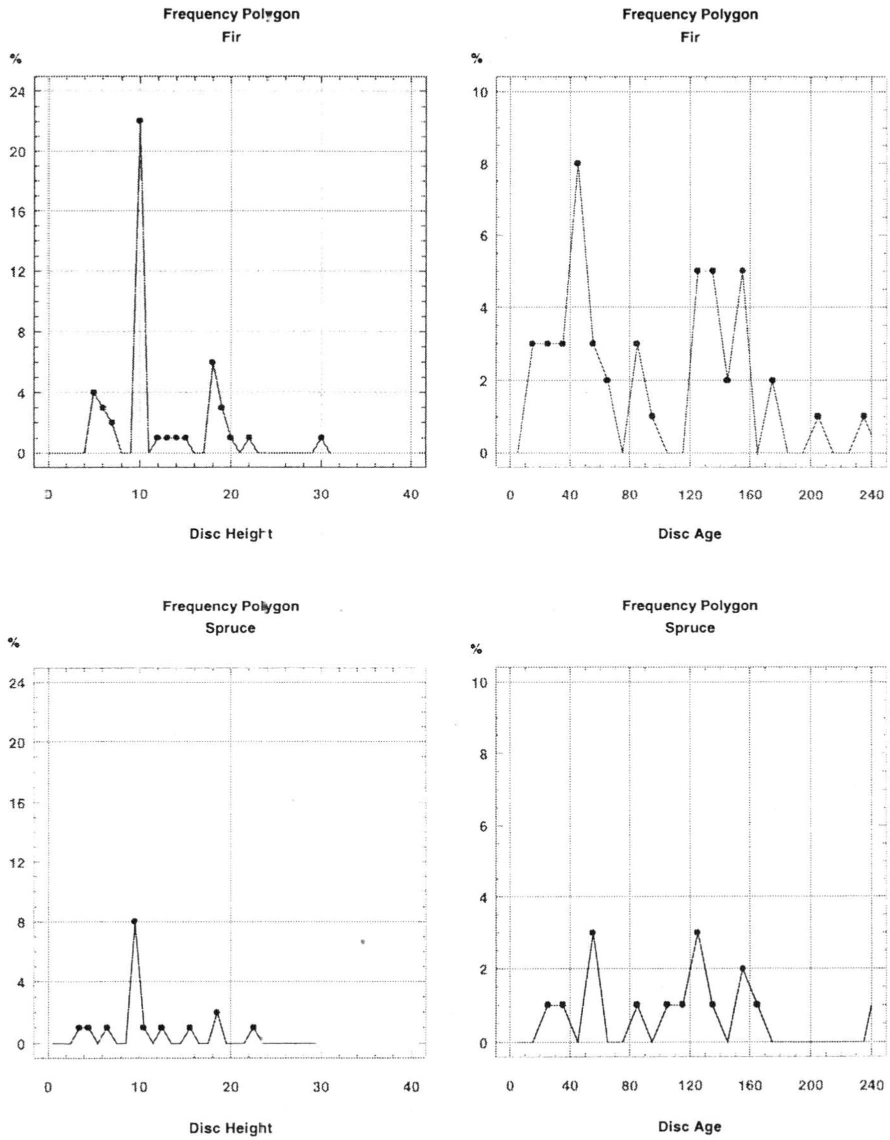

Fig. 1. Distribution of disc heights and age span of data for fir and spruce.

tion and should be eliminated. Several homogenising procedures can be used to eliminate ageing effects, thereby resulting in ageing-free residuals: e.g. relating growth to yield table expectations, relating to other time-dependent functional models or by stratifying the ages.

A time-dependent growth model which assumes ageing as slow changes in physiological processes during a lifetime was applied in order to calculate deviations from expected normal growth patterns. The relative age from the disc's pith (age 1) is referred to as the cambial age (Becker et al. 1990, Picard 1995). The data were averaged with the cambial age to calculate references of age-growth relations. Since the heterogeneous material in disc heights and disc age may influence the averaged age-growth relations, the differences were checked with stratified data. Two subsets both for low and high disc heights as well as for young and old disc ages, for both fir and spruce were used (Table 1). The consequences of the heterogeneous data are discussed in the next two paragraphs.

Table 1. Number of trees for discs age and disc height classes.

Species	Number of trees, old>100 y	Disc age young ≤ 100 y	Disc height low ≤ 11 m	high > 11 m	Total
Fir	21	26	31	16	47
Spruce	11	6	12	5	17

3.1. Stratified Data Averaged According to Cambial Age (Fig. 2A)

Higher stem level disc data show a growth increase of approx. 20% for fir; for spruce the increase is less than 5%. This tendency reflects known radial growth-height relations due to stem form which do not change over time. It seems that the site conditions are at an optimum for fir.

Younger age classes show an increased radial growth for the first 60 years of approx. 10%. In the first 25-30 years, all fir discs start with a slightly suppressed level. Extreme values exist after the age of 55 in the young spruce pattern. These are due to the small number of observations in this class.

3.2. Stratified Data Averaged According to Absolute Age (Fig. 2B)

From 1940 onwards, the number of replications are sufficient. Fir shows a fairly constant and increased parallel growth pattern for younger and higher classes. Similar differences, but more moderate, are shown by spruce. This behaviour over time is proof that for the interpretation of growth shifts and their possible causes, elimination of age-growth relations are necessary. The curves show that silvicultural treatment in the years 1944, 1949, 1954, 1959, 1964, 1969, 1977 and 1985 are followed by short time growth releases as well as declines. Therefore, the felling and changes in tree competition are not the cause of these additional growth shifts.

3.3. Age-Trend Model

The applied Hugershoff model for ageing as a functional relation to age is based on studies by Pollanschütz (1966) and Bräker (1981). Since the data of the growth pattern in Fig. 3 show distortions on a high frequency level, a smoothing function was fitted to the data. The Hugershoff function can be calculated by least-square fitting, by a multiple linear regression (MLR) on log-level or by a non-linear regression approach. The function is normally a combination of a polynomial and a negative exponential part and has, with increasing age, just one turning point:

$$\text{growth increment} = a \times \text{time}^b \times e^{c \times \text{time}}$$

There are three parameters a, b, c to estimate. The resulting parameters for radial increment are e.g.:

	Attribute	a	b	c
Spruce	i_r	363.992	-.257009	.221205E-02
Fir	i_r	286.732	-.0540649	-.403326E-02

The few observations at higher ages disturb the curve and may be suppressed. The time period selected for the fitting influences the shape of the resulting function. In tests, one Hugershoff function for the complete period (>250 years) and, to obtain more than 9 observations, a shortened period (approx. 120 years for all age classes or 55 years for individual age classes) were calculated. A model with approximated parameters (a=360, b=-0.2, c=-0.001) was also tested. The test proved that differences were small when comparing the absolutely dated chronologies. With the function determined for each species, indices as quotients [i(ct)=observed(ct)*1000/fitted(ct)] were calculated for each measurement value according to the cambial age (ct). The weakest fits in the very early growth period of the first 5 years were excluded from further averaging.

To derive the absolutely dated growth levels, the indices were averaged for both fir and spruce (Fig. 4). The values represent (as per mille) the annual growth in relation to an expected age-compensated growth level.

4. Results and Discussion

4.1. Trend Analysis

To estimate growth trends, various methods are proposed in the literature. Among the more modern and complex regression models, response functions introduced by

Ring width, young and old discs, high and low discs

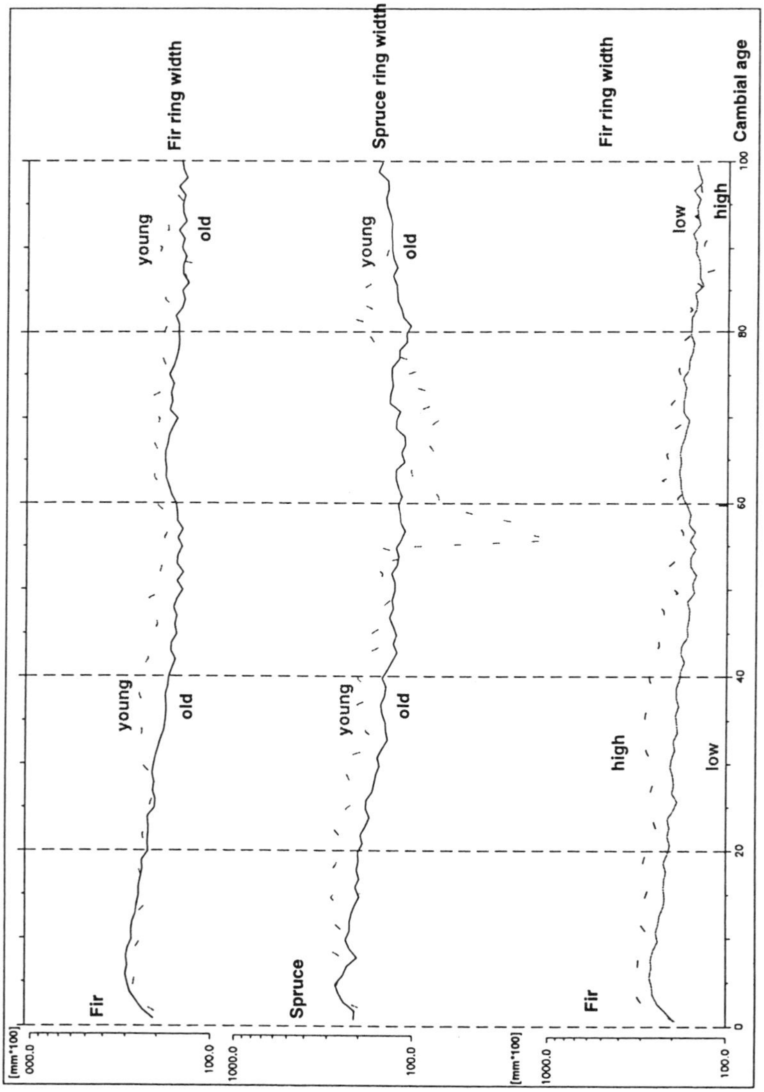

Fig. 2A. Growth patterns of young (dotted line) and old (solid line) discs, high (dotted line) and low (solid line) discs; time scale is cambial age.

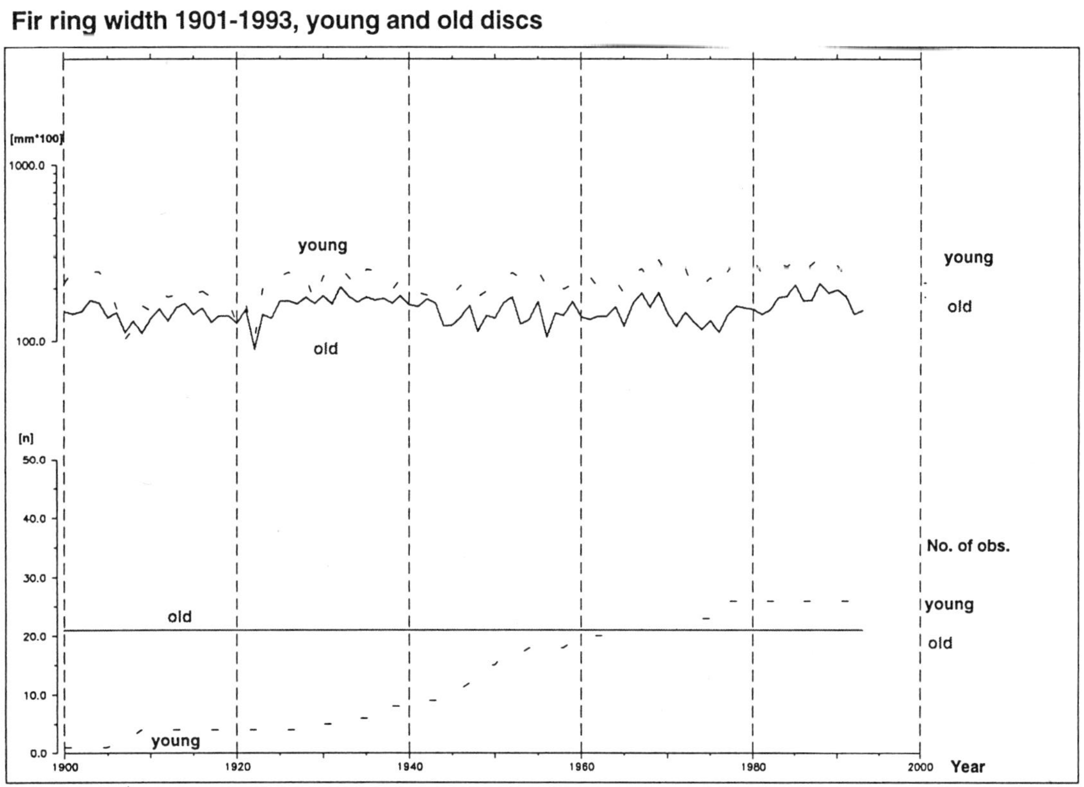

Fig. 2B. Growth patterns of young (dotted line) and old (solid line) discs; time scale is absolute year.

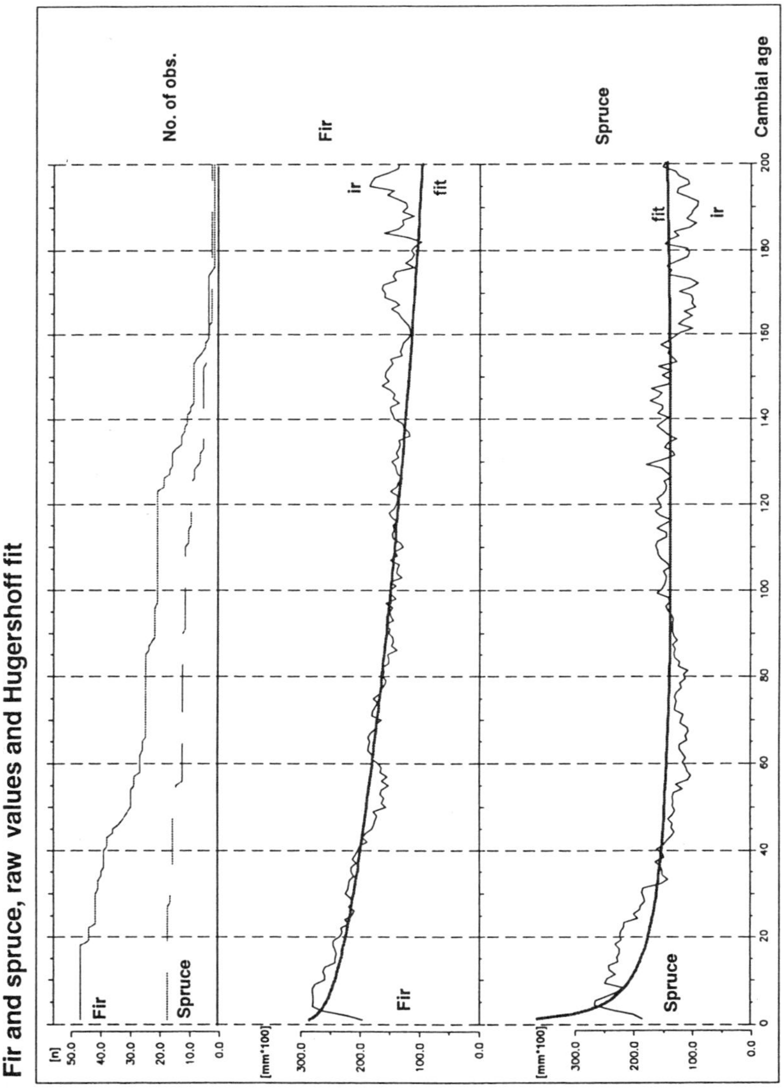

Fig: 3. Ageing pattern of fir and spruce and smoothing functions (Hugershoff).

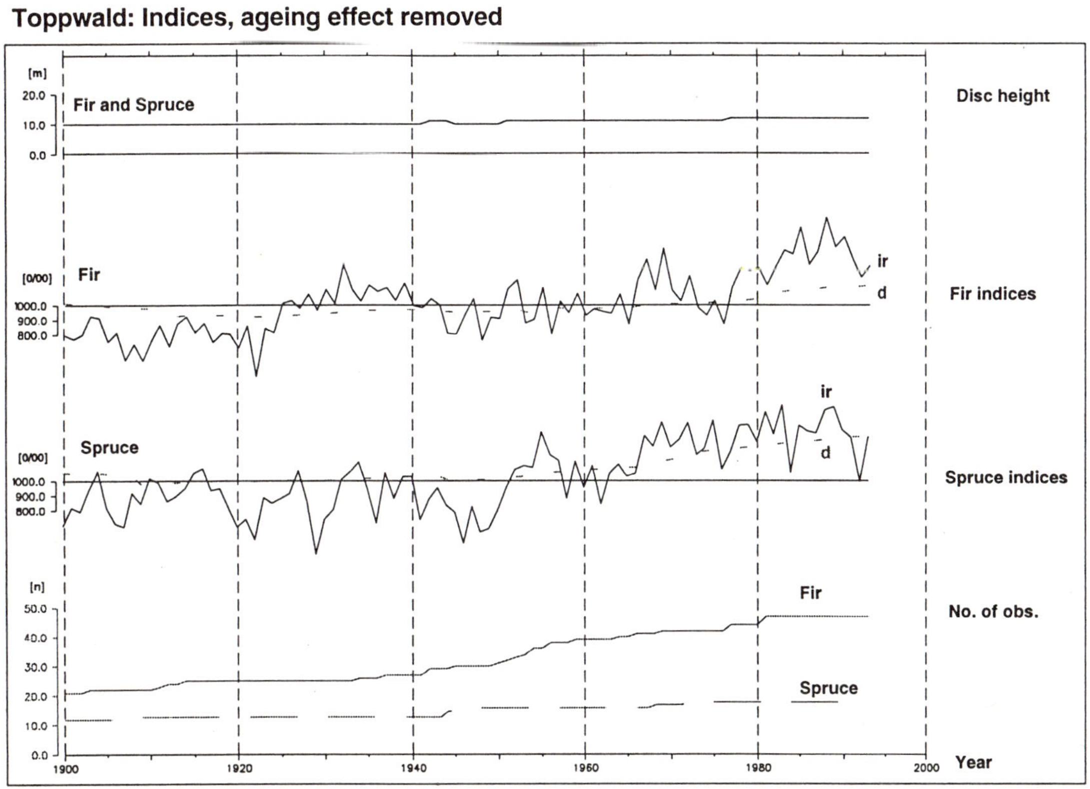

Fig. 4. Growth indices Toppwald for fir and spruce 1900-1993, including disc height and no. of observations. Solid line shows radial increment, dotted line shows diameter.

Fritts (1976) assume relations constant in time while Kalman Filters and Response-Trend models reported by Visser and Molenaar (1992) rely on dynamic relations in time. Kublin (1992) suggested a regression model applying splines for trend changes. In modelling procedures where multiple variables are included, the differentiation between internal and environmental causes can be difficult, i.e. climate trends, time-dependent trends, storage or buffer effects of the plant and serial correlations act as confounded variables. In this context, a simple and stepwise procedure is chosen to reach insight into the separate parts of the explained variance by ageing, climate and time. The ageing effect is modelled by the Hugershoff function. In addition, the effects of changing weather and climate conditions are detected in a climate model search using multiple linear regressions; the same method was used for time effects. A forward stepwise procedure with a standard software package was used for multiple regression modelling, with the F-level set to 4 for variable selection.

Due to the heterogeneous composition in disc height, this variable and other causes may confound the interpretation of trends. For this reason, the height parameter is also averaged. The resulting curves of the mean indices show that, in time, the composition shifts to higher (younger) discs. It is possible to compensate for the height parameter in the applied multiple linear regression model.

The curves for this century show obvious simultaneous increases after 1960 of about 20-30% from the average growth level up to this time (Fig. 4). The radial increment and the basal area increment show high frequency variations and periods at the start of the century where values drop below 100%. The developments of the diameter or basal area are smoother and show nearly horizontal lines before the rise. The recent rise for fir starts after the depression of the mid-seventies (1976); spruce already started to rise in the early 1960s, after the depression of 1962. Although growth has dropped in the last 3 years, the present significance of the trend change is unclear. Similar short-term fallbacks occurred for fir in 1976 and 1986 and for spruce in 1976 and 1984.

4.2. Comparing Results of Toppwald 1993, Switzerland 1984 and Switzerland-Subset of 1992.

A similar procedure for ageing fits and indexing was applied to the data of the fir and spruce of Switzerland collected in the Sanasilva Inventory 1984 (Bräker 1992). The resulting averages cover only the social status of dominant and co-dominant trees (Fig. 5). The ageing fit is a generalised function for all 678 spruce and 255 fir.

Due to similar climatic influences and yearly events at the local site and throughout Switzerland, the comparison of the growth pattern Toppwald and Switzerland documents good correlations for radial growth indices in the high frequency pattern. On a low frequency level the spruce trend is maintained in both time series, while for fir, the rise of Toppwald is not reflected in the Swiss curve. The latest information represented for fir shows the marked drop related to dry years in 1974 and 1976, leading to a "low" which lasted until the beginning of the 1980s. If there is a more recent growth release after 1981, it cannot be seen in the material from 1984.

Preliminary results from a Switzerland-subset of trees recored in 1992 have not proved a further decline of fir (Fig. 6).

Climate models. To detect and eliminate any climatic influences, several models were applied to the indices. As independent variables, temperature and precipitation data averaged for either seasons or vegetation period or whole years were taken (Table 2). Langnau (1931-1994) is the closest local climate station for Toppwald, but Bern (1900-1994) shows a longer reference period (Fig. 7). Appropriate averages of 26 temperature and 45 precipitation stations were taken for Switzerland. The climate data show that a linear climate trend, i.e., greenhouse effect, for this century is weak (+0.5°C / 100 years but not significant at the 5% level). The relevant response coefficients to growth are given below, including a model with a short calibration period 1900 - 1960. Growth of fir is better explained by climatic variables than that of spruce. It is documented that both a warm and wet autumn in the previous year as well as a wet spring in the current year is favourable to the growth of both species, additionally a warm spring or summer benefits fir growth.

Time dependent trend. After climate effects had been determined in a model, the corresponding effect was removed in residual data calculation. To further check a time-dependent trend or other non-climatic effects (i.e. serial correlation) a second-step modelling procedure was applied. In most models with climate residuals, the best input variable was the absolute year. Serial correlation can be eliminated in raw values by "whitening" the time series or using an auto-regressive model. This reduces the signal, which may be a part of climate influence of previous years or an environmental time shift to be looked at. The percentages of explained variance in Table 3 show the amount of signal in common with the confounded variables.

5. Conclusions

The modelled residuals for fir and spruce, on both the local level as well as on the nation-wide pattern, show an above-average growth in the most recent period after 1980 in all cases (Fig. 8). For spruce this rise starts 10 years earlier than for fir. Particularly the fir data of Sanasilva 1984 show the distinct growth depression of the 1980s. The rise is only detectable in the subset of the re-cored trees of 1992. Furthermore, the differences in the growth pattern of fir in the 1970s can be explained: data from Toppwald reflect optimum site conditions for fir, while the average for Switzerland includes poor sites.

This recent, medium-term rise in growth of the final residuals is caused neither by "climate change" in the domain of the seasonal or yearly effect nor by a linear time-dependent trend or ageing effect. These causes were eliminated by the modelling processes. A major cause is associated with changes in site conditions. Changes in forest use and the rescurce management cannot be excluded as causes for the long-term trend of the climate model residuals. The hypothesis of a CO_2, nitrogen or oth-

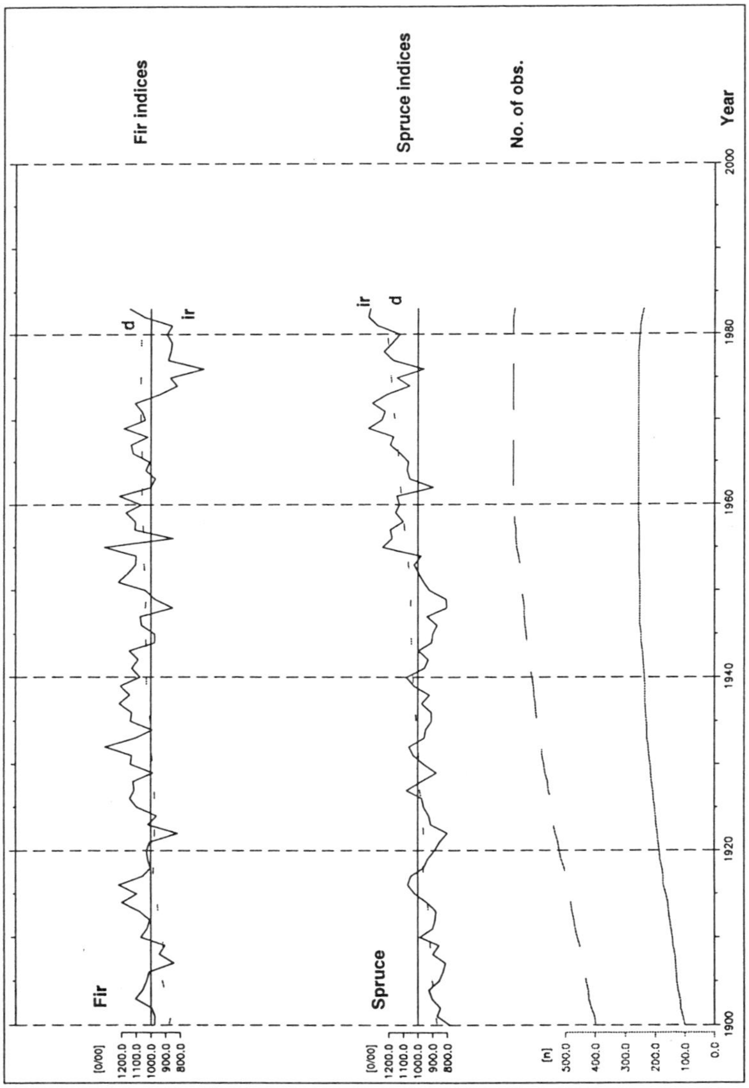

Fig. 5. Growth indices Switzerland for fir and spruce 1900-1983, including no. of observations. Solid line shows radial increment, dotted line shows diameter.

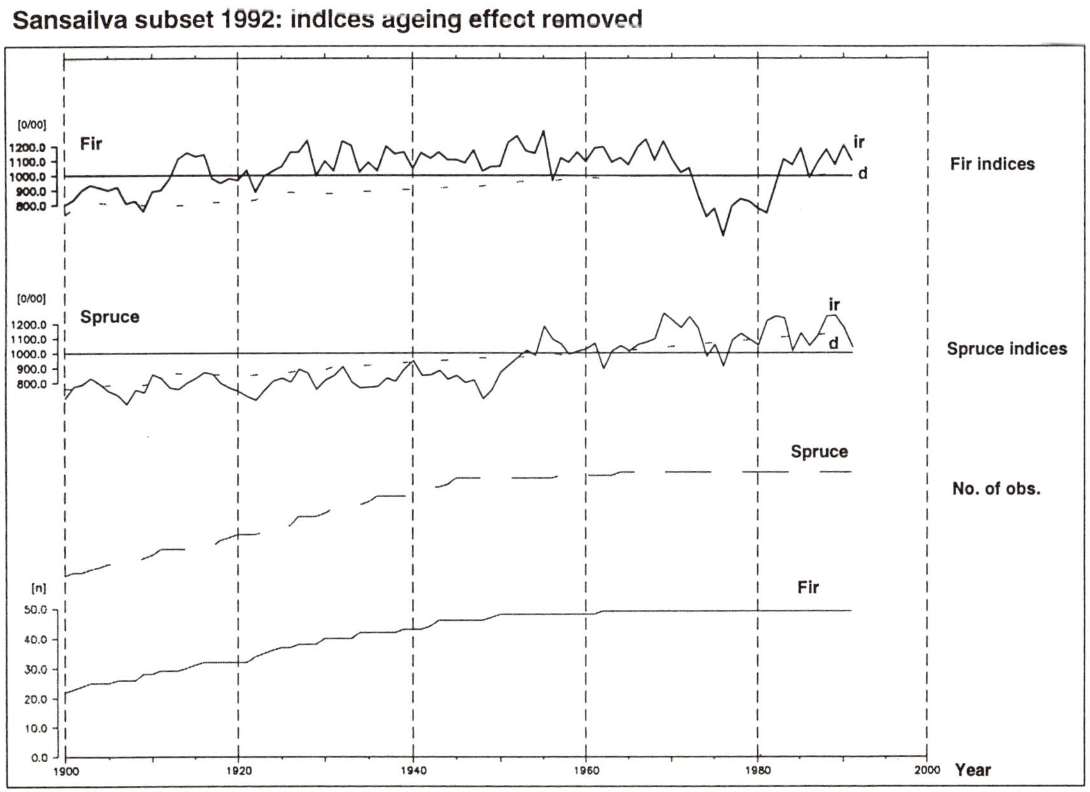

Fig. 6. Growth indices Switzerland-subset for fir and spruce 1900-1991, including no. of observations. Solid line shows radial increment, dotted line shows diameter.

Table 2. Climate model parameters, regression coefficients and explained variance.

Set	Precipitation months Previous year		Current year			Temperature months Previous year		Current year			$r^{2 [\%]}$
Months	7-9	10-12	1-3	4-6	7-9	7-9	10-12	1-3	4-6	7-9	
Spruce											
CH84	0.41			0.74			3.51				15
CH92	0.42			0.70			6.16				17
CH92<1961	0.75				0.47		4.05				26
Toppwald							8.44				9
Toppwald<1961			0.68		-3.76						13
Fir											
CH84	0.61			0.70					2.82		22
CH92	0.64			0.67			3.95		6.33		20
CH92<1961	0.84						4.53	2.39	4.65		30
Toppwald	0.44			0.75			8.98			5.85	31
Toppwald<1961	0.78				0.50		4.34	2.23		5.55	35

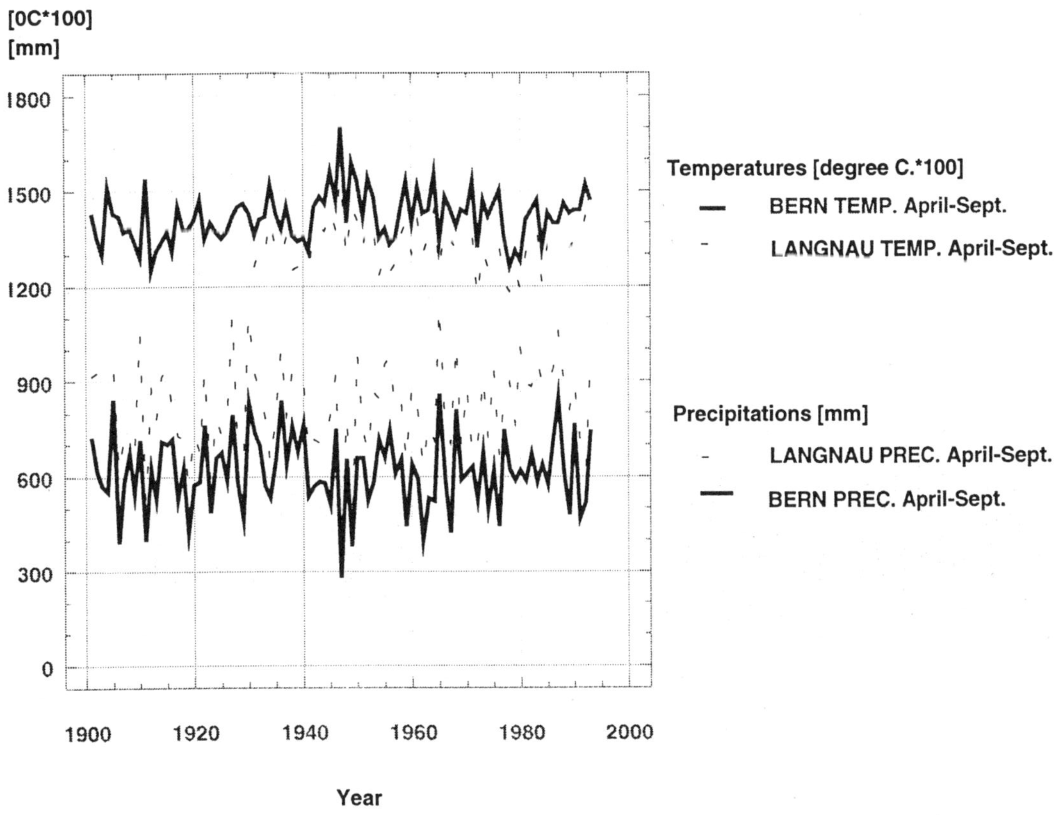

Fig. 7. Variables of nearby climate stations Bern (solid line) and Langnau (dotted line).

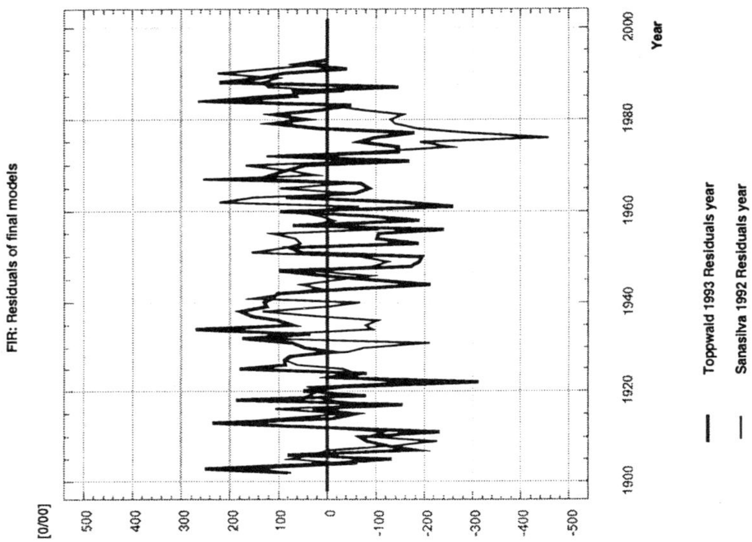
FIR: Residuals of final models
[0/00]
500
400
300
200
100
0
-100
-200
-300
-400
-500
1900
1920
1940
1960
1980
2000
Year
Toppwald 1993 Residuals year
Sanasilva 1992 Residuals year

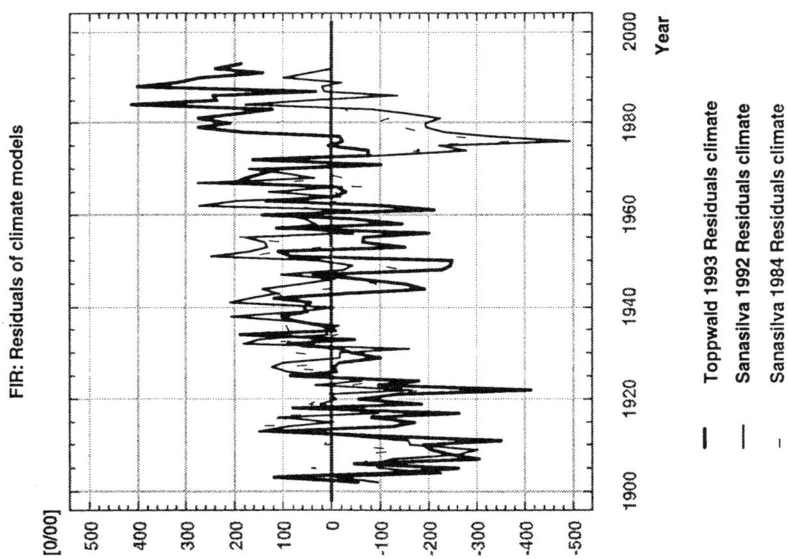
FIR: Residuals of climate models
[0/00]
500
400
300
200
100
0
-100
-200
-300
-400
-500
1900
1920
1940
1960
1980
2000
Year
Toppwald 1993 Residuals climate
Sanasilva 1992 Residuals climate
Sanasilva 1984 Residuals climate

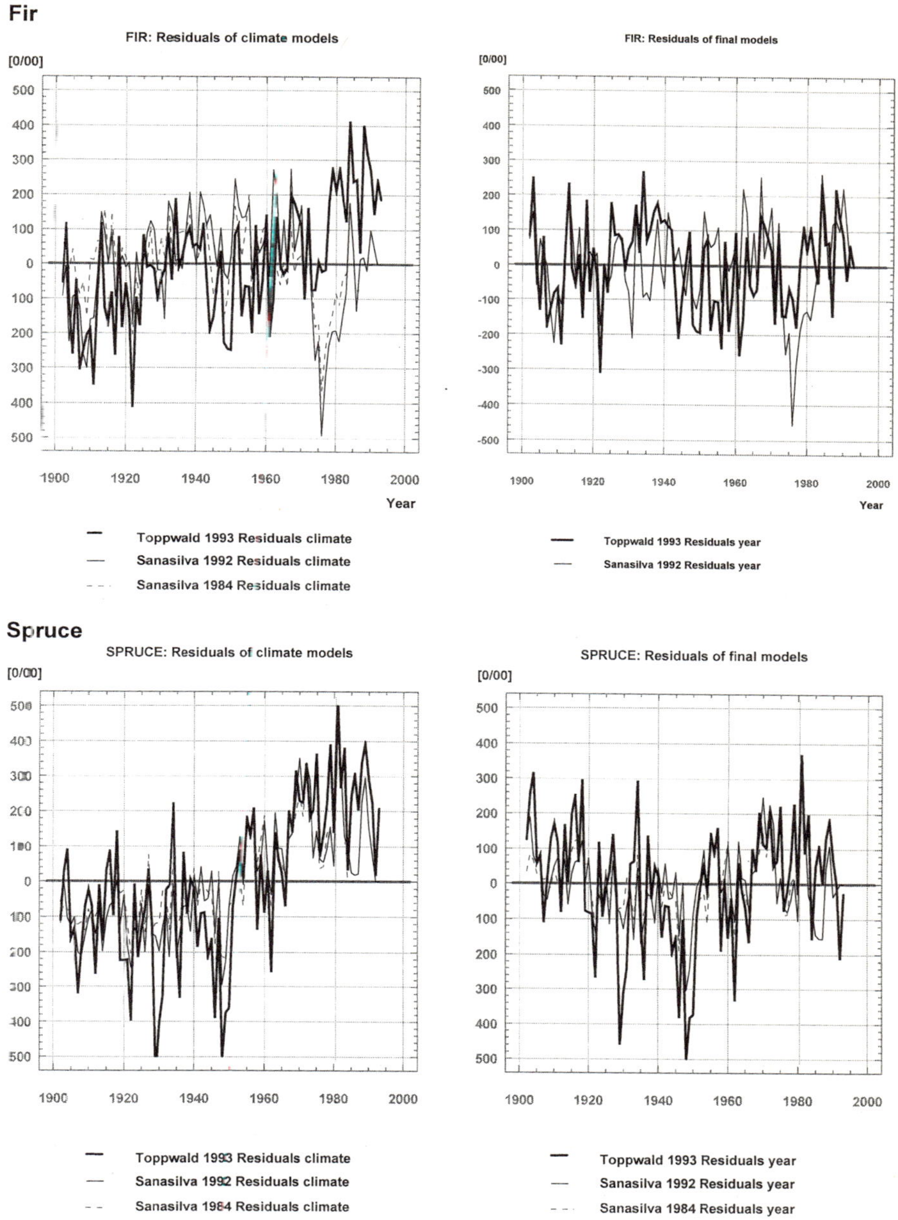

Fig. 8. Residuals of different growth models for fir and spruce. Left side after climate model, right side after climate and time trend model. Thick line for Toppwald, thin line for Sanasilva subset 1992, dotted line for Sanasilva 1984.

Table 3. Explained variances of models for remaining trends in addition to climate effects.

Set	Autoregressive Model, AR (1+2) r^2 [%]	Actual years Regression, Y r^2 [%]	Complete Model AR+Y+CL r^2 [%]	Climate Model CL r^2 [%]	Actual Year on Climate Residuals, RESY r^2 [%]
Spruce					
CH84	63	56	75	15	49
CH92	72	67	81	17	53
CH92<1961	63	41	75	26	-
Toppwald	58	51	65	9	37
Toppwald<1961	25	7	39	13	9
Fir					
CH84	22	-	44	22	5
CH92	57	-	67	20	23
CH92<1961	52	53	68	30	35
Toppwald	62	54	74	31	39
Toppwald<1961	39	25	46	35	4

er fertilising effect still needs to be tested. Analysis of the whole time series reveals neither an exceptional rise nor any growth levels which have not been seen or reached earlier in this century.

Acknowledgement

I would like to thank Christian Matter and Gustav Schneiter for the ring measurements and data handling and Christian Matter and Susan J. Ortloff for correction of the English in the manuscript.

References

Becker M , Bräker OU, Kenk G, Schneider O, Schweingruber FH (1990) Kronenzustand und Wachstum von Waldbäumen im Dreiländereck Deutschland-Frankreich-Schweiz in den letzten Jahrzehnten. München, Allgemeine Forst Zeitschrift 11: 263 - 274.

Bräker OU (1981) Der Alterstrend bei Jahrringdichten und Jahrringbreiten von Nadelhölzern und sein Ausgleich. Mitteilungen der forstlichen Bundesversuchsanstalt Wien, 142: 75 - 102.

Bräker OU (1992) Tree-ring analysis of conifers in the Swiss forest decline study program. In: Tree Rings and Environment. Proceedings of the International Dendrochronological Symposium, Ystad, South Sweden, 3-9 September 1990, LUNDQUA Report 34, Lund University, Deptartment of Quaternary Geology: 56 - 59.

Fritts HC (1976) Tree ring and climate. Academic Press, New York.

Kontic R, Bräker OU, Nizon V, Müller R (1990) Jahrringanalytische Untersuchungen im Sihlwald. Schweizerische Zeitschrift für Forstwesen, 141 (1): 55 - 76.

Kublin E (1992) Ein Verfahren zur Trendanalyse von Zuwachsreihen. In: Sloboda B (ed) Modelle zu automatisierten Zuwachsmess- und Auswertetechniken, klimaorientierte Wachstumsmodelle. Inventurmethoden und ihre Anwendungen. Frankfurt, Sauerländer; Schriften aus der Forstlichen Fakultät der Universität Göttingen und der Niedersächsischen Forstlichen Versuchsanstalt, 106: 103-121.

Pollanschütz J (1966) Verfahren zur objektiven "Abschätzung" (Messung) verminderter Zuwachsleistung von Einzelbäumen und Beständen. Mitt. der Forstlichen Bundesversuchsanstalt Mariabrunn, 73: 129 - 163.

Picard JF (1995) Evolution de la croissance radiale du hêtre (Fagus silvatica L. dans les Vosges. Premiers résultats sur le versant lorrain. Ann.Sci.For. 52: 11-21.

Visser H, Molenaar J (1992) Estimating Trend and Stochastic Response Functions in Dendroecology with an Application to Fir Decline. Forest Science, 38 (2): 221-234.

Growth Patterns in Forests of the Canton of Berne, Switzerland, Based on Inventory Data

Michael Köhl
Inventory Methods Research Group. Swiss Federal Institute for Forest, Snow and Landscape Research. Birmensdorf, Switzerland

1. Introduction

Monitoring forest condition and changes over time has a long tradition in Switzerland. The idea of describing stand development on the basis of permanent observations emerged during the last century. Gurnaud (1878), working in France, elaborated rules for the application of repeated measurements in estimating increment; this procedure is known as the control method. These rules were first applied in Switzerland by Biolley in 1890 to the forest of Couvet in the Jura (Biolley 1921; Kurt, 1987). Schmid (1963, 1967) combined the techniques of continuous forest inventory (CFI) with the control method to establish an inventory system which has now been implemented in many Swiss cantons. Control sampling is based on permanent plots; increment is determined by comparing data taken from successive inventories. In this study, data will be used that have been assessed in forest management planning inventories in the canton of Berne. We are grateful to the Bernese Forest Service, especially to Dr. J. Grilc, who generously made the data for this study available. The findings of this study will be complemented by the reports of Zingg and Bräker (in this volume), who worked with data from permanent growth and yield plots and with tree ring data.

2. General Aspects of Inventory Data

There are some major advantages of forest inventories in comparison to tree-level or stand-level approaches. Inventories are designed on a consistent statistical basis and provide represention for the entire inventory area. All forest conditions as well as the variability of forest stands, sites and management regimes are covered. Very often only a limited range of forest conditions is assessed and conclusions for larger regions cannot be obtained due to the limited number of observations in stand-level and tree-level approaches. In forest surveys, two different approaches are

European Forest Institute Research Report No. 5
Growth Trends in European Forests
Edited by H. Spiecker, K. Mielikäinen, M. Köhl and J. P. Skovsgaard

available for estimating growth: (1) assessment of single trees on sample plots and (2) increment cores.

Two types of plots can be used in forest surveys: permanent plots and temporary plots. Permanent plots have to be marked “invisibly”. However, management actions that occur on a permanent plot can change the degree to which the plot represents a segment of the population. The use of permanent plots is preferred, because the sampling errors are low due to the paired observations. In addition, the history of permanent plots is known, which enables the calculation of growth components (survivor growth, cut, mortality, ingrowth). Sampling errors are larger on temporary plots. Furthermore, as the plot history is not known, it is difficult to assess growth components on temporary plots.

The time period covered by increment cores taken in inventories differs. In some surveys short term series (5 to 15 years) are taken; in others, long-term series are used. As increment boring can cause growth reactions, some authors recommend the taking of increment cores not on but in the neighborhood of plots (e.g. Schöpfer et al. 1994). A study in Swiss forests based on increment cores taken during the Sanasilva forest damage survey in 1984 is presented by Bräker and Köhl in this volume.

3. Growth Trends, Growth Changes and the Mystery of Significant Results

Permanent sample plots were established in the Swiss forests in the late 1960s and have since then been remeasured twice. The inventory period covered does not exceed two decades. The first Swiss National Forest Inventory was conducted between 1983 and 1985 (EAFV 1988) and the first remeasurement is currently taking place. Information on forest growth based on NFI data is not yet available.

The estimation of long-term or seasonal trends is one of the major topics in the analysis of time series. Among many others, Kublin (1992) and van Deusen (1991) described techniques for the analysis of growth trends based on long-term tree ring series. As a rule of thumb, at least 20 to 30 observations in time should be available for time series and trend analysis. In most successive forest surveys, far fewer observations in time are available and for this reason trend analysis techniques are generally not (yet) applicable to survey data. Houllier et al. (1993) used data from the French NFI to assess changes in forest productivity. They concluded that longer chronological series at the regional level are needed and that it is difficult to assess changes in productivity due to the many interacting factors and processes involved in forest growth. The lack of long-term observations on permanent sample plots makes the term growth changes more appropriate than the term "growth trends".

An alternative to trend analysis could be the application of significance tests that allow statistical inference to be made about changes between two successive occasions. However, "significant" results have to be interpreted with caution. Experience shows that the level of significance a of any test is given the value 0.05 or 0.01 and gives the risk of erroneously rejecting a true null hypothesis (Type I error), i.e., a phenomenon such as a difference between the means of two groups may be observed in the data even though it does not actually exist in the population.

The level of significance α considers only the rejection of the true null hypothesis. The error of not rejecting the null hypothesis when it is in fact false is called a Type II error. The probability of committing a Type II error is ß. The probability of rejecting a false null hypothesis is called the power of a statistical test and is defined by 1-ß. For a given sample size n, the value of α is related inversely to the value of ß. A low probability of committing a Type I error results in a higher probability of committing a Type II error unless the sample size n is increased. The power of a statistical test also depends on the "effect size", which may be expressed as the minimum detectable difference between two population parameters.

As the level of significance α, power (1-ß), sample size n, and effect size are directly dependent on each other, effect size can be determined as a function of α, ß, and n. On permanent plots, the significance of changes of tree or plot attributes such as diameter or basal area between two occasions might be tested by paired t-tests. The following example shows the effect size that leads to significant results for various sample sizes. The single tree data from the forest management inventory in the forest domain Toppwald were used to calculate means and standard deviations for the d.b.h. measurements on three occasions (Table 1) and to calculate the correlation matrix (Table 2). Based on these values, the effect size for a paired t-test was determined, assuming that single trees are randomly selected from the target population. α and ß were set to 0.05 and 0.2, respectively. Given these settings, the effect size was plotted over the number of observations, n (Fig. 1). For large n the ef-

Table 1. Single tree data, forest district Toppwald, means and standard deviations

Variable	N	Mean	Std
d.b.h. 1972	6837	23.8653	15.3896
d.b.h. 1982	5739	25.5264	16.3844
d.b.h. 1993	5439	26.3281	17.9310

Table 2. Single tree data, forest district Toppwald, correlation matrix

	d.b.h. 1972	d.b.h. 1982	d.b.h. 1993
d.b.h. 1972	1.00	0.99	0.96
d.b.h. 1982	0.99	1.00	0.99
d.b.h. 1993	0.96	0.99	1.00

fect size becomes very small. It is obvious that even differences between two diameter measurements lead to significant results that are without any practical meaning. Thus, the application of statistical tests and the discussion of "significant" results must be done with great care. Due to the large sample sizes in forest surveys, even marginal (meaningless) differences can result in a significant outcome of a statistical test. Therefore, "no significant" changes in growth patterns will be reported throughout this paper.

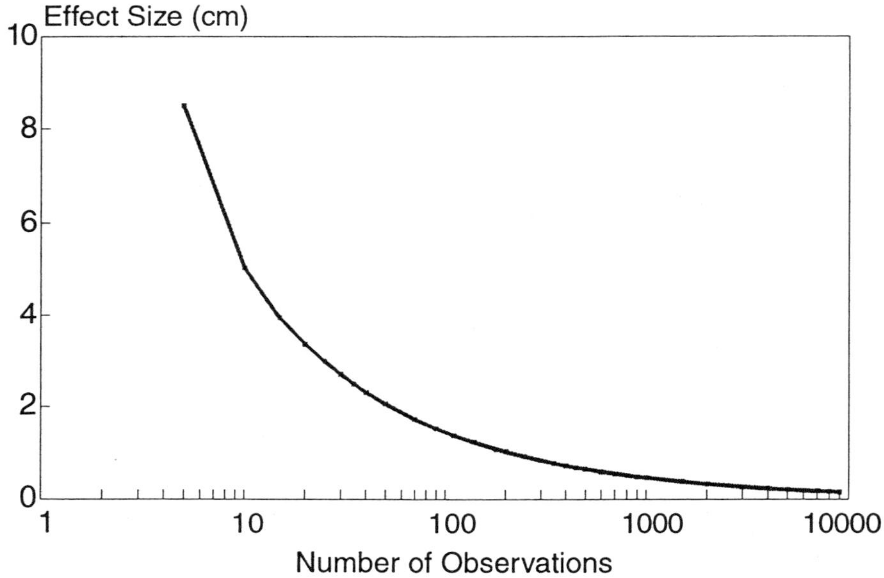

Fig. 1. Effect size versus number of observations for paired t-test

4. Data and Methods

In the canton of Berne, forest management inventories have been conducted on a regular basis since the late 1960s. For this study, the forest service of the canton of Berne provided all data on plots and trees that have been assessed on three successive occasions in the Bernese Mittelland and Oberland. Three productivity regions are thus covered: the plateau region, the Pre-Alps and the Alps. More than 82 000 trees are included in this study, the most common species being spruce, fir and beech (Table 3).

Table 3. Tree species and number of trees included in the study

Deciduous species	Number of trees	Coniferous species	Number of trees
Beech *Fagus sylvatica* L.	17652	Spruce *Picea abies* (L.) Karst.	35 725
Oak *Quercus* L.	1376	Fir *Abies alba* Mill.	22 125
Maple *Acer pseudoplatanus* L.	1268	Pine *Pinus sylvestris* L.	798
Ash *Fraxinus excelsior* L.	1481	Larch *Larix decidua* Mill.	622
other deciduous species	1240	Douglas fir *Pseudotsuga menziesii* (Mirb.) Franco	83
		Eastern white pine *Pinus strobus* L.	89
		Stone pine *Pinus cembra* L.	3
		European mountain pine *Pinus mugo* Turra var. *arborea*	29
		other coniferous species	88
Total	23017	Total	59 563
Grand total			82 580

Thirty nine forest enterprises were sampled, the approximate forested area of each forest enterprise is given in Table 5. The size of the forest domains range from 39 ha (von Roll, Saanen) to 3174 ha (Burgergemeinde Bern). A total forested area of 12 194 ha is covered by the inventory.

The mean time span between the first and the second assessment is 11.3 years and slightly longer than the mean time span between the second and third assessment (10.2 years). Between two successive occasions the minimum time span was 10 years, the maximum being 16 years (Table 4). Since the periods compared are from different time intervals, the d.b.h. growth rate was standardized on the basis of cm per year. For each assessment period, the annual, periodical d.b.h. growth was calculated for individual trees. The expression cm y^{-1} stands for cm per year.

The sampling method applied follows the control sampling design presented by Schmid-Haas et al. (1983, 1993). Fixed area, circular plots with a horizontal area of 300, 400 or 500 m^2 are systematically distributed. Two different grid shapes and sizes are used: rectangular grids with a size of 80×150 m (1 plot per 1.2 ha) in plateau regions and triangular grids with a side length of 150 m (1 plot per 2 ha) in the

Table 4. Assessment cycles

	Number oftrees [Years]	Mean [Years]	Minimum [Years]	Maximum [Years]	Year of first and last measurement
Period 1	82580	11.3	10	16	1966/ 1983
Period 2	82580	10.2	10	12	1977/ 1993

mountainous regions. All trees standing on a sample plot and having a minimum d.b.h of 8 cm are measured, this means that no tree species or social position are omitted. D.b.h. is measured with the caliper (one reading). Azimuth and distance from the plot center of every tree measured on a plot is recorded so that tree location may be identified in further assessments.

No specific forest area definition is applied; all plots falling into forest areas mapped on forest plans are included. The measurement rules and definitions of area-related attributes can be found in Schmid-Haas et al. (1993) and were kept constant during successive occasions. Where results are presented for subsets such as diameter classes, density classes or mixture proportions, the attribute class assessed on the first occasion is taken to form the subsets. As the attributes are not available for all plots, the number of trees in subsets do not necessarily add up to the total number of trees assessed in the inventory. Only those trees which have been assessed on all three occasions have been included in the study.

On each occasion the d.b.h. of trees was recorded. Height measurements were taken for a subset of trees on the first occasion, but height was not remeasured on successive occasions. Therefore, no figures on height growth can be given; the analyses are restricted to d.b.h. growth. The diameter distributions for the first occasion are presented in Fig. 2 for all forest types and separately for selection forests.

As the inventories were not designed to explain growth patterns, only a limited set of attributes is available. Major attributes indicating the initial condition of trees and stands or explaining changes occurring between periods have not been assessed. For this reason, the potential for sophisticated statistical analyses is rather limited: no analysis can be made that could relate a specific period effect to its potential causes.

6. Results

The available inventory data from the 39 forest enterprises have been merged and analyzed for growth changes. To get better insight into growth patterns, subsets have been created and analyzed. Such subsets are mixture proportions, stages of forest

Table. 5. Overview of the forest domains assessed

Forest domain	Forest area [ha]	Plot size [m²]	Grid size [m]	Grid shape
Plateau region				
Gibeleggwaelder	405	400	80x150	rectangular
EG Lyss	127	400	80x150	rectangular
BG Kallnach	177	400	80x150	rectangular
BG Grossaffoltern	66	400	80x150	rectangular
BG Diessbach	201	400	80x150	rectangular
BG Leuzigen	405	400	80x150	rectangular
BG Arch	167	400	80x150	rectangular
BG Oberwil	202	400	80x150	rectangular
BG Rüti b. Büren	198	400	80x150	rectangular
BG Meinisberg	190	400	80x150	rectangular
BG Büren a. A.	477	400	80x150	rectangular
BG Roggwil	534	400	80x150	rectangular
BG Langenthal	392	400	80x150	rectangular
BG Wynau	185	400	80x150	rectangular
FK 8 Bern	1102	400	80x150	rectangular
FK 6 Bern Region Sued	304	400	80x150	rectangular
Burgergut Wahlern	499	400	80x150	rectangular
Guggisberg	372	400	80x150	rectangular
Rueschegg	415	400	80x150	rectangular
Burgergemeinde Bern	3174			
Mountainous region				
Bäuert Betelried	148	300	150	triangular
Gem. Weiden Betelried	93	300	150	triangular
Bäuert Mannried	197	300	150	triangular
Bäuert Grubenwald	184	300	150	triangular
von Roll (Saanen)	39	300	150	triangular
EG Saanen	165	300	150	triangular
Staatswald FK 4	304	300	150	triangular
Bäuert Zweisimmen	127	300	150	triangular
Oberlangenegg Lindenwald	64	500	100x150	rectangular
Unterlangenegg Hängenegg	77	500	100x100	rectangular
Burgerbäuert Spiez	113	300	150	triangular
Burgerbäuert Faulensee	114	300	150	triangular
Burgerbäuert Hondrich	48	300	150	triangular
Burgerbäuert Spiezwiler	43	300	150	triangular
EG Hofstetten	61	300	150	triangular
Gem. Gemeinde Schwanden	66	300	150	triangular
FK 19Diemtigtal	217	400	200x200	rectangular
FK 19 Simmenwälder	106	300	150	triangular
EG Brienz	100	300	150	triangular
Grindelwald BS Holzmatten	100	400	100x200	rectangular
Grindelwald BS Bach	236	400	100x200	rectangular

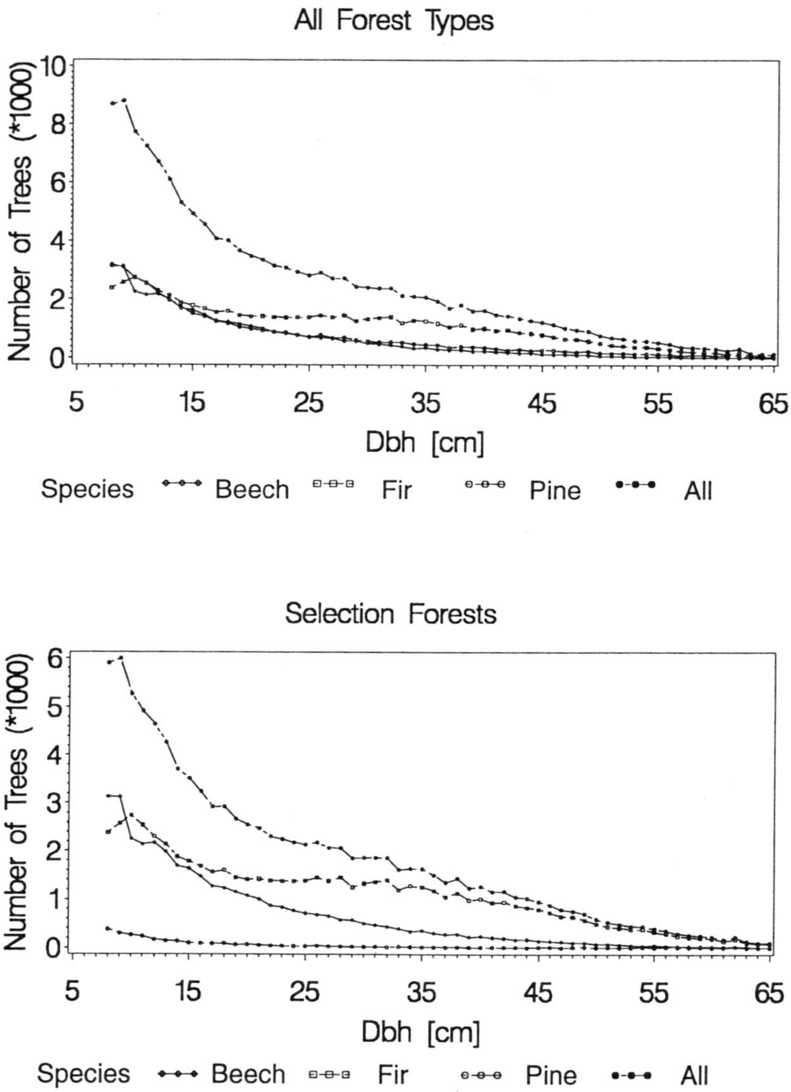

Fig. 2. Diameter distribution for the total data set and for selection forests, truncated at 65 cm

development, diameter classes and density classes. Results will be presented for all trees as well as separately for spruce, fir and beech. Growth patterns in selection forests are less sensitive to age effects, as all diameter classes are present on the same site and are kept in balance. Therefore, inventory data will be presented separately for selection forests.

6.1. Results for 39 Forest Enterprises in the Canton of Berne

Periodic annual d.b.h. growth for all trees and for beech, spruce and fir are given in Fig. 3. In the second observation period, which ranges from the beginning of the 1980s to the beginning of the 1990s, growth increases for all trees as well as for beech and fir. For all trees, d.b.h. growth of the second period is about 108% of the d.b.h. growth of the first period; beech has a slightly higher increase (109%). The growth increase of spruce is marginal. Only fir shows a substantial increase in d.b.h. growth (127%).

Results for the subsets of density classes, stages of forest development and diameter classes are presented in Tables 6 to 8. The density of forest cover has been divided into three classes: dense (the tree crowns are in close contact), normal (the crowns touch each other, spatial distribution of the trees is regular, the tree crowns of the main stand cover between 50 and 80% of the area) and open (tree crowns do not touch each other, the tree crowns of the main stand cover less than 50% of the area). Beech, fir and the set of all species have the highest d.b.h. growth in open stands. The highest (0.436 mm y^{-1}) and the lowest (0.291 cm y^{-1}) d.b.h. growth amongst all the subsets can be observed for fir. For beech, d.b.h. growth increases consistently with decreasing density. Spruce and fir grow best in open stands; spruce grows least well in dense stands. Beech and spruce maintain the ranking of growth patterns in relation to stand density in the first and second period, but the ranking for fir changes. In the first period fir grows least well in stands with normal density, in

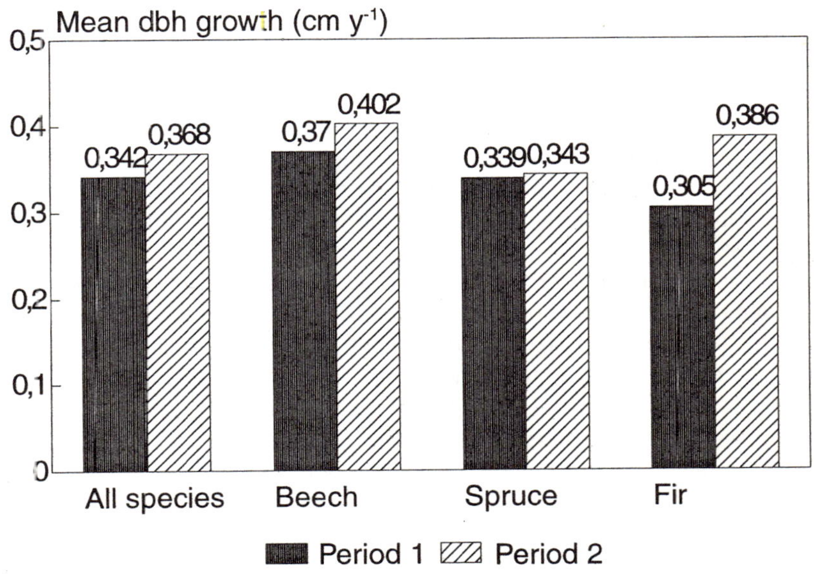

Fig. 3. Mean d.b.h. growth [cm/a]

Table 6. D.b.h. growth for density classes

Species	Density class	Number of trees	Mean d.b.h. growth [cm y^{-1}] period 1	Mean d.b.h. growth [cm y^{-1}] period 2	Difference[1]
All	dense	24267	0.340	0.375	110.2
	normal	22194	0.346	0.385	111.3
	open	10878	0.361	0.395	109.5
Beech	dense	4883	0.333	0.383	115.0
	normal	5235	0.379	0.416	110.0
	open	2380	0.379	0.424	111.9
Spruce	dense	10860	0.365	0.379	103.8
	normal	8272	0.356	0.373	104.8
	open	4530	0.349	0.361	103.7
Fir	dense	6907	0.298	0.369	123.5
	normal	7037	0.291	0.378	129.7
	open	2909	0.349	0.436	124.9

[1] D.b.h. growth of period 2 in percentage of d.b.h. growth of period 1

the second period in dense stands. The difference in d.b.h. growth between period 1 and period 2 is smallest for spruce with minor differences between the density classes. Beech shows the highest difference in dense stands. As the stand density is given for the first assessment, it is very likely that the number of stems in dense stands was reduced during the observation period. Thus beech shows a growth pattern that is probably consistent with the ability of reacting very quickly in closing openings. Almost 40% of all beeches assessed in the inventory grow in dense stands. Thus, the relatively strong increase in the growth of beech could be linked to a growth behavior caused by forest management practices. The highest overall increase in d.b.h. growth can be observed for fir. Differences in growth between the first and the second period are high for all density classes and thus are probably not affected by forest management.

As age data on the sampled trees are not available, the view of stages of forest development and diameter classes provides some insight into the growth patterns in relation to age. In closed, even-aged stands, diameter growth culminates at an early stage and decreases with advancing years (Kramer 1988). According to Erteld and Hengst (1966), diameter growth reaches its maximum between the ages of 8 and 12 years for spruce, between 10 and 20 years for beech and between 25 and 30 years for fir. However, diameter growth is known for deviations from a "standard" pattern. This holds true especially for even-aged and uneven-aged mixed stands. Changes in environmental conditions have a stronger effect on diameter growth than on height growth (Mitscherlich 1978).

The figures for stages of forest development and diameter classes are given in Tables 7 and 8. Diameter classes are based on the d.b.h. values taken at the first meas-

Table 7. D.b.h.- growth for stages of forest development

Species	Stage of forest development	Number of trees	Mean d.b.h. growth [cm y^{-1}] period 1	Mean d.b.h. growth [cm y^{-1}] period 2	Difference[1]
All	Young stands	682	0.656	0.626	95.4
	Polewood	3479	0.455	0.421	92.6
	Timber I	5800	0.312	0.339	108.7
	Timber II	15 981	0.297	0.349	117.6
	Timber III	9553	0.319	0.392	123.1
	Mixed	2575	0.353	0.412	116.7
	Select. forests	3237	0.315	0.417	132.4
Beech	Young stands	90	0.669	0.545	81.4
	Polewood	1037	0.385	0.384	99.8
	Timber I	1156	0.305	0.353	115.7
	Timber II	1739	0.306	0.375	122.7
	Timber III	1355	0.378	0.448	118.5
	Mixed	458	0.364	0.442	121.3
	Select. forests	723	0.272	0.369	135.8
Spruce	Young stands	306	0.677	0.639	94.4
	Polewood	1005	0.602	0.497	82.4
	Timber I	2179	0.329	0.328	99.6
	Timber II	7679	0.301	0.336	111.8
	Timber III	3597	0.297	0.355	119.6
	Mixed	852	0.359	0.390	108.7
	Select. forests	717	0.405	0.510	125.8
Fir	Young stands	171	0.663	0.717	108.2
	Polewood	1015	0.375	0.404	107.7
	Timber I	2159	0.289	0.344	119.0
	Timber II	6141	0.285	0.358	125.7
	Timber III	4272	0.313	0.407	130.2
	Mixed	1047	0.346	0.435	126.0
	Select. forests	1754	0.296	0.401	135.6

[1] D.b.h. growth of period 2 in percentage of d.b.h. growth of period 1

uring. Four diameter classes were calculated: ≤ 10 cm, 11-30 cm, 31-50 cm and > 50 cm. Seven stages of forest development are presented: young stands, polewood (mean d.b.h. below 20 cm), timber 1 (mean diameter of the main stand 20-35 cm), timber 2 (mean diameter of the main stand 35-50 cm), timber 3 (mean diameter of the main stand over 50 cm), mixed stands (several stages of forest development on the same area, multiple crown layers) and selection forests. All trees are grouped into both the diameter classes and the development classes which were recorded in the first assessment. However, there is one important difference between assigning trees to diameter classes and stages of forest development. Trees are grouped into

Table 8. D.b.h. growth for diameter classes (related to d.b.h. on first occasion)

Species	Diameter class	Number of trees	Mean d.b.h. growth [cm y^{-1}] period 1	Mean d.b.h. growth [cm y^{-1}] period 2	Difference[1]
All	> 50 cm	4094	0.422	0.479	113.5
	31-50 cm	20203	0.386	0.426	110.3
	11-30 cm	45269	0.333	0.357	107.1
	≤ 10 cm	13014	0.279	0.284	102.0
Beech	> 50 cm	272	0.544	0.558	102.6
	31-50 cm	2747	0.472	0.511	108.2
	11-30 cm	11087	0.357	0.397	111.2
	≤ 10 cm	3546	0.320	0.324	101.1
Spruce	> 50 cm	2310	0.412	0.456	110.7
	31-50 cm	11892	0.360	0.381	105.9
	11-30 cm	17929	0.324	0.318	97.9
	≤ 10 cm	3594	0.295	0.272	92.4
Fir	> 50 cm	1071	0.432	0.536	124.1
	31-50 cm	4104	0.414	0.532	128.6
	11-30 cm	12299	0.294	0.373	126.8
	≤ 10 cm	4651	0.206	0.256	123.9

[1] D.b.h. growth of period 2 in percentage of d.b.h. growth of period 1

diameter classes according to the individual diameter at the first occasion. Thus the diameter classes can be directly linked to individual trees. The stages of forest development are assigned to stands and do not necessarily reflect the age status of single trees.

In the first observation period, diameter growth shows an expected development only for spruce: it is highest in young stands, decreases continuously and reaches a minimum in mature timber (timber III). For the collective of all trees as well as for fir and beech, d.b.h. growth increases again for mature timber. In the second period, d.b.h. growth reaches a minimum for all four species categories in timber I and increases for timber II and timber III. Assuming that the stage of forest development reflects the age of trees, the observed growth pattern in the second period departs from the expected one. Except for fir, d.b.h. growth is lower in the second period for young stands and polewood. Spruce shows lower d.b.h. growth in timber I as well. The highest increase of d.b.h. growth can be found for selection forests. Growth increases by 125% for spruce and by more than 135% for fir and beech. Fir shows the highest overall changes. In stands with mixed stages of forest development, the increase of all tree species is 117%. It is lower for spruce (108%) and clearly higher for fir and beech.

D.b.h. growth for all four species categories is highest in the largest diameter class and decreases continuously. The highest overall value (0.558 cm y^{-1}) can be found for beech in the largest diameter class in the second period; the lowest overall value (0.206 cm y^{-1}) is observed for fir in the smallest diameter class in the first period. In the first period, diameter growth is higher for beech than for fir and spruce. Fir shows a clear decrease in d.b.h. growth in the two lower diameter classes. As the diameter classes can be linked directly to individual trees, the figures presented in Table 8 show that trees with large diameters (old trees) show a higher d.b.h. growth than trees with low diameters (young trees). The differences in diameter growth between the first and second period are all positive with two exceptions: spruce in the diameter classes ≤ 10 cm and 11-30 cm. The highest differences are found for fir and these are substantially higher than those for the other tree species. On the basis of former research on growth and yield plots, from growth and yield tables and from the widely accepted growth patterns of trees and stands, it could be expected that the diameter growth decreases when trees become older than 60 to 100 years. The findings of this study lead to a contradictory result. In spite of the expected growth pattern, the diameter growth of the trees being studied increases with increasing diameter. As diameter is an indicator of age, it can be assumed that there is a positive shift (trend?) in the diameter-growth/age relationship.

In Table 9, d.b.h. growth is presented for three mixture proportions: more than 80 percent coniferous species, 20 to 80% coniferous species and more than 80 percent deciduous species. With an increasing proportion of deciduous trees, the d.b.h. growth rates increase. In all but one category (spruce, >80% deciduous), the growth of the second period is higher than the growth of the first period. However, the increase is only marginal for all trees and beech in stands with more than 80% deciduous trees. The highest d.b.h. growth values can be found for spruce in deciduous

Table 9. D.b.h.- growth for mixture proportions

Species	Mixture proportion	Number of trees	Mean d.b.h. growth [cm y^{-1}] period 1	Mean d.b.h. growth [cm y^{-1}] period 2	Difference[1]
all	>80% conif.	33063	0.308	0.351	114.1
	20=80% conif.	13569	0.367	0.410	111.9
	>80% decid.	2721	0.444	0.445	100.3
Beech	>80% conif.	4179	0.349	0.384	110.1
	20=80% conif.	3641	0.341	0.395	115.8
	>80% decid.	1306	0.414	0.419	101.1
Spruce	>80% conif.	15565	0.300	0.326	108.8
	20=80% conif.	3582	0.449	0.457	101.7
	>80% decid.	485	0.566	0.560	98.8
Fir	>80% conif.	11697	0.297	0.377	127.1
	20=80% conif.	5413	0.321	0.398	124.1
	>80% decid.	480	0.406	0.465	114.4

[1] D.b.h. growth of period 2 in percentage of d.b.h. growth of period 1

stands (0.566 cm y^{-1} and 0.560 cm y^{-1}) the lowest figure is obtained for fir in coniferous stands (0.297 cm y^{-1}). The highest growth increase between the two periods is realized by fir. This might be due to the fact that fir has the smallest dissemination and is located on rich sites. The same holds true for deciduous species that are found less frequently on poor sites. Thus the growth patterns found for mixture proportions are very likely confounded with site quality.

The growth figures for the total of all forests covered by the inventory have been broken down to the individual forest enterprises. This breakdown shows a wide spread of growth patterns in relation to the forest enterprises. The minimum and maximum growth rates are presented in Table 10. For the collective of all tree species, the minimum difference in growth between the two periods is 76%, the maximum is 135%. Slightly wider ranges can be observed for spruce (71 vs. 140%) and beech (80 vs. 175%). Fir shows the widest spread; the maximum difference (204 percent) is almost three times as big as the minimum difference (73%). These figures reflect the heterogeneity of growth conditions, which can be influenced by many factors such as site quality, management regimes or genetic variability. However, they do not show a uniform trend that could be interpreted as an effect of transboundary air pollution or global climate change.

Table 10. Forest enterprises with minimum and maximum growth figures

Species	Minimum/ Maximum	Forest domain	Number of trees	Difference[1]
All	Minimum	Mannried	887	76
		von Roll AG, Saanen	249	
	Maximum	Wahlern	6321	135
Beech	Minimum	Oberlangenegg	13	80
	Maximum	Zweisimmen	29	175
Spruce	Minimum	von Roll AG, Saanen	146	71
		Lyss	299	
	Maximum	Wahlern	3643	140
Fir	Minimum	Spiez	24	73
	Maximum	Meinisberg	203	204

[1] D.b.h. growth of period 2 in percentage of d.b.h. growth of period 1

6.2. Selection Forests

Selection forests are a special type of uneven-aged, mixed forests with shade tolerant species. In the region under consideration, they are composed mainly of spruce, fir and beech (Table 8). All age classes, diameter classes and height classes can be found within small areas. In selection forests, the ideal diameter distribution is constant over large periods and can be expressed by the so-called Plentergleichgewicht

Table 11. Tree species and number of trees per species in selection forests

Species	Number of observations
Beech	723
Maple	13
Ash	18
other decid.	5
Spruce	717
Fir	1754
Pine	5
Larch	1
White pine	1

(balance of selection forests) (Prodan 1946; Mitscherlich 1978; Schütz 1981). This balance is realized if trees which die or grow out of one diameter class are replaced by trees from the next smallest diameter class. The amount of cuts must be kept constant over long periods to preserve the balance of the stem distribution in selection forest. Especially in small, privately owned forests, however, heavy cuts are likely to occur periodically and can be seen as "weaves" that move through the diameter classes in the course of time. Biolley (1923) and Flury (1929) described those shifts in the diameter distribution of selection forests. Diameter growth of single trees culminates much later and lasts much longer in selection forests than in even-aged pure stands. It has to be kept in mind that the following discussions do not deal with single selective cut stands but are related to a composite of several stands.

In Table 12, d.b.h. growth for all species and for spruce, fir and beech are presented. The highest growth figures can be observed for spruce, the lowest for beech. The rank of growth rates between tree species is retained in both periods. There is a marked increase of growth between the first and the second period. The level of increase is about the same for beech (135.8%) and fir (135.6%) and lower for spruce (125.8%). Beech in selection forests does not reach the growth figures of beech in the collective of all forest types (0.370 cm y^{-1} for period 1 and 0.402 cm y^{-1} for period 2, Fig. 3), but the increase in growth is much higher in selection forests. Fir shows slightly lower d.b.h. growth in the first period in selection forests and higher d.b.h. growth in the second period. In both periods, the growth rates of spruce in selection forests is higher than those in the collective of all stands. The dif-

Table 12. D.b.h. growth for selection forests

Species	Number of trees	Mean d.b.h. growth [cm y^{-1}] period 1	Mean d.b.h. growth [cm y^{-1}] period 2	Difference[1]
All	3237	0.315	0.417	132.4
Beech	723	0.272	0.369	135.8
Spruce	717	0.405	0.510	125.8
Fir	1754	0.296	0.401	135.6

[1] D.b.h. growth of period 2 in percentage of d.b.h. growth of period 1

ference of growth between the first and second period is obviously higher for all species. Even fir, which showed the highest difference in the total collective of forests (127%), has a larger difference in selection forests.

These findings are complemented by the analysis of growth patterns in diameter classes (Table 13). Highest d.b.h. growth rates are found in the largest diameter classes. The highest (0.595 cm y^{-1}) and lowest (0.198 cm y^{-1}) growth rates are found for spruce in diameter class 31-50 cm and diameter class <10 cm respectively. Beech shows the highest difference in growth between the two periods in the smallest diameter class; spruce and fir have their largest difference in the larger diameter classes.

Table 13. D.b.h. growth for diameter classes (related to d.b.h. at first occasions) in selection forests

Species	Diameter class	Number of trees	Mean d.b.h. growth [cm y^{-1}] period 1	Mean d.b.h. growth [cm y^{-1}] period 2	Difference[1]
All	>50 cm	289	0.413	0.559	135.3
	31-50 cm	414	0.416	0.551	132.4
	11-30 cm	1746	0.319	0.422	132.3
	<10 cm	788	0.218	0.285	130.5
Beech	> 50 cm	13	0.392	0.505	128.7
	31-50 cm	68	0.360	0.457	127.0
	11-30 cm	468	0.285	0.387	135.7
	<10 cm	174	0.191	0.276	144.0
Spruce	> 50 cm	90	0.438	0.562	128.4
	31-50 cm	153	0.416	0.549	131.9
	11-30 cm	340	0.424	0.521	123.1
	<10 cm	134	0.325	0.407	123.1
Fir	> 50 cm	184	0.405	0.566	139.8
	31-50 cm	184	0.442	0.595	134.5
	11-30 cm	912	0.295	0.403	136.6
	<10 cm	474	0.198	0.257	129.9

[1] D.b.h. growth of period 2 in percentage of d.b.h. growth of period 1

7. Discussion

Data from forest management inventories in the canton of Berne were used to analyze growth trends. The inventories were conducted in 39 forest enterprises which total forested area of 12 194 ha. The forests assessed by forest management inventories cover a multitude of site conditions, management regimes and genetic variation. Thus the observed, consistent changes in growth patterns can hardly be caused by those factors; instead, they are likely to be linked to other factors such as transboundary air pollution, climate changes, change of primary forest functions and changes in site productivity.

The growth patterns presented in this paper are related to diameter growth. Unfortunately, height measurements were taken only at the first measurement of the plots and were not repeated on successive occasions. Height growth would be more suitable for the interpretation of growth patterns and changes in site productivity, as it is not as sensitive to management operations as diameter growth. The data presented are not linked to single stands, and growth changes are caused not only by forest management activities but reflect a wide range of effects that influence growth. All trees have been assessed three times over a period of more than 20 years. The earliest measurements were completed in 1966, the last remeasurement was finished in 1993.

The data are presented for the collective of all tree species assessed on sample plots and separately for beech, spruce and fir. Subsets are formed according to density, stage of forest development, mixture proportion and diameters at the beginning of the period. Additionally, results are presented for selection forests. An obvious increase in diameter growth can be observed for fir and beech. Spruce shows only marginal increases in the total collective of forests, but a clear increase in selection forests. The growth rates for all tree species are substantially higher in selection forests than in the other forest types. This increase between the first and the second observation period may be confounded with site quality, as selection forests are typically stocked on better sites. Spruce has a much wider area of distribution than fir and beech and has to cope with a wider range of growth factors, amongst which climate is not the only limiting factor. Fir and beech grow on richer soils and thus are more sensitive to climate changes than spruce. However, where spruce grows on rich soils, it grows under optimum conditions and, like the other species, demonstrates a clear increase in growth. These findings are supported by the analysis of individual forest enterprises. The difference in growth between forest enterprises is fairly large for all tree species. The widest range can be observed for fir, but both beech and spruce show a clear variation.

The contradictory findings of growth patterns for individual forest enterprises and the differences in d.b.h. growth of spruce shows that growth patterns are subject to high variability even in small areas. Forest inventories provide representative figures of growth patterns and their variations in large areas. In the past, growth and yield studies were based mainly on permanent observation plots. Köhl et al. (1995) compared data from permanent growth and yield plots in Switzerland with data from the national forest survey and found remarkable differences in factors indicating site quality. From a statistical point of view, not all permanent observation plots can be regarded as experiments but must be considered as case studies due to the lack of repetitions. The figures from those case studies can show growth patterns but must be interpreted with great care. The results must not be generalized. However, if data from permanent observation plots are compared to inventory data, the frame of forest conditions covered can be derived. A proof for growth trends in forests can be obtained if growth trends found in many independent growth and yield studies point in the same direction.

References

Biolley HE (1921) L'aménagement des forêts par la méthode expérimentale et spécialement la méthode du contrôle. Paris, Neuchatel, 90 p.

Biolley HE (1923) Die Forsteinrichtung auf der Grundlage der Erfahrung und insbesondere das Kontrollverfahren, Karlsruhe, Müller Verlag, 72 S.

EAFV (1988) Schweizerisches Landesforstinventar: Ergebnisse der Erstaufnahme 1982-1986. Eidgen. Anst. Forstl. Versuchswes., Bericht Nr. 305, 375 p.

Erteld W, Hengst E (1966) Waldertragslehre, Radebeul, Neumann Verlag, 332 S.

Flury Ph (1929) Ueber den Aufbau des Plenterwaldes. Mitteilungen der Schweizerischen Centralanstalt für das Forstliche Versuchswesen, Band 15, Heft 2, 53 S.

Gurnaud A (1878) Cahier l'aménagement pour l'application de la méthode par contenance exposée sur la forêt des Eperous. Paris, 160 p.

Houllier F, Pignard G, Schmitt F, Bertrand P (1993) Medium-term evolution of forest productivity: the use of national forest inventory data. Unpubl. manuscript, 17 p.

Köhl M, Scott CT Zingg A (1995) Evaluation of permanent sample surveys for growth and yield studies. Forest Ecology and Management, Vol. 71 (3): 187-194

Kramer H (1988) Waldwachstumslehre. Verlag Paul Parey, Hamburg und Berlin, 374 S.

Kublin E (1992) Ein Verfahren zur Analyse von Zuwachsreihen. In Sloboda, B. (Hrsg.): Modelle zu automatisierten Zuwachsmess- und Auswertetechniken, klimaorientierte Wachstumsmodelle, Inventurmethoden und ihre Anwendungen, Schriften aus der Forstlichen Fakultät der Universität Göttingen und der Niedersächsischen Forstlichen Versuchsanstalt, Band 106: 103-121

Kurt A (1987) Die Schweizer Kontrollstichprobe, Allg. Forst- u. J. Ztg. 158(5/6): 101-105

Mitscherlich G (1978) Wald, Wachstum und Umwelt - Form und Wachstum von Baum und Bestand (Band 1), 2. Auflage, Frankfurt

Prodan M (1944) Zuwachs- und Ertragsuntersuchungen im Plenterwald. Dissertation, Univ. Freiburg i. Br.

Schmid P (1963) Vergleich von Vollkluppung und Stichprobenaufnahme. Schweizerische Zeitschrift für Forstwesen 114 (7): 412-425

Schmid P (1967) Die Weiterentwicklung der Leistungskontrolle in der Schweiz, Wiss. Z. Techn. Univ. Dresden, 16(2): 545-549

Schmid-Haas P (1983) Swiss continuous forest inventory: twenty years' experience. In: Bell JF, Atterbury T (ed) Renewable resource inventories for monitoring changes and trends. Corvallis, OSU, College of Forestry: 389-393

Schmid-Haas P (1993) Forest inventories by unmarked permanent plots: instructions. Birmensdorf, Swiss Federal Institute for Forest, Snow and Landscape Research, Bericht 186: 143 pp.

Schöpfer W, Hradetzky J, Kublin E (1994) Wachstumsveränderungen der Fichte in Baden-Württemberg. Forst und Holz 49. Jg, Nr. 21: 633-644

Schütz JPh (1981) Que peut apporter le jardinage à notre sylviculture? Schweizerische Zeitschrift für Forstwesen, 132: 747-760

van Deusen PC (1991) Trend monitoring with varying coefficient models. Forest Science, Vol. 37, No. 5: 1365-1375

Diameter and Basal Area Increment in Permanent Growth and Yield Plots in Switzerland

Andreas Zingg
Swiss Federal Institute for Forest, Snow and Landscape Research
Birmensdorf, Switzerland

Abstract

The development trend of the diameter increment and the basal area increment of single trees and of stands is examined by means of 25 permanent trial plots of the growth and yield research. A number of these plots have been periodically measured for more than 100 years. The plots are not systematically distributed over Switzerland. The periodic resolution of the measurements lies between 4 and 10 years. In the study, the diameter increment is used for single trees and tree classes, the basal area increment for stands. These two quantities are preferred over other quantities such as height and volume because of their simple and reliable measurability. The assessment of a possible trend is done by means of a comparison of the growth with either the mean of the whole observation period, or, in those cases where it is permissible, with yield tables. Despite increasing age, a constant or, as is the case in the most recent observation period, an increase of the diameter and basal area increment can be shown for various tree species and for various forest structures in different areas of Switzerland. Relationships between the observed basal area increment and the development of the climate parameters or the silvicultural treatment can be ascertained only in some cases. Apparently the site productivity has changed or is in the process of changing. On the basis of the available data, none of the possible hypotheses – climate change, change of the CO_2 content in the air, influx of substances, extension of silvicultural management etc. – can be either accepted or rejected.

1. Introduction

Studies concerning the development of forest growth in recent decades have been published for some areas in Switzerland. Schmid-Haas (1989) found that damaged fir showed a continuous decrease in increment until the end of the 1970s. This was

European Forest Institute Research Report No. 5
Growth Trends in European Forests
Edited by H. Spiecker, K. Mielikäinen, M. Köhl and J. P. Skovsgaard

followed by a rapid increase in increment. Buffi (1989) calculated a higher site index than was to be expected on the basis of the site class for larch in southern Switzerland. Hartmann and Schneider (1989) showed that spruce in western Switzerland grows better today than it did at the beginning of the century. Bräker (1989) could establish an increment peak since the 1950s for spruce and, following on an increment low, a recovery since the beginning of the 1980s for fir. Becker et al. (1990), Kontic et al. (1990) and Bräker (1992) showed from tree ring analyses that various tree species grow better today than is to be expected on the basis of the age trend. Keller (1992) studied different-aged spruce on former growth and yield plots and determined that they are now higher than the previous generation at the same age and on the same plot. Hartmann et al. (1992) found out that in western Switzerland the increment of most of the studied spruce stands lies above the expected values of the yield tables.

Most of the studies carried out in Switzerland are based on tree ring analyses of single trees. No studies have been made on the changes of basal area or on the volume increment in reference to a given area. It is therefore obvious to use data from growth and yield studies to get insight into the increment development.

2. Data

Data from 25 long-term permanent plots were used in this study (Table 1, Fig. 1.) The first measurements were made between 70 and 105 years ago; the most recent growth and yield measurements were carried out between 1990 and 1994.

The present ages of the stands are between 104 and 217 years with an average age of 136 years. Five plots are stocked with an uneven-aged stand (selection forest). The choice of the plots at the time of installation and the measurement intervals depended on experimental objectives (see type of experiment Table 1). The questions formulated in this paper did not stand to discussion when the plots were laid out and data from these experimental plots are not representative for Swiss forests (see Köhl et al. 1995).

The following parameters have been assessed since the beginning of the permanent growth and yield experiments:

- tree species, diameter at 1.3 m callipered crosswise (d.b.h.), IUFRO code or social class for all trees above the recording limit,
- tree height, diameter at 7 m (D7), height of the crown base and crown radii (generally for only a part of the collective).
- The time of growing above the recording limit and the removal (utilisation or mortality) of each individual tree is known.

Climate data (precipitation, temperature) from the Swiss Meteorological Institute (SMA) were used for this study. Precipitation data from the climate stations at Lang-

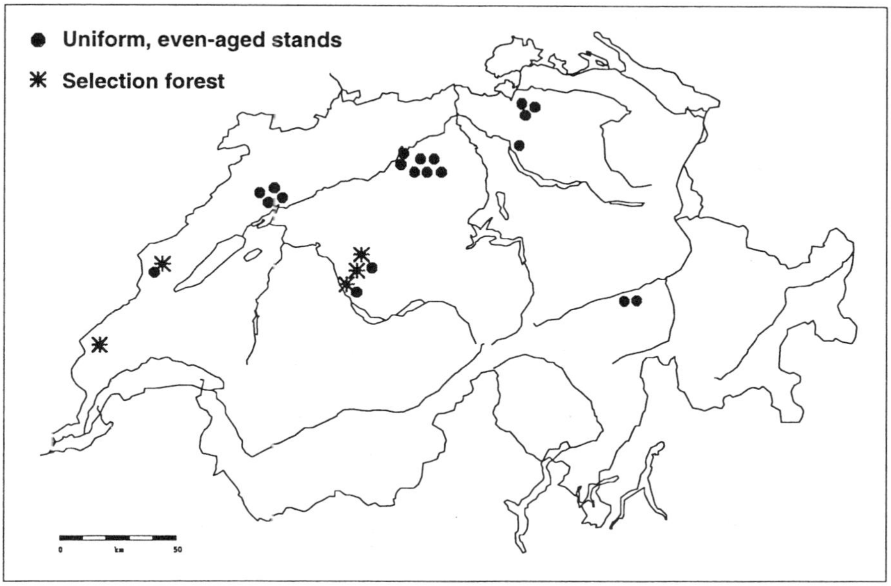

Fig. 1. Distribution of the plots

nau im Emmental for the plots Toppwald (01-015 and 01-021), Thun for Hasliwald (01-019 and 21-276), Couvet for 02-034 and Bois du Pays (02-035) as well as Le Sentier for La Rolat (21-294) have been collected since 1900, i.e. before the mentioned trial plots were laid out. The long temperature data series from the Berne climate station was used for all these examples.

3. Methods

In this study, the diameter increment of individual trees and the basal area increment per hectare and year of the main tree species, spruce (*Picea abies* [L.] Karst.), fir (*Abies alba* Mill.) and beech (*Fagus sylvatica* L.) are shown. In Switzerland, 83.4% of the growing stock consists of these three tree species (EAFV 1988).

It is well known that the top height is a good indicator of site classes and their changes (Assmann 1961). The stand treatment does not influence it very much. On the other hand, the measurement of the tree heights is subject to errors (Schmid-Haas et al. 1971; Winzeler 1986; Falkengren-Grerup et al. 1990; Kaufmann 1995). Volume and volume increment are derived magnitudes which are influenced by the measurement errors of the input variables. Basal area and diameter develop in pro-

Table 1. Trial plots

Plots	Location	Altitude	Aspect	Gradient	Relief	Area	Type of Experiment	Tree Species *Yield Table* (Badoux 1983)	h_{dom}*	Number of Samplings since **	Age 1995
		m.a.s.l		%		ha			m		
01-012	Morissen GR, Escherwald	1630	E	30	slope	0,50	Growth and yield in mountains	*Spruce* Cembra pine	(14) (12)	13 (1905)	121
01-013	Morissen GR, Escherwald	1640	N	30	slope	0,50	Growth and yield in mountains	*Spruce* Cembra pine	13 11	13 (1905)	119
01-021	Oberhünigen BE, Bannholzegg, Wildenei	1050	N	20	slope	0,40	Thinning B-degree, check plot of 01-015,	*Spruce* *Fir*	(18) (13)	15 (1912)	151
02-013	Biel-Bienne BE, Frinvillier Est	570	N	40	slope	0,20	Thinning B-degree	*Beech* *Spruce* Ash	18 20 18	14 (1906)	119
02-015	Biel-Bienne BE, Frinvillier Ouest	570	N	50	slope	0,20	Thinning C-degree	*Beech* Scots pine Ash	19	13 (1921)	119
02-034	Buttes NE, Crêt rond	970	N	30	slope	1,00	Thinning L-degree, check plot of 02-035	Spruce Fir Beech		13 (1913)	217
02-039	Zollikon ZH, Breitbirch	540	W	10	slope	0,30	Growth and yield	Larch *Beech*	(28) (24)	11 (1919)	166
02-042	Biel-Bienne BE, ob dem Neuhaus Ost	670	NW	30	slope	0,50	Thinning B-degree	*Beech* Fir	20 19	13 (1921)	104
02-043	Biel-Bienne BE, ob dem Neuhaus West	670	NW	30	slope	0,50	Thinning H-degree	*Beech* Fir	19 20	13 (1921)	104
21-276	Oppligen BE, Hasliwald	570	-	0	plain	0,27	Thinning B-degree, check plot of 01-019	*Spruce*	(25)	16 (1908)	142
41-007	Embrach ZH, Tisch, Blauen	590	W	10	slope	0,25	Thinning C-degree	*Beech*	(18)	16 (1891)	146

41-193	Embrach ZH, Tisch, Blauen	585	W	10	slope	0,17	Thinning B-degree	*Beech*	(18)	13 (1905)	146
41-194	Embrach ZH, Tisch	585	W	10	slope	0,25	Thinning H-degree	*Beech*	(17)	13 (1905)	146
41-014	Zofingen AG, Mühletaler Halde, Baan	510	NW	50	slope	0,25	Thinning B-degree	*Beech*	19	16 (1890)	128
41-118	Zofingen AG, Mühletaler Halde, Baan	490	NW	30	slope	0,20	Thinning D-degree	*Beech*	20	14 (1892)	123
41-155	Zofingen AG, Mühletaler Halde	510	NW	40	slope	0,25	Thinning H-degree	*Beech*	20	13 (1898)	124
41-017	Zofingen AG, am Kohlerweg, Baan	560	W	10	slope	0,25	Thinning C-degree	*Beech*	(19)	16 (1890)	134
41-018	Zofingen AG, am Kohlerweg, Baan	560	W	10	slope	0,25	Thinning B-degree, since 1954 H-degree	*Beech*	(18)	16 (1890)	134
41-022	Aarburg AG, Brüschholz	475	S	20	slope	0,25	Thinning H-degree	Beech	19	18 (1890)	140
41-024	Aarburg AG, Brüschholz	485	W	10	slope	0,25	Thinning B-degree	*Beech*	(19)	18 (1890)	142
Mixed, uneven-aged stands, selection forests											
01-015.001	Niederhünigen BE, Obere Moosmatten, *Toppwald 1*	947	W	30	slope	1,80	Selection forest	Fir Spruce Beech	22	16 (1905)	-
01-015.002	Niederhünigen BE, Obere Moosmatten, *Toppwald 2*	931	NW	10	hollow	1,20	Selection forest	Fir Spruce Beech	22	16 (1905)	-
01-019	Oppligen BE, *Hasliwald*	575	-	0	plain	2,00	Selection forest	Spruce Fir	22	16 (1908)	-
02-035	Buttes NE, *Bois du Pays*	983	NW	40	slope	2,00	Selection forest	Spruce Fir Beech	22	13 (1913) -	-
21-294	Le Chenit VD, *La Rolat*	1340	W	20	slope	2,00	Selection forest	Spruce	12	9 (1925)	-

* top height h_{dom50} (mean height of the thickest 100 trees/ha at 50 years age) interpolated or extrapolated (values in brackets) on the basis of measurements. For selection forests site index for spruce.
** the first measurement is in brackets

portion to the volume increment and they are affected only by the d.b.h measurement error and are thus applied in this study. The diameter increment is shown for individual trees and tree collectives of the trial plots and the basal area is shown for stands.

The periodical diameter development of each tree is known. The diameter increment was determined for each tree species as a mean per year of an observation period. This is illustrated graphically for every mid-period. In even-aged stands dominant and co-dominant trees are considered for the evaluation, since they produce on the average 94% of the basal area increment. These trees correspond to the tree classes 1, 2 and 3 according to Kraft (according to Assmann 1961).

In selection forest plots, only the trees with more than 30 cm d.b.h. are combined for the evaluation which, with the exception of beech, have a diameter increment double that of the small diameters, and achieve on the average 65 % of the basal area increment. The assignment occurs according to the corresponding diameter at the beginning of a measuring period (Spiecker 1986) and thus the combination of these collectives changes. Of the trees which had a d.b.h. greater than 30 cm at the beginning of the observation, only 35% of the spruce and 12% of the fir remained till the last observation. Analyses with various selection methods (30 cm d.b.h. at the beginning of the experiment, at the end of the experiment or at the beginning of each observation period) resulted in small differences for the corresponding increment value. They did not show different increment patterns.

Over the years, the maximum diameter increases on all plots. The number of trees having large diameters (> 60 cm) is always relatively small. The following was done to prove that growth changes are not attributed to a change in structure or in diameter distribution of the stands: the average diameter increments per year of each observation period were computed separately for spruce and fir for two diameter classes (30 cm ≤ d.b.h. ≤ 60 cm and d.b.h > 60 cm). A comparison of averages (z-test according to Riedwyl (1978) with a confidence limit of 5%) revealed that less than a third of the pair comparisons showed significant differences in the averages of the diameter increment. In 40% of these cases, the diameter increment of the trees with a d.b.h. > 60 cm was less than that of the trees with a d.b.h. between 30 and 60 cm. Slight differences result in their absolute values; the pattern, however, presents parallel increment developments.

The basal area increment per hectare and year of an observation period is calculated according to the formula

$$I_G = \frac{G_{t+p} - G_t}{p}, \qquad \text{where} \quad G = \sum \pi \left(\frac{\text{d.b.h}}{2} \right)^2$$

G is basal area/ha of a stand or of a tree species in a stand,
t is time, year,
p is period length.

This is also illustrated graphically for the mid-point of every period. As a rule, the basal area increment I_G is shown as a total value for a stand. In mixed stands, to enable references to yield tables, it is shown for each tree species for the period t to t+p as a percentage of the basal area at the beginning of the period, Gt, as $I_{Gt\text{-}t+p}$ (t = year, p = period length).

Diameter and basal area increment culminate fairly early in uniform, even-aged stands and decrease with advancing physiological age (Assmann 1961). Diameter increment can be compared with the development of previous years, e.g. by means of yield tables. The latter reflect the growth of the experimental stands in the first half of the 20th century. It cannot be said with certainty whether or not this was a period of high or low increment. Another possibility is the comparison of the increment in the reference period with the same or with several stands.

The Swiss yield tables were developed for high-thinned pure stands. According to Assmann (1961), yield tables based on top-height site indices may be used for the estimation and assessment of the increment in stands provided that the top height development corresponds to that of the yield tables. Whenever this is the case, the used yield table is mentioned in Table 1.

The basal area increments of pure stands are compared directly with the corresponding yield tables. In the case of mixed stands, they are compared relatively for each tree species by means of the basal area increment percentage.

As age is not given in selection forests, yield tables cannot be applied to these. A comparison with a second generation is also not possible. A comparison with a reference, e.g. the mean value of the whole observation period, is applied to these forests.

On selection forest plots as well as on the corresponding even-aged check plots, the influence of the weather and the stand treatment on the basal area increment was tested with regression analysis. Average annual temperature, precipitation sum values (mean annual precipitation from May to September in the observation period, i.e. 5 years before the first measurement until the end of the measurement period) and the thinning, expressed as removed basal area in percentage of the basal area before silvicultural operations, were identified as independent variables. The influence of thinning only will be examined on all trial plots in this study.

4. Results

4.1. Diameter Increment

Diameter Increment in Uniform, Even-Aged Stands. The mean diameter increment of the dominant and co-dominant trees is described. The increment development of spruce, fir, and beech in pure stands as well as in mixed stands is shown in Fig. 2 in percentage of the mean increment for the whole period, I_d [%].

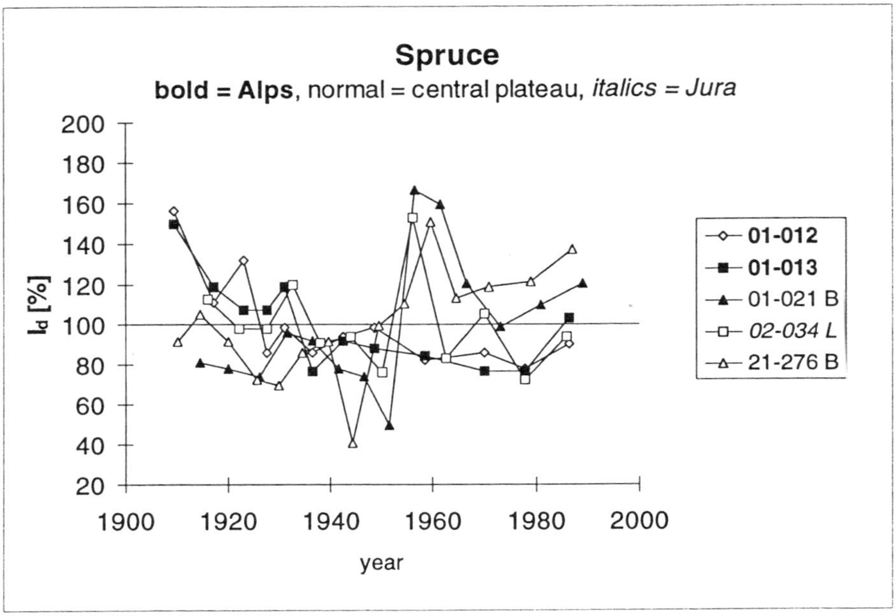

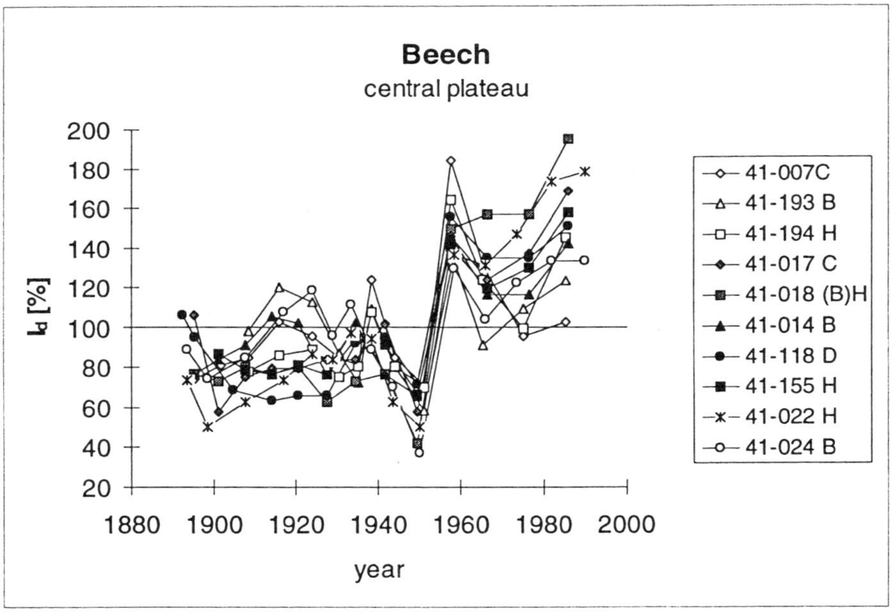

Fig. 2. Diameter increment I_d [mm y^{-1}] for dominant and co-dominant spruce, fir and beech, in percentage of the mean for the whole period

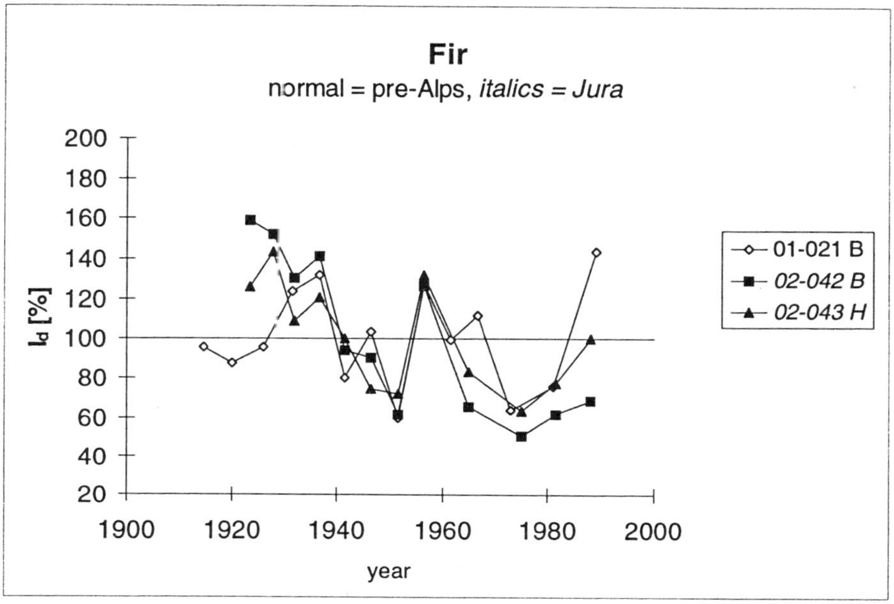

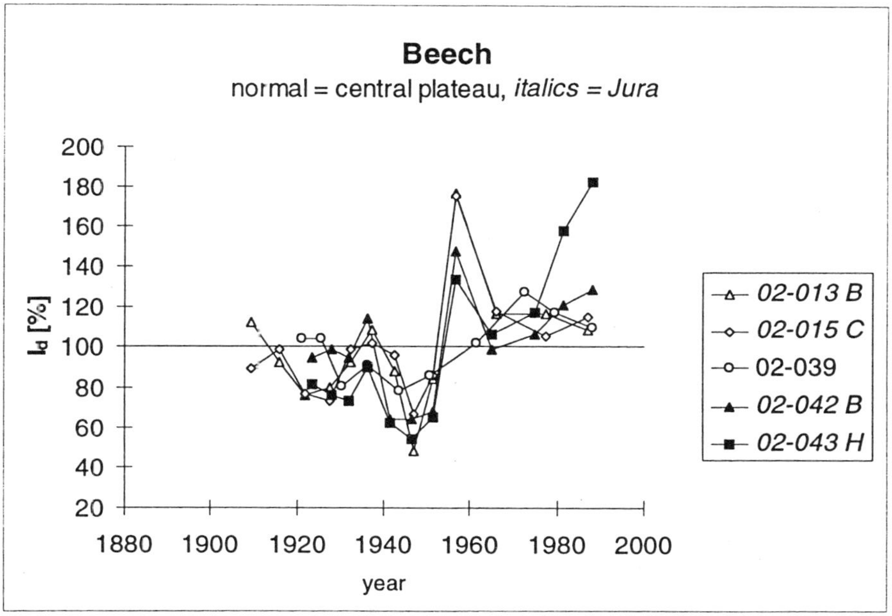

Fig. 2. Continued

In the even-aged stands, spruce (*Picea abies* [L.] Karst.) shows a fairly consistent decline in the diameter increment from 3 to 2 mm with a range of approx. 2 mm. The mean for the whole period and for all the plots is approx. 3 mm. In the second half of the 1940s, there was a marked reduction to 1.5 mm. This was followed by a marked increase of up to 5.5 mm in the first half of the 1950s on the trial plots in the region central plateau and pre-Alps. The diameter increment of the spruce on the plots situated in the Alps does not show these fluctuations. Instead, it has had a markedly sinking tendency since the beginning of the experimental studies, and thus shows a typical growth pattern. The spruce of the mixed stand plot in the Jura has an intermediate position. The increment reduction of the 1940s - the spruce were 160 years old at the time - is not visible on the diameter increment curve. The diameter increment rises from 2 to 4 mm and, contrary to the development on the other plots, it falls back to the previous level. The last measurements in the 1980s indicate an increase of the diameter increment on all plots, including those of the Alps. The diameter increment on all plots in all regions now lies between 90 and 135% of the mean diameter increment for the whole period.

Fir (*Abies alba* Mill.) is represented in a sufficient number (at least ten trees in the most recent measurement) as an associated species on three trial plots (01-021, 02-042, 02-043) and shows the same pattern observed in spruce: after a sinking trend of the same magnitude as that of spruce (mean diameter increment over the whole period approximately 3 mm), there is a reduction in increment which is followed by a sharp increase. After this diameter increment increase, fir again falls back to the increment level of the 1940s. However, the increase after 1980 is much clearer for fir than it is for spruce. The diameter increment now lies between 70 and 120% of the mean diameter increment for the whole period, that of fir on the Jura plots lies on or below the mean diameter increment for the whole period.

In beech (*Fagus sylvatica* L.) stands, a pattern similar to that of spruce can be determined with diameter increments between 2 and 3 mm (mean for the whole period is 3 mm); however, there is no clear decrease of the diameter increment until 1940. The reduction of the diameter increment in the measurements of the second half of the 1940s falls to 1 mm. The following marked increase in the first half of the 1950s rises to 6 mm. The practically identical development on all beech trial plots is conspicuous. After 1960, two groups whose diameter increments differ, can be formed: today the plots with the high thinned treatment variants H, (B)H and D show a diameter increment of 5 to 7 mm per year whereas the low thinned plots show an increment of between 3 and 4 mm. The differing stand treatments are clearly responsible for this difference. In comparison to the mean diameter increment for the whole period, the diameter increment now lies between 100 and 195%.

The pattern of diameter increment of beech in mixed stands is similar to that of beech in pure stands (mean for the whole period is 3 mm). On all plots in the Jura, the increment reduction and the following increase occurred to the same extent and direction as that on the pure stand plots. The diameter increment in these stands also lies between 110 and 130%, in one case even 180%, in comparison to the mean increment for the whole period.

One exception to the above are beech on the mixed larch-beech plot 02-039: the diameter increment increased from the 1940s up to the first half of the 1970s when they reached the age of approx. 150 years. Since then it has been decreasing but, with 4.2 mm per year, it is still 10 % higher than the mean for the whole period and higher than at the beginning of the experiment in 1919 at the age of 90 years. The increment reduction of the 1940s and the following increase is only faintly noticeable. This may be due to the fact that beech is mixed with 50% larch on this plot. The latter are slightly thicker and higher, so that they served as a shelter for the beech in the dry years of 1947 and 1949.

Diameter Increment Development in Selection Forests. The development of the mean diameter increments of spruce and fir in selection forests with diameters of 30 cm and greater are shown in Fig. 3.

Spruce on the plots situated in the Emmental, 01-015 Toppwald and 01-019 Hasliwald, show in the course of time, compared both absolutely as well as relatively to the mean diameter increment for the whole period (approx. 4 mm), an increment increase of up to between 6 and 7.5 mm. On the two plots in the Toppwald, the reduction and the following increase of increment in the 1940s, are clearly visible, whereas in the Hasliwald they are only faintly noticeable. The diameter increment of spruce has been increasing on these plots since the 1950s. The trial plots in the Jura are in contrast to this. On the lower plot, 02-035 Bois du Pays, a decrease is noticeable whereas on the higher plot, 21-294 La Rolat, there is no change.

This is less evident with fir: depending on the trial plot, there is either a slightly increasing or a slightly decreasing development throughout the whole period. An increase in the diameter increment can be determined only in the 1980s. Compared to the whole period (on the average approx. 5 mm), the diameter increment in the 01-019 Hasliwald plot is 118% and on the two adjacent plots in 01-015 Toppwald it amounts to 177 and 152%. Both locations are situated in the Emmental. An increase in the diameter increment in comparison to the whole period cannot be established on the Bois du Pays plot, even though, in the most recent measurement, the increment again increased by 10 % above the mean for the whole period. This is in contrast to the previous years.

In Table 2 the comparison of the mean periodic diameter increment of spruce, fir and beech at present, $I_d lp$ (the latest observation period), with the mean diameter increment for the whole period, I_d, is shown. The table is sorted according to decreasing altitude above sea level of the trial plots.

On the higher altitude plots of the Alps and the Jura the diameter increments of spruce remain the same at present as they did in previous years. At present fir and beech below 1000 m.a.s.l. grow clearly better than they did in the past. Spruce have a higher diameter increment in the pre-Alps whereas this applies to fir and beech on the central plateau. Spruce grow on the average 12% better in even-aged stands and in selection forests the growth has increased by 8%. The increment increase for fir in selection forests amounts to 31% on the average. The marked diameter increment increase of beech is conspicuous: on the average the increase amounts to 60% with a peak of 150%. The high thinned, pure beech stands show the highest increase despite their age of more than 120 years: on the average 88%.

Table 2. Comparison of the diameter increment of all trees: growth in the whole period (I_d) with growth in the latest period (I_d lp), *italics = selection forest*

Plot	Region	Altitude	Spruce			Fir			Beech		
			$\bar{I}_d$ whole period	$\bar{I}_d$ lp latest period	$\bar{I}_d / \bar{I}_d lp$	$\bar{I}_d$ whole period	$\bar{I}_d$ lp latest period	$\bar{I}_d / \bar{I}_d lp$	$\bar{I}_d$ whole period	$\bar{I}_d$ lp latest period	$\bar{I}_d / \bar{I}_d lp$
		m.a.s.l.	mm/a	mm/a	%	mm/a	mm/a	%	mm/a	mm/a	%
01-013	Alps	1640	2.2	2.6	117						
01-012	Alps	1630	2.1	2.0	95						
21-294	*Jura*	*1340*	*1.6*	*1.6*	*100*	*1.6*	*1.6*	*99*	*1.4*	*1.2*	*87*
01-021 B	pre-Alps	1050	2.7	3.3	124	2.1	2.4	114			
02-035	*Jura*	*983*	*3.0*	*3.3*	*108*	*3.2*	*3.3*	*102*	2.6	*1.9*	*74*
02-034 L	Jura	970	2.7	2.6	95						
01-015.001	*pre-Alps*	*947*	*2.5*	*3.3*	*132*	*2.6*	*3.7*	*141*	*2.3*	*3.3*	*145*
01-015.002	*pre-Alps*	*931*	*2.9*	*3.9*	*137*	*2.7*	*4.1*	*150*	*2.0*	*3.2*	*160*
02-042 B	Jura	670				2.3	1.4	61	2.1	2.8	132
02-043 H	Jura	670				2.1	2.7	129	2.0	5.1	250
41-007 C	plateau	590							2.7	2.9	108
41-193 B	plateau	585							2.3	3.1	133
41-194 H	plateau	585							1.7	3.8	219
01-019	*plateau*	*575*	*3.3*	*2.3*	*69*	*3.5*	*5.2*	*149*	*2.5*	*3.6*	*142*
02-013 B	Jura	570	2.0	1.5	76				2.1	2.5	119

02-015 C	Jura	570							3.1	3.6	115
21-276 B	plateau	570	3.6	5.1	142						
41-017 C	plateau	560							2.2	3.7	170
41-018 B/H	plateau	560							2.3	4.9	216
02-039	plateau	540							3.5	3.8	110
41-014 B	plateau	510							2.2	3.3	149
41-155 D	plateau	510							2.1	4.5	216
41-118 H	plateau	490							3.8	5.7	152
41-024 B	plateau	485							2.3	3.5	150
41-022 H	plateau	475							2.7	5.8	214
even-aged			2.5	2.9	112	2.2	2.2	100	2.5	3.9	159
Selection forest			*2.7*	*2.9*	*108*	*2.7*	*3.6*	*131*	*2.2*	*2.6*	*123*

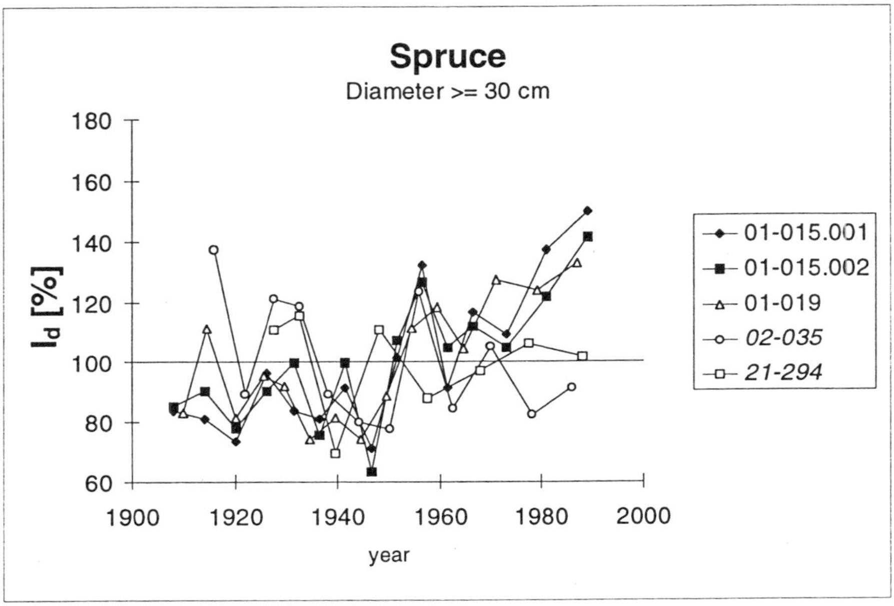

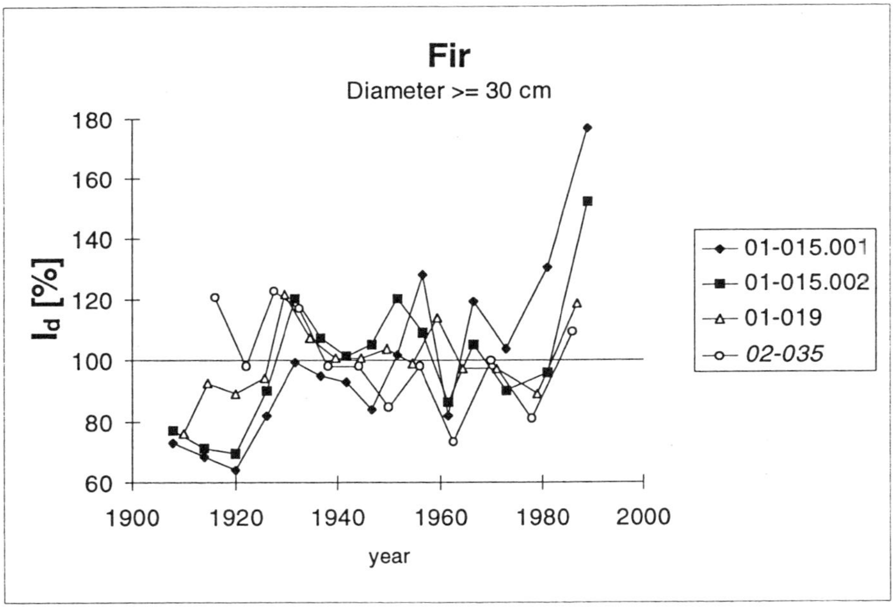

Fig. 3. Diameter increment in selection forests I_d [mm y^{-1}] for spruce and fir, in percentage of the mean for the whole period

4.2. Basal Area Increment

Basal Area Increment in Even-Aged Stands. The development of the basal area increment on the studied plots mirrors to a large extent the same conditions already established with the diameter increment. In the even-aged stands spruce and fir are, with the exception of plot 21-276, associated species. As a result of this, the basal area increment per hectare and year in terms of a percentage of the basal area per hectare is used for the comparison with the yield tables.

In all cases, from the age of 50 years, the basal area increment developed below that of the yield tables. From the age of approx. 90 years, the increment of spruce on plot 21-276 exceeds the corresponding values of the yield tables and continues on a higher level. The spruce from plot 01-021 and the two mountain plots 01-012 and 01-013 do this from the age of approx. 100 years. On the other hand, the increment of fir on plot 01-012 lies, with the exception of one period, more or less clearly below the increment of the yield table. Despite the fact that the top height development of these stands corresponds fairly well to that of the yield tables, these stands show a different behaviour in the pattern of the basal area increment. Different silvicultural treatment of the stands as opposed to those of the yield tables or different growth dynamics in mixed stands could be mirrored in these discrepancies.

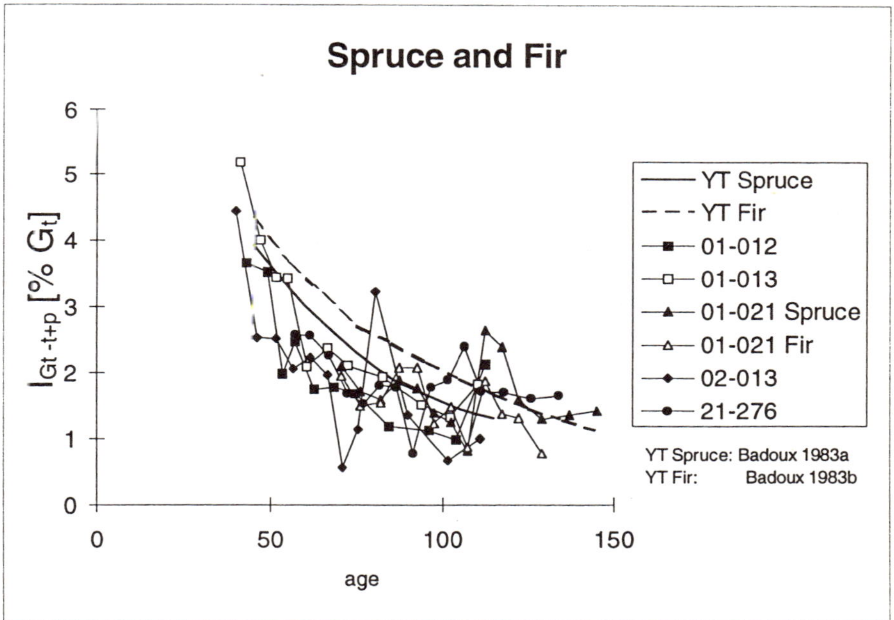

Fig. 4. Basal area increment of spruce and fir: Comparison with the yield table

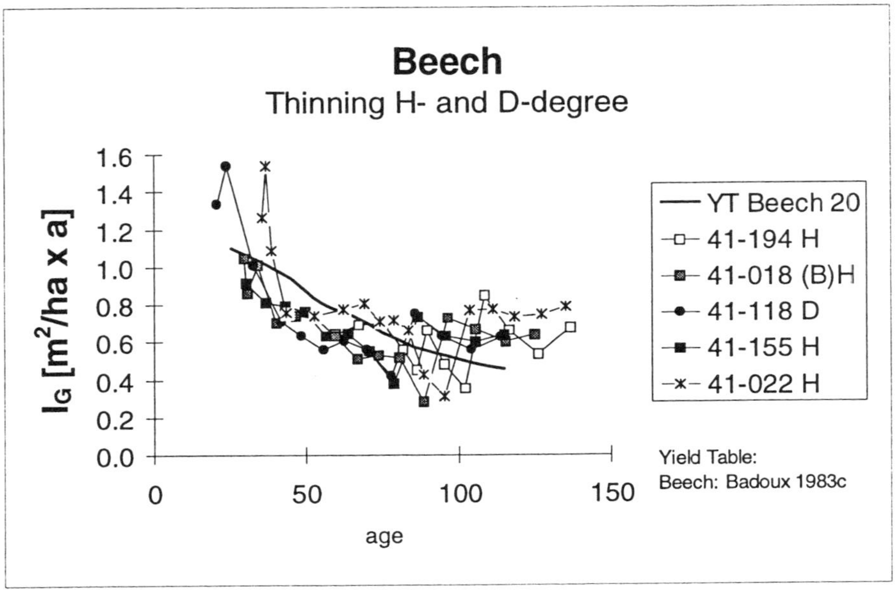

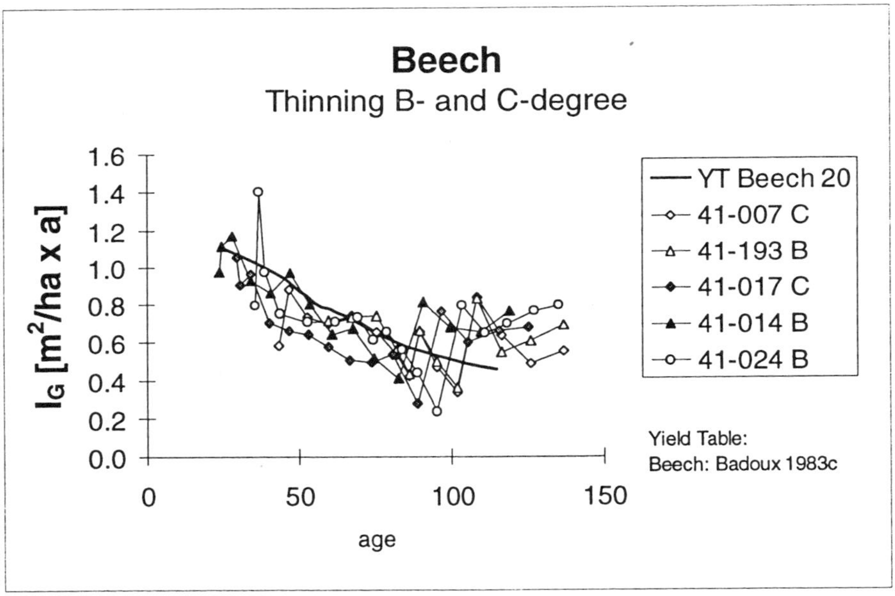

Fig. 5. Basal area increment of beech: comparison with the yield table

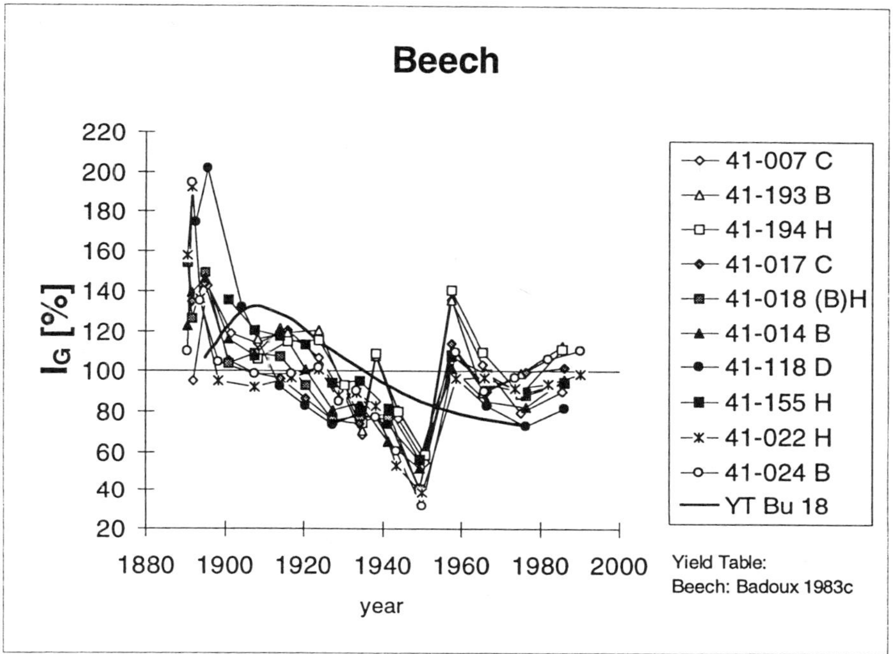

Fig. 6. Basal area increment of the pure beech stands in comparison with the mean for the whole period.

Regardless of thinning treatment, the basal area increment of all pure beech stands moves clearly above that of the yield tables from the age of 90 to 110 years. This is in contrast to the development of spruce and fir. The jump to a higher basal area increment level in all the stands occurred after the increment reduction in the second half of the 1940s which can be clearly observed in the diameter increment.

When the development is compared to the mean basal area increment for the whole period, the following picture is revealed (Fig. 6): the pattern corresponds to that of the diameter increment. The expected sinking of the basal area increment with the advancement of age, is clearly visible here. Until the reduction of the 1940s, the course of the basal area increment corresponds to the expectations (yield tables). The steep rise after the marked increment reduction is also visible here. In spite of the following sinking, the level remains clearly higher than before the reduction and in all cases shows an increase in the 1980s. In Fig. 6 the pattern of the basal area increment for a hypothetical yield table stand is shown (stand establishment in 1855, age 1995 = 140 years). From this, an increment increase of up to 50 % can be estimated for the stands on the research plots.

Beech in mixed stands show, with the exception of two plots, a similar basal area increment development as spruce and fir: a lower increment pattern than the yield tables from the age of 50 years (no figure). On plot 02-043 beech make an exception to this. In the last two measurement periods, aged approx. 80 to 100 years, an increase of the basal area increment of up to double the value of that of the yield table has been determined. The other exception are the beech mixed with larch on plot 02-039. The basal area increment of these beech stands corresponds to the yield tables from age 90 (at beginning of experiment in 1919) until 118 years. Since then there has been a slight increase in the basal area increment percentage.

Basal Area Increment in Selection Forests. The basal area increment in selection forests shows a similar but less clear pattern to that of the diameter increment. (Fig. 7, Table 3).

The basal area increment of spruce on the plots in the Emmental, 01-015 Toppwald and 01-019 Hasliwald have, compared to the whole period, increased in the mean by 60 - 90% since the 1960s. In the Jura the basal area increment of spruce on the 02-035 Bois du Pays plots has stagnated, while that on the 21-294 La Rolat plot has fluctuated around the mean value for the whole period.

The basal area increment of fir had a slightly sinking tendency on all of the selection forest plots until the beginning of the 1980s. Increment reductions can be discovered between 1940 and 1950, around 1960 and in the second half of the 1970s. The basal area increment reduction did not reach the same magnitude as that of the beech plots. A basal area increment increase has been observed on all plots in the last measurement period, which with 50% on the Toppwald plot, was clearly above the mean.

The basal area increment for all tree species of the five selection forest plots can be characterised as follows: on the Toppwald 1 plot, (slope location) there is a clear increase of 0.67 to 1.21 m^2 ha^{-1} and year or 70%. On the neighbouring plot Toppwald 2 (hollow location) the increase amounts to 28 % only, i.e. from 0.65 to 0.85 m^2/ha and year. The 01-019 Hasliwald plot, at 575 m.a.s.l. the lowest of the selection forest plots, has the highest average increment level of all the selection plots described here. The increase in comparison to the whole period amounts to 41%, which is from 0.78 to 1.15 m^2 ha^{-1} and year. This increase resulted only in the most recent period.

The lower altitude plot of the Jura, 02-035 Bois du Pays, shows a strong fluctuation with a sinking trend: but the basal area increment of the most recent measurement for the entire stand lies with 0.87 m^2 ha^{-1} and year slightly above the 0.83 m^2 ha^{-1} and year of the mean for the whole period. This value is reached only due to the high increment of beech. Spruce and fir lie below the mean for the whole observation period. The higher altitude spruce selection forest of the Jura, 21-294 La Rolat, situated on karstic Jura lime stone, shows consistent growth over the entire observation period. The basal area increment does not deviate more than 18% or 0.07 m^2 ha^{-1} and year from the mean for the whole period. In the last three periods it lay between 1 and 4% above the mean.

In Table 3 the basal area increment for all plots is shown as the mean for the whole period, the last period and the deviation of the latter from the whole period expressed in percentage. The corresponding values for two spruce and one beech

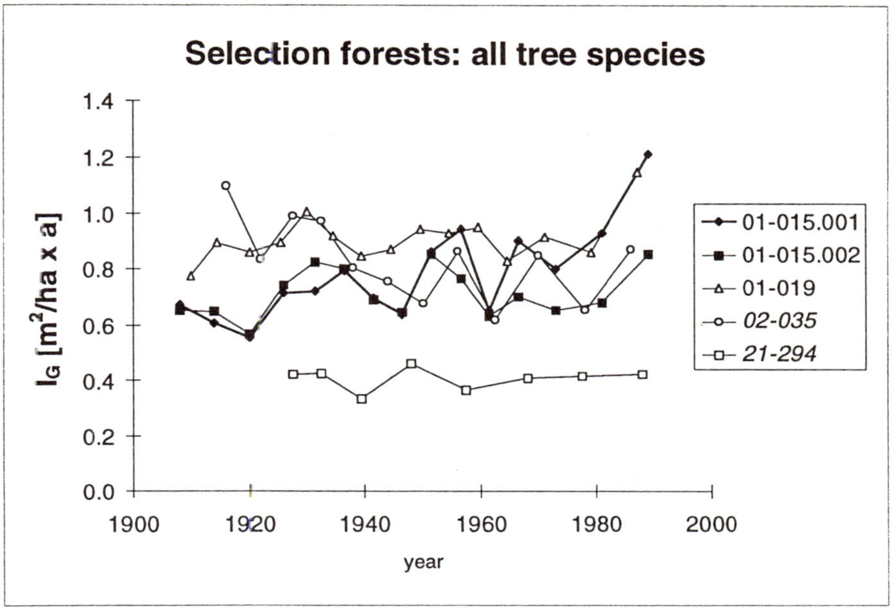

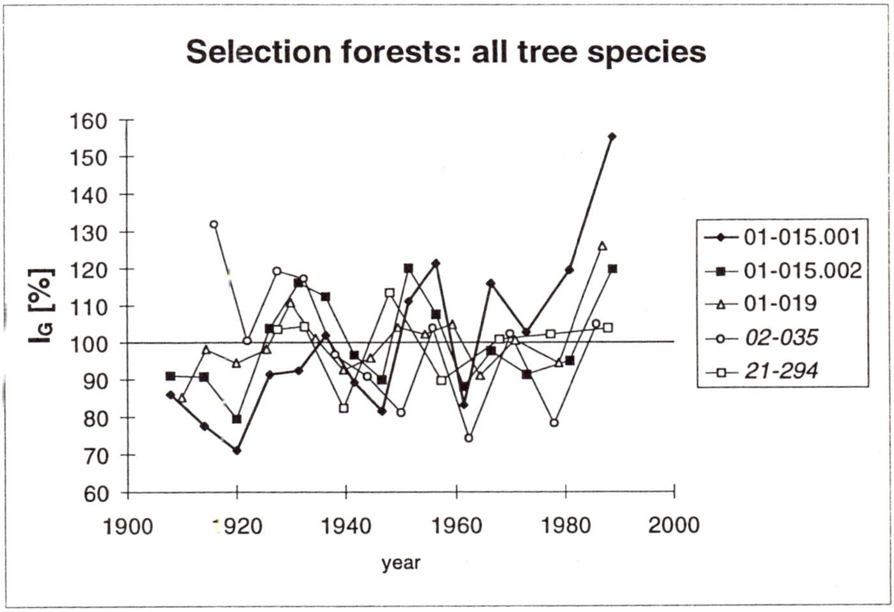

Fig. 7. Basal area increment in selection forests

yield table stands are presented as a reference. In most cases where there is a lower basal area increment than in the whole period, this is generally higher than the values that were to be expected (yield tables). The values of the spruce-cembran pine plots in the Alps (01-012 and 01-013) are the only ones which lie within the framework of the yield tables. The low value of the plot 02-034 (Jura) may be explained by the high age of spruce.

Apart from the plot Bois du Pays in the Jura, all selection forest plots show a slight increase. It is only on the plot Toppwald 1 that a clear rise of the basal area increment is established.

Table 3. Basal area increment

Plot	$\bar{I}_G$ whole period	$\bar{I}_{G^{lp}}$ latest period	$\bar{I}_G / \bar{I}_{G^{lp}}$
	[m2/ha x a]	[m2/ha x a]	%
01-012	1.17	1.00	85
01-013	1.15	0.69	60
01-021 B	0.70	0.93	132
02-013 B	0.75	0.60	80
02-015 C	0.74	0.57	77
02-034 L *	0.37	0.18	48
02-039 H	0.58	0.69	120
02-042 B	0.96	0.80	84
02-043 H	1.00	0.84	84
21-276 B	0.61	0.30	49
41-007 C	0.62	0.55	89
41-193 B	0.62	0.69	111
41-194 H	0.61	0.67	110
41-014 B	0.81	0.76	95
41-118 D	0.77	0.62	82
41-155 H	0.71	0.65	91
41-017 C	0.70	0.69	98
41-018 B(H)	0.69	0.64	93
41-022 H	0.80	0.76	95
41-024 B	0.72	0.80	110
even-aged	0.75	0.67	89
YT Spruce 14			*63*
YT Spruce 18			*59*
YT Beech 20			*68*
01-015.001	0.78	1.21	155
01-015.002	0.71	0.85	120
01-019	0.91	1.15	126
02-035	0.83	0.87	105
21-294	0.40	0.42	104

* Spruce

4.3. Modelling of Basal Area Increment

The development of the increment, as shown here, can depend on various factors. The weather is probably the quantity with the greatest influence. Regression models were calculated for the selection forest plots and their corresponding check plots that are described here. The periodic basal area increment of the stand was the dependent variable. The year in the middle of the measurement period, the precipitation sum values (see Sec. 3, Methods), average annual temperature and the defined exploitation were used as variables. The model was tested with a F-test and the significance level of the model as well as the coefficient of determination, R^2, was given.

Suitable models could be derived for five plots with these values. In the case of 01-015.001 Toppwald 1 plot and the 21-294 La Rolat plot the explanation is satisfactory. The Toppwald 1 plot already shows clear positive tendencies. The precipitation explains the largest part of the variation (72%) in this case. In contrast, the

Table 4. Regression models for basal area increment

Plot	Thinning type	Level of significance for the variables					Significance of the model	R^2
		year	temperature	precipitation	precipitation + 5	Cut		%
01-015.001	P	0.0018	-	0.0279	-	-	0.0004	71
01-015.002	P	-	-	*0.0751*	-	-	0.0751	17
01-021	B	-	-	-	0.0148	-	0.0148	38
01-019	P	-	-	-	*0.1795*	-	0.1795	7
21-276.001	B	0.0042	-	-	-	-	0.0042	47
02-035	P	-	-	0.0202	-	-	0.0202	41
02-034	(H) L	-	-	-	*0.1321*	-	0.1321	15
21-294	P	-	0.0253	-	-	-	0.0253	60

italics = variable with largest F-value forced into regression

average annual temperature plays the decisive role on the higher altitude plot of La Rolat in the Jura. It explains 60% of the variation. This may be partly due to the diminishing influence precipitation has on tree growth with increasing altitude and with the increasing importance of temperature to tree growth with increasing altitude. The variation of the basal area increment of the lower altitude plots can be related to the model, to the low periodic resolution of the increment or to the application method of the climate information.

Thinning was not selected as an independent variable by any of the models on any of the permanent plots. In four of the thinning plots there are significant correlations with coefficient of determination between 30 and 65%, in three others it ranges from 25 to 28% and this is not significant. No correlation was found in selection forest plots.

5. Discussion

If stands today have a greater diameter and basal area increment than in previous years, even though they have grown older and the stand treatment has not changed, the hypothesis that site productivity is changing can be supported. Although the diameter and the basal area increments are influenced more strongly by the stand treatment than are, e.g. height and volume increments, these variables present over the years a pattern which lies higher than the expectations.

For a number of decades, beech has grown better than is to be expected on the basis of present experience (e.g. yield tables) on all the examined plots. Spruce also shows an increase in increment in some of the experimental plots. In most of the plots, both even-aged as well as uneven-aged, the increment development corresponds to the expectations. For fir, after a decrease in increment in the late 1940s and the early 70s, a clear recovery since the beginning of the 1980s can be determined. These deviations from the expected development and the differences between the various plots cannot be explained for any of the three fir plots. The treatment does not contribute to an explanation of the increment development. The causes remain unknown.

Nevertheless, these investigations permit a number of conclusions. It can be assumed on the basis of available data that climate factors, although statistically not proved, play an important role. Other possible reasons are site changes caused by utilisation changes in a broad sense. Unfortunately, there are no data available concerning either the CO_2 content in the air or the influx of substances with a fertilising effect. Differences in the growth behaviour of different research plots may be traced back to site differences.

The influence of climate can be confirmed as in the even-aged stands in higher altitudes an age trend is visible right up to the 1970s and as in the selection forest there is a balanced diameter and basal area increment. The strong fluctuations caused by climate (e.g. precipitation in the lower altitude plots) have not been observed in the growth development of the stands on these plots. On the lower selection forest plot in the Jura, the fluctuations from the 1940s and the 1950s, clearly visible on the other plots, are unclear.

The selection forest plots of the Emmental, where a lightly to strongly rising trend is established, are a contrast to those of the Jura. The diameter increment as well as the basal area increment of spruce in all three plots increases slightly over the whole observation period. The same can be said of fir in the case of one plot only. In these plots the increment reduction of the 1940s with the following increase is visible only in a weakened form. These plots attained the highest basal area increment since the start of the experiment in the most recent period. Spiecker (1991) made similar observations concerning volume increment in selection forest plots in the Black Forest. Differences in diameter increment between spruce and fir in the Emmental are not uniform; however, in none of the cases do they attain those values mentioned by Spiecker (1991). This may be related to site differences and treatment variants.

An important difference between the selection forest plots of the Jura and those of the Emmental can be the water balance. The soils on the Jura plots are appreciably more permeable and their water storage capacity is not as good as those of the Emmental. The hypothesis of the influence of the water storage capacity is supported by the fact that no direct connection with the precipitation can be proved. Investigations of the water balance in the various experimental plots must be undertaken to verify this hypothesis. Abetz (1988) gave an indication that the water balance may play an important role in connection with growth changes. He indicates that through entry of nitrogen or through the raising of the CO_2 content in the air, site improvements in beech stands become more apparent on sites well supplied with water. In high altitudes the temperature is decisive to the increment development.

The increment pattern on all other plots is marked by the reduction in the second half of the 1940s followed by the sharp, above average rise. The reduction is a result of the two dry and hot years in 1947 and 1949. These two summer precipitation minima that followed one another in such a short period of time had a pronounced negative influence on growth, especially in beech stands. On the other hand, the causes for the strong increment increase that was observed on most of the plots in the following observation period are not obvious.

In almost all of the plots, a more or less clear increase in the increment has been determined in the most recent observation period. Falkengren-Grerup et al. (1990) came to similar significant results for beech and oak. They showed that the increment increase observed for beech was not caused by favourable weather conditions. They interpreted it as a result of the increased nitrogen depositions. These results were supported by soil and vegetation parameters. In spite of the lack of such data, the results presented here indicate an influence by the weather.

A possible cause for the increment changes could be due to the changes in utilisation intensities: this can be the partial utilisation of the phytomass of the trees or the abstention from secondary yields (e.g. forest litter), which leads to soil enrichment with nutrients. The forests were intensively exploited before the observations started. The complete exploitation of the biomass of the trees and of the ground vegetation led, through nutrient withdrawal, to an impoverishment of the sites before the start of the experiments. This withdrawal could be compensated in recent decades since this type of exploitation (extensive yield) was stopped. Amongst a number of other studies, Sterba (1988, 1993) shows the large influence of the intensity of exploitation on growth. However, this effect could not be proved in this study due to the lack of information on utilisation intensities.

Last but not least, a number of methodical questions remain open: the problems involved in the comparison with the yield tables have already been mentioned. The yield tables seem to reflect the growth of old beech stands only partially correctly. Pretzsch (1992) reports observations similar to the ones made here, amongst others, in even-aged beech thinning experiments.

Nevertheless, the increment development of the stands described here remains, in a long-term comparison, i.e. 70 to over 100 years, on a high level. A clearly rising tendency is shown for the past few years and under various conditions for nearly all the stands included in this study.

6. Conclusions

On the Swiss forest growth experimental plots, which have been under observation for between 70 and 100 years, the development of the diameter and basal area increments indicates changes. Both values show a rising trend in the course of time. In the cases where there is a decline in increment, this is smaller than was to be expected on the basis of previous experience (yield tables).

Apart from cores taken during the Sanasilva Survey (Bräker 1989), there are no extensive or representative data. There are no additional data concerning soil, climate, CO_2 content, depositions, etc. for the long-term observed experimental plots or for the tree-ring data mentioned. These would be necessary for the development of an explanatory model. This leads to the conclusion that long-term trends can be neither clearly verified nor can they be explained.

The long-term observed experimental plots are not representative of Swiss forests (Köhl et al. 1995). The research questions and the research planning did not specifically consider the questions posed here at the design of the experimental plots. As these are the only available data at present, additional surveys must be conducted to close the knowledge gaps. Supplementary attributes must be assessed on existing and on new permanent observation plots to enable the development and influence of explanatory models.

There are two approaches in attempting to pool the representative data from the National Forest Inventory or from regional sampling inventories with data from long-term growth research and tree-ring analyses:

- retrospective view: this enables representative data and results, broadens knowledge but remains inadequate for explanatory models.
- prospective view: in the future improved information will be available if research planning is focused on current questions. This will result in improved explanatory possibilities.

The expansion of the present experimental plot network must be clearly linked to the problems that are to be investigated: experimental plots on extreme sites are suited to answer ecological questions, experimental plots on good and average sites are interesting for questions pertaining to forestry. The latter, however, should be implemented with extreme treatment variants. At present, the continuation and supplementation of forest growth research finds itself in this conflicting area. Finally, the fact must be mentioned that in the near as well as in the far future, research should have experimental plots with long data sets at its disposal. In this way, similar future research questions may be answered using data that were not captured for the particular question.

Acknowledgements

I thank my two fellow researchers PD Dr. Michael Köhl and Dr. O.U. Bräker for their co-operative collaboration in the EFI-project Growth Trends of European Forests, my colleague, Christian Matter, for the swift and critical translation into English and PD Dr. Paul Schmid-Haas for critical and encouraging comments on this manuscript. I would like to extend my thanks to the three reviewers of the manuscript, Alain Franc, Pelle Gemmel and an unknown reviewer, who made considerable contributions to its improvement.

References

Abetz P (1988) Untersuchungen zum Wachstum von Buchen auf der Schwäbischen Alb. Allgemeine Forst- und Jagdzeitung 159, 11/12: 215-223.

Assmann E (1961) Waldertragskunde. 490 S. München Bonn Wien, BLV.

Badoux E (1983a) Ertragstafeln für die Fichte in der Schweiz. Eidg. Anst. forstl. Versuchswes., Birmensdorf, 3. Auflage.

Badoux E (1983b) Ertragstafeln für die Tanne in der Schweiz. Eidg. Anst. forstl. Versuchswes., Birmensdorf, 3. Auflage.

Badoux E (1983c) Ertragstafeln für die Buche in der Schweiz. Eidg. Anst. forstl. Versuchswes., Birmensdorf, 3. Auflage.

Becker M, Bräker OU, Kenk G, Schneider O, Schweingruber FH (1990) Aspect des houppiers et croissance des arbres au cours des dernières décennies dans les régions frontalières d'Allemagne, de France et de Suisse. Revue Forestière Française, XLII, 3: 284-300.

Bräker OU (1989) Ergebnisse zum Baumwachstum aufgrund von Jahrringanalysen an Bohrkernen aus der SANASILVA-Waldschadeninventur 1984. "Waldwachstum und Waldschäden", Tagung an der ETH-Zürich vom 14. Juni 1989. Sanasilva-Tagungsbericht: 32-40.

Bräker OU (1992) Erkenntnisse aus Zuwachsuntersuchungen. Forum für Wissen 1992, Waldschadenforschung in der Schweiz: Stand der Kenntnisse, Eidgenössische Forschungsanstalt für Wald, Schnee und Landschaft: 55-58.

Buffi R (1989) La crescita giovanile del larice europeo e del larice del Giappone al Sud delle Alpi, negli orizzonti submontano e montano. Zürich, Dissertation Eidg. Technische Hochschule Nr. 9074, 159 p. Also: Mitteilungen der Eidgenössischen Forschungsanstalt für Wald, Schnee und Landschaft, 66 (1990/91), 2: 155-329.

EAFV (Hrsg.) (1988) Schweizerisches Landesforstinventar: Ergebnisse der Erstaufnahme 1982-1986. Berichte der Eidgenössischen Forschungsanstalt für Wald, Schnee und Landschaft, 305: 375 p.

Falkengren-Grerup U, Eriksson H (1990) Changes in soil, vegetation and forest yield between 1947 and 1988 in beech and oak sites of southern Sweden. Forest Ecology and Management 38 (1990): 37-53.

Hartmann Ph, Schneider O (1989) Etudes régionales sur l'accroissement de l'arbre et du peuplement. "Waldwachstum und Waldschäden", Tagung an der ETH-Zürich vom 14. Juni 1989. Sanasilva-Tagungsbericht: 46-50.

Hartmann Ph, Schneider O, Petter D, Zimmermann D (1992) L'accroissement en diamètre des arbres. Evolution et connaissances. Influence du climat, de la sylviculture et de la pollution. Bulletin de la Société Vaudoise des Sciences Naturelles 81: 131-151.

Kaufmann E (1995) Kontrollaufnahme LFI2. Gesamtübersicht 1994. Internes Arbeitspapier. Birmensdorf, Eidgenössischen Forschungsanstalt für Wald, Schnee und Landschaft, 96 p.

Keller W (1992) Bonität in Fichten-Folgebeständen ehemaliger Fichten-Versuchsflächen der WSL. Deutscher Verband Forstlicher Versuchsanstalten, Sektion Ertragskunde, Tagungsbericht: 123-129.

Kontic R, Bräker OU, Nizon V, Müller R (1990) Jahrringanalytische Untersuchungen im Sihlwald. Schweizerische Zeitschrift für Forstwesen 141, 1: 55-76.

Köhl M, Scott CT, Zingg A (1995) Evaluation of permanent sample surveys for growth and yield studies: a Swiss example. Forest Ecology and Management 71 (1995): 187-194.

Pretzsch H (1992) Zunehmende Unstimmigkeit zwischen erwartetem und wirklichem Wachstum unserer Waldbstände. Konsequenzen für zukünftige ertragskundliche Informationssysteme. Forstwissenschafliches Centralblatt 111 (1992): 366-382.

Riedwyl H (1978) Angewandte mathematische Statistik in Wissenschaft, Administration und Technik. 243 S. Bern Stuttgart, Paul Haupt.

Schmid-Haas P, Roiko-Jokela P, Mingard P, Zobeiry M (1971) Optimal determination of the volume of standing trees. Mitteilungen der forstlichen Bundes-Versuchsanstalt Wien 91 (1971): 33-54.

Schmid-Haas P (1989) Der Nachweis der Ursache des Tannensterbens an einem Beispiel. Forstwissenschaftliches Centralblatt 108 (1989): 244-254.

Spiecker H (1986) Das Wachstum der Tannen und Fichten auf Plenterwald-Versuchsflächen des Schwarzwaldes in der Zeit von 1950 bis 1984. Allgemeine Forst- und Jagdzeitung 157, 8: 152-164.

Spiecker H (1991) Zur Dynamik des Wachstums von Tannen und Fichten auf Plenterwald-Versuchsflächen des Schwarzwaldes. Allgemeine Forstzeitung 46, 21: 1076-1080.

Sterba H (1988) Increment Losses by Full-Tree Harvesting in Norway Spruce (*Picea abies*). Forest Ecology and Management, 24: 283-292.

Sterba H (1993) Zuwachsreaktionen auf Biomassenentzug durch Vollernte bei Stammzahlreduktionen und Auslesedurchforstungen. Forschungsauftrag des Bundesministeriums für Land und Forstwirtschaft, Abschlussbericht. 10 p. Wien, Institut für Waldwachstumsforschung, Universität für Bodenkultur.

Winzeler K (1986) Volumenbestimmung durch Messung am stehenden Baum. Mitteilungen der Eidgenössischen Forschungsanstalt für Wald, Schnee und Landschaft, 62 (1986), 1: 3-99.

A Comparison of Analytical Approaches for Detecting Diameter Growth Trends: Case Studies from Switzerland

Michael Köhl, Andreas Zingg and Otto Ulrich Bräker
Swiss Federal Institute for Forest, Snow and Landscape Research
Birmensdorf, Switzerland

Abstract

Growth patterns in forests can be monitored using three approaches: assessments on permanent growth and yield plots, tree ring and stem analyses and forest inventories. Each of the three approaches has distinct advantages and disadvantages. To evaluate the effects of the three approaches on the detection and description of growth trends, they are applied to three identical areas: Switzerland, the Toppwald region in the Canton of Berne and the selection forests in the Toppwald region. The results of the case studies are discussed and the specific contribution of each approach for monitoring forest growth is highlighted. The data presented show a clear increase in growth rates in the last decades. However, these growth rates cannot be related to site productivity alone; confounding factors are very likely.

1. Introduction

Different approaches are available to describe growth patterns in forests. There is a long history of growth assessments upon permanent growth and yield plots and tree ring analyses. At the end of the last century, sampling procedures had been introduced into forestry by Zetsche (Schmidt 1891) and have since become a valuable tool for monitoring forest growth (Schmid-Haas 1983). All three approaches have distinct advantages and disadvantages, which will be discussed. The three methods were applied to three units of reference: Switzerland, the Toppwald region in the Canton of Berne and the selection forests in the Toppwald region. Based on the results obtained by the three methods which will be presented, the specific benefits of each of the approaches for the detection of changes in forest growth will be discussed. The findings of each approach have been discussed separately by Bräker, Köhl and Zingg (in this volume). This chapter attempts to summarize the findings in the light of the specific advantages and disadvantages of each approach and gives recommendations for further long-term observation studies of forest growth.

European Forest Institute Research Report No. 5
Growth Trends in European Forests
Edited by H. Spiecker, K. Mielikäinen, M. Köhl and J. P. Skovsgaard

2. Advantages and Disadvantages of the Three Approaches

Observations on permanent growth and yield plots give insight into the development of single trees in a stand and also into stand dynamics. As most long-running growth and yield experiments were not planned according to statistical considerations and experimental designs, they lack replication and have to be considered as case studies which do not allow statistical inference. The lack of replication is partly due to the cost of intensive maintenance. Rarely do they represent a large variety of forests. Information based on observations from permanent growth and yield plots is sensitive but the results can be transferred only to a very small part of forest conditions.

Extensive forest inventories cover a multitude of site conditions and forest characteristics and provide representative results for regions. To obtain information on forest growth, plots have to be remeasured on successive occasions or increment cores have to be taken. As increment boring causes growth reactions and infection by fungus, it is widely accepted that this practice should not be applied on permanent sample plots. There are some major disadvantages connected to sample surveys regarding the assessment of growth patterns. Extensive forest surveys cover areas of more than 50,000 ha and have sample plot fractions of much less than 1%. This sampling fraction is too small to enable useful conclusions for single stands (Pelz, 1980). Only in specially designed stand inventories (Lund et al. 1994), which have much higher sampling fractions, can stand dynamics, concurrence situation and growth patterns of individual stands be captured. Another shortcoming of forest surveys is the periodicity of assessments which partially applies to growth and yield plots as well. Even if an inventory has a permanent set-up with annual assessments, e.g. Austria and Sweden, only a subset of the total number of sample plots assessed in a survey cycle are visited each year. Thus individual plots are remeasured after several years. The results of the assessments during the inventory cycle, e.g. in Austria, Sweden or Finland, are aggregated and do not reflect the situation at one common point in time. In national or regional forest surveys and in forest management planning surveys, the time interval between successive occasions can be too long to detect short-term changes. Only mid- or long-term changes with large amplitudes can be detected. If the number of observations needed to conduct time series analysis (approximately 30 points in time) is compared with the time period between successive samplings in forest surveys (5 to 15 years), it becomes clear that forest survey systems have to be conducted for more than one century until enough observations in time are available to conduct time series analysis. In summarizing the advantages and disadvantages, it can be said that extensive forest surveys provide representative but not sensitive results. Stand inventories with annual assessments provide sensitive results, but - as they are generally too costly to be conducted for large areas - they lack representativity for larger regions.

Tree ring or stem analyses are ideal tools for retrospective analyses because they allow one to follow growth patterns of single trees for long periods. However, the sampling schemes and selection rules for trees where cores or discs are taken can have a significant effect on the outcome and influence of the transferability of results. As tree ring information can be taken from (all) trees in a stand or from a subset of trees selected on sample plots, the advantages and disadvantages cover a wide span ranging from sensitive to representative data.

3. Data

Details on data assessment and the applied methods can be found in the contribution of Bräker, Köhl and Zingg (in this volume) and will not be repeated. The synthesis focuses on the results presented for the whole of Switzerland and for the «Toppwald» region and is restricted to spruce (*Picea abies* [L.] Karst.), fir (*Abies alba* Mill.) and beech (*Fagus sylvatica* L.).

The results for Switzerland are based on permanent growth and yield plots and tree ring analyses and are complemented by results based on the forest management inventories of the Bernese forest service. The permanent growth and yield plots included in this study are not systematically distributed over Switzerland but cover a wide range of forest characteristics and sites. They were established with a distinct research objective in mind and are thus not representative of Swiss forests (Köhl et al. 1995). The data available cover a time period of 70 to 105 years.

The tree ring data analyzed were gathered in 1992 from a subsample of the 1984 Sanasilva forest health survey. The locations where the core samples were taken from are systematically distributed, but the sampling frame was restricted to publicly owned and accessible stands. The number of observations is large enough to give representative results within the sampling frame dating back to 1900. The inventory data cover only a subset of the geographical scale and time period. The data were collected in the Bernese Oberland and Mittelland between 1967 and 1993. As the plots have been remeasured only twice, no long-term time series is available and only changes occurring in the last two decades can be reported.

Data from the Toppwald region were gathered from the permanent growth and yield plots 01-015.001, 01-015.002 and 01-021, which cover an area of approximately 3.4 ha. Stem disks were taken from 67 trees on plots 01-015.001 and 01-015.002 (selection forest), which grew on these two plots. The permanent growth and yield plots have been monitored since 1905; the last sampling was taken in 1993. Sufficient observations dating to 1900 were available for tree ring data. The inventory data originates from the Toppwald forest district, which is part of the state forest district number 8 (KFA 8) in the canton of Berne and covers approximately 800 ha.

4. Growth Patterns in the Three Regions

The results for Switzerland are presented in Table 1. The periodic annual d.b.h. growth (mm y^{-1}) is given for three periods (1900 to 1970, 1970 to 1980 and 1980 to 1990). Inventory data are available only for parts of the canton of Berne and do not represent growth patterns for all of Switzerland. In addition, the periods covered by inventories are slightly different from the periods covered by growth and yield plots and tree ring analyses. The inventories cover the years between 1973 and 1983 and 1983 and 1993 respectively. No inventories were conducted before 1973.

Although the absolute values for d.b.h. growth differ considerably among the three data sources, the direction and size of the percentage changes between the two periods 1970-1980 and 1980-1990 are similar. Lower absolute values can be found for growth and yield plots because the tree collectives are mostly old growth stands. The other data sources cover all age classes. Spruce shows only a slight increase in d.b.h. growth; the highest values can be observed for fir. Tree ring analyses and inventory data show almost equal results for changes in beech with a higher increase for trees assessed on growth and yield plots. This might be due to the fact that most of the stands under investigation in growth and yield studies are of pure, mature timber. Inventory data and tree ring analyses cover mature timber in pure and mixed stands and at various stages of forest development. In relation to the growth of the period 1900 to 1970, a clear decrease can be observed for fir in the period 1970 to 1980. However, fir recovers in the last observed period (1980 to 1990) and shows even higher values on growth and yield plots than in the first period (1900 to 1970).

Table 1. D.b.h. growth for Switzerland

Species	Data source	d.b.h. growth 1900-1970 [mm y^{-1}]	d.b.h. growth 1970-1980 [mm y^{-1}]	d.b.h. growth 1970-1980 [% of period 1900-1970]	d.b.h. growth 1980-1990 [mm y^{-1}]	d.b.h. growth 1980-1990 [% of period 1970-1980]
Spruce	G&Y plots	2.60	2.54	98	2.67	105
	tree ring analyses	3.40	3.54	104	3.54	100
	inventory data[1]	-	3.38	-	3.43	101
Fir	G&Y plots	2.56	2.09	82	2.65	127
	tree ring analyses	4.14	2.86	69	3.48	122
	inventory data[1]	-	3.05	-	3.86	127
Beech	G&Y plots	2.21	2.89	131	3.51	121
	tree ring analyses	3.16	3.62	115	3.92	108
	inventory data[1]	-	3.70	-	4.02	109

[1]Canton Bern, Mittelland and Oberland region

Similar growth signals can be found in all data sources. A majority of conditions of Swiss forests are covered by both the inventory data and the tree ring data taken in the Sanasilva forest damage survey. Therefore, these findings can be treated as a common pattern in Swiss forests. The pattern is not only related to specific site conditions, stand structures or management regimes. Some of the changes in tree growth can be explained by climate and climatic succession. Although the overall pattern can be observed in all three data sets, the findings are subject to large local variations.

Results for the Toppwald region are summarized in Table 2. No data originating from tree ring analyses are available for this unit of reference. Pronounced increases among the three periods can be seen for spruce and beech. Fir clearly shows a decrease in the period from 1970 to 1980 with recovery in the last period (1980 to 1990). Higher absolute growth values can be found for inventory data. Compared to the results for Switzerland (Table 1), the absolute and relative values are higher for spruce, which might be due to the fact that spruce grows under close to optimal growth conditions in the Toppwald region. Beech shows a considerably higher rate of change in inventory data in the Toppwald region between the two last observation periods. This reveals specific growth conditions in the Toppwald region. Spruce and beech show a consistent increase in growth among the three periods. The decrease for fir in the years between 1970 and 1980 can clearly be seen in growth and yield plots. The level of change between the second and the last period is pronounced in the inventory data set (145%).

In Table 3, results for selection forests in the Toppwald region are shown. Pronounced differences in the absolute levels of d.b.h. growth can be found in the three data sets, with growth and yield plots showing the lowest values (with one exception) and inventory data showing the highest values. In all data sets, a consistent increase in growth can be found for spruce. The change in growth patterns between the second and the last period is lowest for the tree ring analyses and indicates no obvious change. Growth and yield plots indicate a clear growth increase between

Table 2. D.b.h. growth for Toppwald region

Species	Data source	d.b.h. growth 1900-1970 [mm y^{-1}]	d.b.h. growth 1970-1980 [mm y^{-1}]	d.b.h. growth 1970-1980 [% of period 1900-1970]	d.b.h. growth 1980-1990 [mm y^{-1}]	d.b.h. growth 1980-1990 [% of period 1970-1980]
Spruce	G&Y plots	2.57	2.98	116	3.32	111
	inventory data	-	4.02	-	4.85	121
Fir	G&Y plots	2.44	2.28	94	2.95	129
	inventory data	-	2.89	-	4.12	145
Beech	G&Y plots	2.03	2.40	118	2.85	119
	inventory data	-	2.91	-	3.74	129

the last two periods, while inventory data show a marked increase (126%). The decrease in growth of fir between 1970 and 1980 is not large, and tree ring analyses even show an increase. However, between the second and the last period, an obvious increase can be found in all data sources, with growth and yield plots and inventory data showing a marked increase (130 and 139%). Consistent increase in growth can be found in beech. Tree ring analyses show only a slight increase between the first and second period. This may be related to the fact that the stems for the tree ring analyses were not randomly selected. The discs were taken from trees that were harvested in connection with regular management operations and may thus lack representativity. All three data sources indicate increasing growth rates between the second and the last period. The highest change rates can be found for inventory data.

Table 3. D.b.h. growth for Toppwald region, selection forests

Species	Data source	d.b.h. growth 1900-1970 [mm y^{-1}]	d.b.h. growth 1970-1980 [mm y^{-1}]	d.b.h. growth 1970-1980 [% of period 1900-1970]	d.b.h. growth 1980-1990 [mm y^{-1}]	d.b.h. growth 1980-1990 [% of period 1970-1980]
Spruce	G&Y plots	2.56	3.05	119	3.40	111
	tree ring analyses	2.82	3.98	141	4.08	103
	inventory data	-	4.68	-	5.92	126
Fir	G&Y plots	2.58	2.55	99	3.33	130
	tree ring analyses	3.32	3.86	116	4.56	118
	inventory data	-	3.34	-	4.63	139
Beech	G&Y plots	2.03	2.40	118	2.85	119
	tree ring analyses	2.86	2.96	104	3.80	128
	inventory data	-	3.33	-	4.37	131

5. Discussion

A comparison of the results from growth and yield plots, tree ring analyses and inventory data reveals similar growth patterns for Swiss forests. Spruce, fir and beech show an increase in d.b.h. growth in recent decades. An increase in d.b.h. growth can be observed for spruce in the 1960s, beech shows an increase since the 1970s and fir is obviously growing better since the early 1980s after a decrease between 1970 and 1980.

Growth patterns observed in Swiss forests are similar to those found in the Toppwald region. Spruce and beech consistently grew better and had higher growth increases between the second and the last period than between the first and second period. A decrease in fir growth can be found between 1970 and 1980 on growth and yield plots. In the selection forest of the Toppwald region, beech and fir grew considerably better as time progressed. A slight decrease for fir between 1970 and 1980 can only be observed on growth and yield plots. The growth decrease that can be observed for fir in Switzerland is much less pronounced in the Toppwald region. However, the increase in growth between the second and last period is higher than the one observed in Switzerland. Considering only data from growth and yield plots and inventory data, the findings for fir reveal that they are growing on sites close to optimal growth conditions in the Toppwald region.

In growth and yield tables, decreasing d.b.h. growth is presented for higher age classes. This growth pattern can only partially be observed in the data sets used in this study. If the growth pattern indicated by growth and yield tables is taken as a reference, above-average d.b.h. growth can be assumed for recent years. It is increasingly recognized that only in some cases can growth and yield tables be a suitable tool to describe growth (Köhl et al. 1995). Long term growth patterns vary considerably over time and it is not always clear in which phase of the long-term pattern the data used to construct growth and yield tables were assessed. They may originate from "low-growth" periods and thus positive growth trends may be overestimated.

One of the major objectives of EFI's project was to verify changes in site productivity. The data presented show a clear increase in growth rates in the last decades. However, these growth rates cannot be related to site productivity alone; confounding factors are very likely. The climatic situation in recent years, changes in forest management practices, the abandonment of grazing and littering with resulting fertilization effects might lead to positive growth changes. Based on the available data, no single factor can clearly be identified as being responsible for growth increases.

To summarize the findings for growth changes in Swiss forests, it has to be emphasized that the data available for the analyses of large-scale, long-term growth patterns are still limited. Data have been available for approximately 100 years. The time resolution is not sufficient to explain the mid-term variation in growth patterns by time series analysis. Beside the resolution in time, the spatial resolution has to be considered carefully. As Swiss forests are characterized by small-scale heterogeneity in both site conditions as well as stand structures, the spatial variation is confounded by a multitude of factors that affect growth. It is doubtful whether the spatial and temporal resolution of the available data and the knowledge about the interaction of confounding factors and their relative weight are sufficient to develop models that take into consideration both growth patterns and potential cause-effect relationships. It is extremely difficult to model the high adaptation rate of trees to changing environmental and stand conditions. A decrease in growth is very likely to be compensated for in succeeding years.

The growth patterns in Swiss forests are subject to large local variation. The arbitrary (non random) selection of test sites can lead to contradictory results. Under the heterogeneous forest conditions which can be found in Switzerland, permanent monitoring plots will hardly allow representative conclusions for all forests and sites. Quite to the contrary, they cover only a very limited scope of specific site, stand, climatic and genetic conditions. Due to the lack of repetition (under similar constraints) permanent monitoring plots must therefore be regarded as case studies. However, they have the advantage of providing information on stand dynamics, which cannot be obtained by (representative) forest surveys. Tree ring and stem analyses are a suitable tool for retrospective analyses of single tree growth, but these are rather costly. If cores are taken in sample based surveys by a sound statistical selection rule, long-term growth patterns for large areas and various forest conditions can be described in a representative and reliable way. If methods are provided to combine the three approaches utilizing their specific advantages, they will become a powerful tool for monitoring forest growth on large representative areas and on long-term, representative time frames.

Acknowledgments

We thank Dr. Anton Bürgi, Swiss Federal Institute for Forest, Snow and Landscape Research, Birmensdorf, Switzerland, Dr. Francis A. Roesch, Champion International Corporation, Jacksonville, FL, USA, and a third, anonymous reviewer for carefully reading the manuscript and for their critical but rewarding comments. We highly appreciate their efforts and their advice, which facilitated the revision of our manuscript. We are grateful to Christian Matter, Swiss Federal Institute for Forest, Snow and Landscape Research, Birmensdorf, Switzerland, for language corrections. The Cantonal Forest Service of Berne generously made its inventory data available for this study. Dr. Janez Grilc facilitated our work by compiling the inventory data and giving helpful comments on the use of the data sets.

References

Bräker OU (in this volume) Growth trends of Swiss Forests: Tree Ring Data. Case Study Toppwald

Köhl M (in this volume) Growth patterns in forests of the canton of Berne based on inventory data

Köhl M, Scott CT, Zingg A (1995) Evaluation of permanent sample surveys for growth and yield studies. Forest Ecology and Management, Vol. 71 (3): 187-194

Lund G, Landis E, Atterbury T (1994) Stand Inventory Technologies, World Forestry Center, Oregon, USA.

Pelz DR (1980) Stichproben in kleinflächigen Beständen, Forstw. Cbl 90, pp 55-60

Schmid-Haas P (1983) Swiss continuous forest inventory. Twenty years' experience. In. Bell, J.F., T. Atterbury (eds.): Renewable resource inventories for monitoring changes and trends. Corvallis, OSU, College of Forestry, pp 389-393

Schmidt K (1891) Das "Kreisprobeflächen-Aufnahmeverfahren" des Herrn Oberforstrath Zetsche, Allg. Forst-u. Jagdztg. 67 (11): 73-76

Zingg A (in this volume) Diameter and Basal Area Increment in Permanent Growth and Yield Plots in Switzerland

Growth Trends in Austria

K. Schadauer
Federal Forest Research Centre
Vienna, Austria

Abstract

Different increment parameters (radial increment, top height increment and current annual increment (CAI) per hectar for volume and for basal area), which were derived from data sampled by the Austrian Forest Inventory are presented for different time spans. Special emphasis is laid on the comparability of inventory results from different time periods for basal area increment. Since 1961 inventory results show an increase in CAI per hectare for basal area of about 24% with the maximum increase from the 1970s to 1980s. Long increment core data reveal a long-term trend in radial increment for the last 100 years. Some causes like climatic changes, increasing nitrogen supply, and CO_2 are discussed, but no clear cause relationships can be presented.

1. Introduction

1.1. Forest in Austria

Austria, lying in the centre of Europe, has about 3.8 million ha of forest land (46% of the whole area). The Austrian forest belongs to 214 000 owners, beginning with a size from one to 600 000 hectares owned by the state. Forest in Austria provides income for 125 000 people, who produce 3.7% of the gross national product; 20% of the forest area has protection character in higher elevations of the Alps up to 2300 m. The most important tree species are Norway spruce (*Picea abies*, 56%) and beech (*Fagus sylvatica*) covering 9% of the forest area.

1.2. Aims

The main objective is to detect growth trends and to differentiate between trends and medium frequency changes in growth. Four increment parameters (radial incre-

European Forest Institute Research Report No. 5
Growth Trends in European Forests
Edited by H. Spiecker, K. Mielikäinen, M. Köhl and J. P. Skovsgaard

ment, top height increment, CAI/ha for volume and basal area), which were all derived from data sampled by the Austrian Forest Inventory, are used in order to reach this aim. A main aspect is to check comparability of the results of different inventory periods.

2. Data and Methods

2.1. Inventory Sampling Design

The first forest inventory on a systematic grid started in 1961 with an assessment scheme continuous to the present. The Austrian Inventories have always used a cluster sampling design (tract system). One cluster is in the form of a square. Before 1971 we used different numbers of satellites; since 1971 there have been four plots for tree sampling at the corners of each tract with a distance of 200 m. The first two inventories had a period length of 10 years and a temporary design (Table 1). From 1981 to 1985 permanent tracts were established and the period length was reduced to 5 years. Currently we are carrying out the second reassessment of the permanent plots.

Table 1. Austrian Forest Inventories from 1961 to 1990

period	design	radial increment
1961 - 1970	temporary plots	increment bored
1971 - 1980	temporary plots	increment bored
1981 - 1985	permanent plots	
		2 DBH measures
1986 - 1990	permanent plots	

2.2. Inventory Measurements and Increment Derivation

The Austrian Forest Inventory has always used horizontal point sampling with basal area factor 4, the minimum DBH was 10.5 cm until 1980 and 5 cm from that time on. DBH was measured with caliper in one direction downhill. Tree height was measured with the Blume-Leiss until 1980 and from that time on with the Mirror Relascope, which was also used to measure the diameter in 3/10 of the height (D03) and for angle count sampling. The volume functions include DBH, D03 and the height of the trees.The methods of increment estimation changed over time. In the first two temporary assessment periods radial increment of 5 years was assessed by boring. Height increment for five year periods for conifers was estimated, in the field using binoculars for the same time span, but height increment for broad leaves was neglected. Since 1981, increment estimation has changed. On the permanent plots, DBH, D03 and height were measured twice. On the second, occasion D03 and height were measured only on a subsample and the missing values were replaced by regression model values derived from the measured subsample (Knieling 1994).

In the next section, results for CAI for volume and for basal area will be given. The following study of comparability emphasizes the results for basal area increment as a first step. The influence of the models for D03 and height for the comparison of volume increment will not be studied in this chapter. Nevertheless, the cited formulas will use the term volume, so that original notation is used. Table 2 gives the number of sample trees used for the calculations of CAI for basal area in the three inventory periods.

To prove comparability, the different methods of increment estimation have to be analysed carefully. As described above, we had two periods with different increment measures and estimates: until 1980 and afterwards. The first step towards comparability is to assure the constancy of growth definition over time. According to Beers (1962), the Austrian Inventory always used Gross growth including ingrowth:

$$V_2 + M + C - V_1 = G_s + I, \qquad (1)$$

V_1 is the volume of trees measured at the first inventory,
V_2 is the volume of trees measured at the second inventory,
M is the initial volume of trees dying during the period between inventories,
C is the initial volume of trees which were cut during the period between inventories,
I is the volume of trees at the second inventory which were below merchantile size at the first inventory.

The inventories used the right side of (1) but the estimates for G_s and I changed over time. In the following formulas four categories of trees concerning successive horizontal point sampling are differentiated. They are defined according to Martin (1982):

s is "in" and above minimum DBH at both measurements (survivor trees),
i is "in" at both measurements and below minimum DBH at the first measurement and above minimum DBH at the second measurement (ingrowth trees),
o is "out" and below minimum DBH at the first measurement and "in" and above minimum DBH at the second measurement (ongrowth trees),
n is "out" and above minimum DBH at the first measurement and "in" at the second measurement (nongrowth trees).

Table 2. Number of sample trees for the calculation of CIA/ha for basal area for the different inventory periods

Period	1961/70	1971/80	1981/90
Number of trees	239 631	106 623	54 071

Until 1980 the temporary design with increment boring led to an estimate for survivor growth which was published later on by Roesch et al. (1989):

$$S^{**} = s_2 - s_1' + n_2 - n_1' , \tag{2}$$

s_2 is volume obtained from the final volumes and the final basal areas of survivor trees,
s_1' is volume obtained from the initial volumes and the final basal areas of survivor trees,
n_2 is volume obtained from the final volumes and the final basal areas of nongrowth trees,
n_1' is volume obtained from the initial volumes and the final basal areas of nongrowth trees,

combined with an ingrowth estimate near to the so-called Purdue Estimator (Martin 1982):

$$I_p = i + o , \tag{3}$$

i is volume obtained from the final volumes and the final basal areas of ingrowth trees,
o is volume obtained from the final volumes and the final basal areas of ongrowth trees.

The Purdue Estimator was reduced to the 5 year increment of i and o. For the survivor growth (2) bark increment was not taken into account, according to increment boring.

The described method had the advantage that for basal area only increment measured values were used. But there are two underestimating factors: the omission of bark increment and ΔI with

$$\Delta I = I_v - I_5 , \tag{4}$$

I_v is total volume of ingrowth trees,
I_5 is 5 years increment of ingrowth trees.

Since 1981, a permanent system has been established. Therefore increment boring was replaced by two measures of DBH. The minimum DBH was reduced to 5 cm, but the stems between 5 and 10.4 cm were sampled on a fixed area plot R = 2.6 m, which is the plot limit using horizontal plot sampling with basal area factor 4 for a tree with 10.4 cm. Of course, for the comparison of CAI over time, minimum DBH was constant at 10.5 cm, but the measurements of the smaller trees under 10.5 cm on the first occasion gave information to separate ingrowth trees from ongrowth and nongrowth trees without any modelling for DBH on the first occasion. This led to the use of the so-called Furnival Estimator for ingrowth:

$$I_f = i', \tag{5}$$

i' is volume obtained from the final volumes and the initial basal areas of ingrowth trees.

Survivor growth was estimated according to Grosenbough (1958; cf. Martin 1982):

$$\hat{S} = s_2' - s_1 , \tag{6}$$

s_2' is volume obtained from the final volumes and the initial basal areas of survivor trees,

s_1 is volume obtained from the initial volumes and the initial basal areas of survivor trees.

This method uses only initial basal area, whereas the method until 1980 used the basal area of the second occasion.

Three steps can be taken to prove comparability:

1. The underestimating factors (bark increment and DI) for the method used until 1980 have to be regarded and basal area increment has to be recalculated.
2. The two different increment measures have to be compared (boring versus two DBH measures).
3. The different estimates for survivor growth and for ingrowth should be compared empirically.

To accomplish the first point, bark functions for the different tree species were calculated using all sample trees in the inventory 1971/80.

The empirical comparisons referred to in point (2) need special investigations. A fifth additional sample plot was therefore established on each permanent tract of the inventory 1981/85. DBH was measured on both occasions and all sample trees were bored on the second occasion, 1986/90. After this, the plots were abandoned. We have 6831 sample trees to carry out the comparison. The results are shown in Table 3 (Schieler, personal communication). The difference is small, but highly significant due to the large size of the sample. Assuming that the increment measurements us-

Table 3. Result of the t-Test for paired samples for wood increment of two successive DBH measures versus increment cores

Type of increment mesure	mean (cm)	2 - tailed Sig
two DBH measures	1.66	
increment cores	1.63	
Paired differences	0.03	0.009

ing cores were biased with -0.03 cm, the CAI for basal area increment would be underestimated by 2%. The third suggestion to prove comparability has not yet been completed, but we can say that survivor growth S** (2) as well as $\hat{s}$ (6) are unbiased estimates (Roesch et al. 1989; Martin 1982). This may also be stated for the two ingrowth estimates (3) and (5), because there is no possibility of trees in Austria reaching DBH of 10.5 cm within 5 years starting with a tree height below 1.3 m. This fact would bias the Furnival Estimator for ingrowth (Martin 1982). Of course, there is some uncertainty because of the missing empirical comparisons.

2.3. Data and Methods of Special Investigations Concerning Top Height and Long Increment Cores

Between 1971 and 1990 top height was estimated for the main tree species in even-aged and (nearly) pure stands as the mean height of the three largest stems of the angle count sample (Pollanschütz 1973). The distribution of the top height plots according to age is given in Table 4. The age was assessed in 5-year classes. Top height increment was calculated by averaging the height increment for the last 5 years of the top height stems. As described above, height increment measure changed over time. Field estimation with binoculars was replaced after 1980 by successive measuremens of tree height. As for DBH increment, both methods were carried out on the fifth plot of each tract. The result of the comparison (4283 spruces) is shown in Table 5. The difference for the 5-year height increment is significant, even if only 6 cm.

Long increment cores, which hit the pith or were close to it, had been taken on two occasions, in 1977 (Sample A) and between 1986 and 1990 (Sample B) (Neumann and Schadauer 1995). Table 6 shows the distribution of the sample trees according to age class for Norway spruce (*Picea abies*). A single core was taken at breast height and on the right side according to downhill measurement from each sample tree, which was chosen out of the inventory sample (normally one per plot) out of dominant or predominant trees, avoiding trees with heavy mechanical damage. Sample B includes data for the whole country, whereas for sample A the data of the eastern parts are not available. In sample A about 60% of the cores met the center, while for sample B this fact was not recorded. The measurements were car-

Table 4. Distribution of top height plots according to age for the inventories 1971/80 and 1981/90 (grouped in 20 years classes)

	period	
age	1971/80	1981/90
15 - 20	516	94
21 - 40	871	305
41 - 60	785	264
61 - 80	849	274
81 -100	526	220

Table 5. Result of the t-Test for paired samples for five year height increment values of two successive height measures versus binocular estimate

Type of increment measure	mean (cm)	2-tailed Sig.
two height measures	135.6	
binocular estimates	129.8	
Paired differences	5.8	0.001

Table 6. Number of increment cores according to age classes of sample A and B

	sample	
age class	a	b
- 20	58	-
- 40	118	97
- 60	176	108
- 80	196	110
-100	156	96
-120	99	67
>120	201	92

ried out on the Ecklund measuring equipment (sample A) and on the Digitalpositiometer (Johann 1977; sample B). The series were cross-checked in order to reveal measurement errors or missing tree rings.

The evaluations can be divided into two sections. The first part deals with analysis methods comparable to Becker's (1989), where a special form of age detrending leads to the detection of long-term trends. Becker's method of age detrending is rather simple. He constructs one general mean curve of annual radial growth according to the current ring age (i.e. cambial age) for all ages and sites together. This mean curve for annual radial growth for Norway spruce in Austria is given in Figure 1. After that, the deviations from the general curve are calculated for age detrending. The classical method of using separate curves for each core is not useful in this case, because it may remove possible time trends.

The second part deals with analyses of mean tree ring width over time without detrending (Neumann and Schadauer 1995). The mean tree ring width is synchronised over cambial age for different age classes. This comparison of annual radial growth for different age classes representing different time periods of growth enables the detection of time trends.

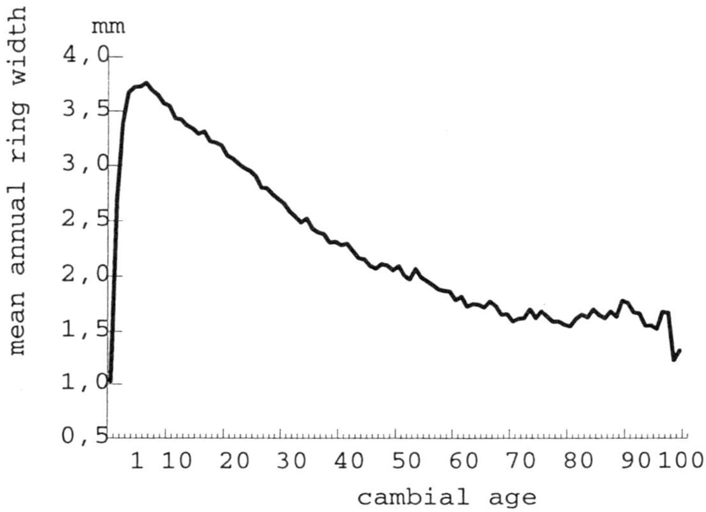

Fig. 1. Mean curve for annual radial growth for Norway spruce

3. Results

3.1. Inventory Results

The results of the mean CAI for volume (minimum DBH 10.5 cm) show a clear increase, especially since 1981 (Table 7). Equally, the values of the growing stock and of the stems ha^{-1} show an increase with time. The stable stem numbers per ha of the last period is a hint for intensified thinning activities. The volume harvested accounts only for two thirds of total increment.

Volume increment is very important when dealing with economic topics, but it is burdened with several assumptions (model for form factor, height growth and D03). Therefore the development of DBH increment and top height increment gives better evidence for ecological questions like changes in site productivity. So the mean

Table 7. Main results for three periods of the forest inventory (minimum DBH 10.5 cm)

	1961/70	1971/80	1986/90
CIA ha^{-1} for volume	5.8 m^3	6.3 m^3	8.7 m^3
growing stock ha^{-1}	239 m^3	262 m^3	288 m^3
stems ha^{-1}	577	599	602
ann. Cutting ha^{-1}	–	–	5.6 m^3

CAI for basal area per hectare was calculated for the different inventory periods with regard to the discussed underestimating factors until 1980 (Fig. 2). The results show the same trend as for volume increment: initially only a small increase from the 1960s to the 1970s and then a large increase in the following decade.

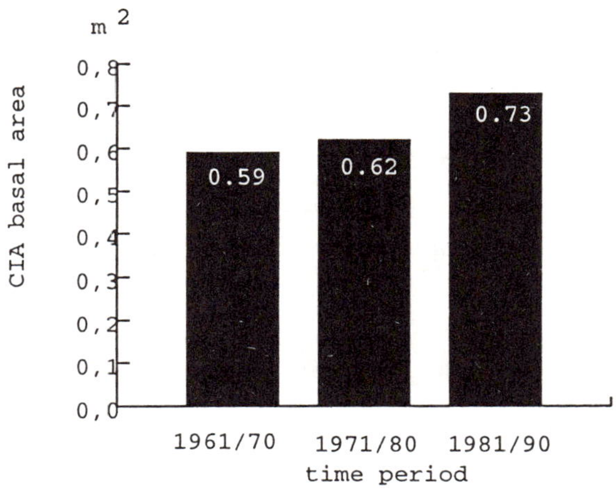

Fig. 2. Mean CAI/ha for basal area for three periods of the forest inventory

3.2. Top Height Increment

For the development of top height increment, only two inventory periods are available. Figure 3 shows the mean top height increment for Norway spruce (*Picea abies*) for the periods 1971/80 and 1981/90 over age. The mean increments were calculated for each 5-year age classes separately. The lines between the 5-year steps are only for better visualisation. Top height increment also increased in the 1980s. The largest differences occur in young stands aged up to 40 years. If we assume an

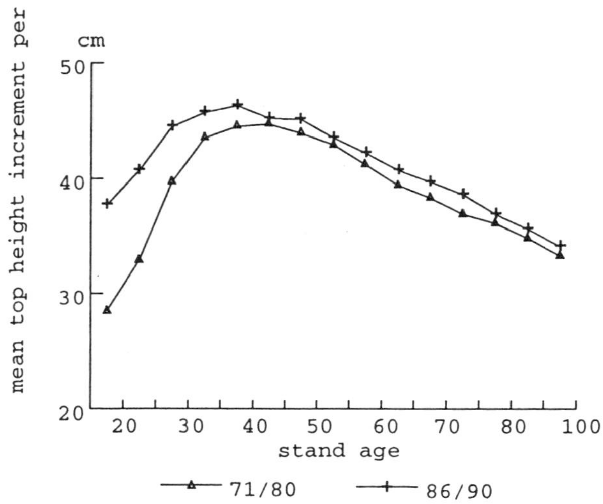

Fig. 3. Comparison of top height increment for two periods (1971/80 - 1981/90) calculated for Norway spruce

underestimation of the values for 1971/80 for 1.2 cm for the annual height increment (5.8 cm in 5 years, see sect. 2.3), the difference for the older stands becomes rather small.

3.3. Radial Increment

The results from the long increment cores using Becker's method show a continuous increase in the deviations from the mean curve of annual radial growth from 60% to 120% (Fig. 4). The results are similar to the ones found by Becker in the Vosges Mountains for silver fir (*Abies alba*).

Using the method of Neumann and Schadauer (1995) the general results are quite the same. Radial increment has increased steadily, at least since the end of the last century. Figure 5 shows the mean annual ring width over age for different age classes for sample A. The differences in juvenile growth of different time periods are very clear. The juvenile growth of the youngest age class (representing the time period from 1958-1977) increases steeply and later decreases equally fast. The juvenile growth for former time periods (corresponding to higher age classes) becomes flatter in a very regular way. A second remarkable point is that the annual ring width of the old trees does not show a "normal ageing trend". The values remain at the same level for up to 100 years and more.

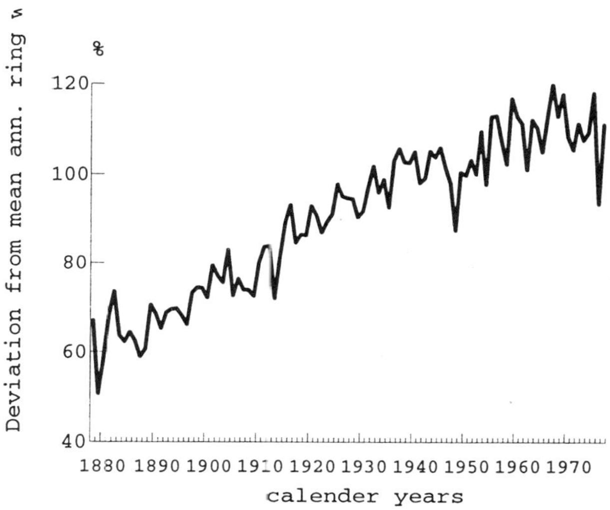

Fig. 4. Average deviation from the mean curve for annual radial growth for Norway spruce

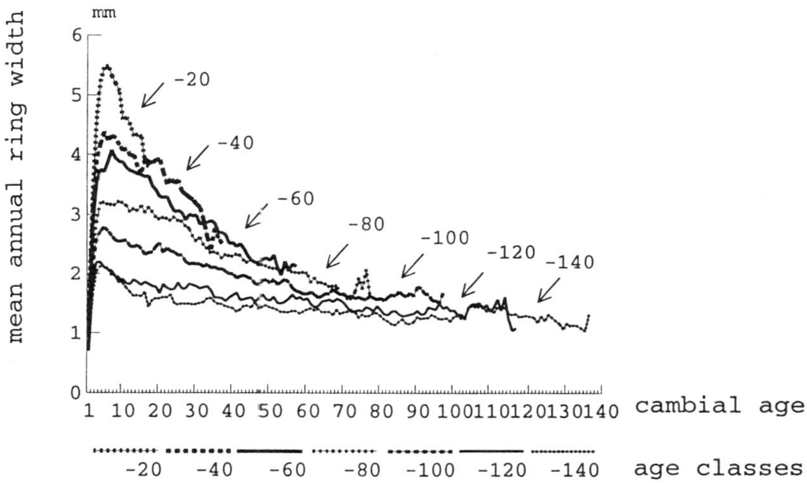

Fig. 5. Mean annual ring width versus tree age for Norway spruce of sample A stratified by age classes

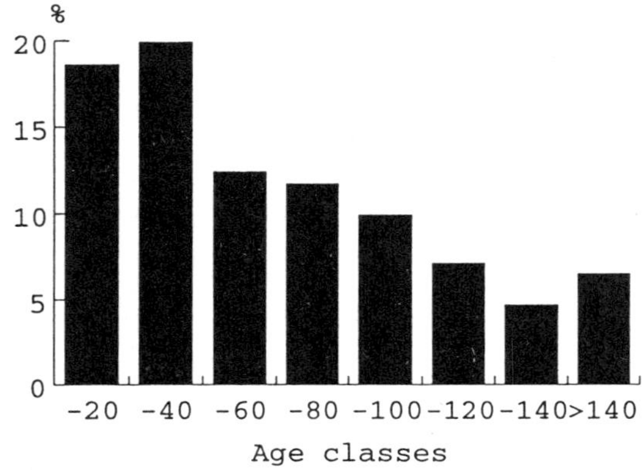

Fig. 6. Age class distribution according to the forest inventory 1986/90

4. Interpretation

4.1. Age Class Distribution and Afforestations

The age class distribution of the Austrian forests (Fig. 6) shows the highest values for the young stands up to 40 years old. This maximum is caused by afforestations of marginal agricultural sites and by the reestablishement of stands exploited after World War II. The age classes from 20 to 60 years produce 50% of the volume increment. So on the one hand, many stands reached a very productive age class in the 1980s, and on the other, many of them stocked on more productive sites than the average Austrian forest.

4.2. Climatic Changes on Medium Frequency Level

The mean radial increment for Norway spruce over calendar year for sample B (Fig. 7) shows a depression in the 1970s with a minimum in 1976 (very dry year). According to Böhm (1992), the 1970s was the driest period of the century in Austria. Therefore the large increase in mean CAI for basal area per hectar from the 1970s with 0.62 m^2 to 0.73 m^2 in the 1980s is probably influenced by changes in weather conditions. This becomes even more evident if this increase in basal area increment is stratified according to soil moisture (Fig. 8). The moderately fresh group shows the largest relative increase compared to the fresh and very fresh sites for Norway spruce older than 40 years (in order to exclude the effects of afforestations, see above).

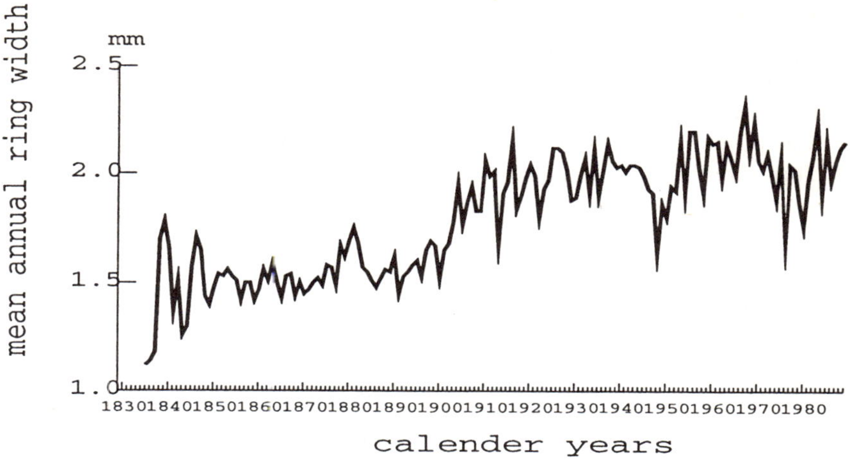

Fig. 7. Mean annual ring width versus calendar year for Norway spruce from sample B

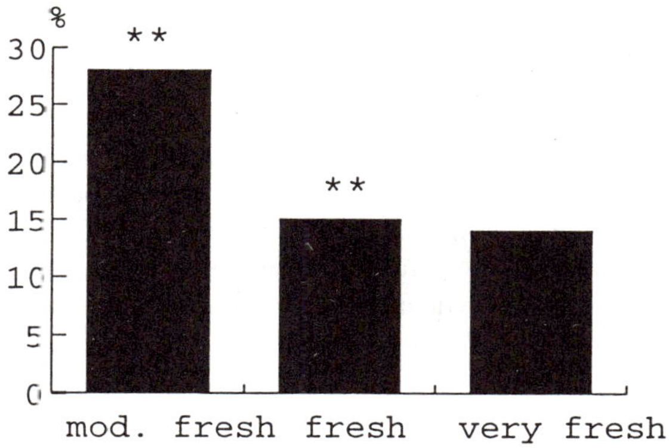

Fig. 8. Relative increase in CAI per hectar for basal area for different soil moistures (** significant for 95% level)

4.3. Long-Term Trend in Radial Increment

The long-term trend showing nearly double tree ring width since the end of the last century can only be explained by trends of environmental conditions lasting for at least the same time span.

The attempt to find correlations between radial growth indices (see Fig. 4) and climatic variables (Becker 1989) using mean monthly temperature and water balance (monthly precipitation minus potential evapotranspiration) was not successful. There has been an increase in mean temperature of about 1°C (Böhm 1992) since the middle of the last century for Austria as a whole but precipitation showed no trend. Therefore the multiple linear regression shows some correlations with temperature and none with the water balance.

4.4. The Role of Changes in Silvicultural Treatment, Nitrogen Supply and CO_2

The changes in silvicultural methods (decreasing number of plants for the establishment of stands and more intensive thinning activities) over the last 30 years are likely to have an influence, especially on radial growth of juvenile single trees; but changes in silvicultural methods, continuously effecting radial growth of single trees and lasting for 100 years at least, are not recorded for Austria.

Increasing nitrogen supply caused by soil regeneration, after litter raking or by higher atmospheric nitrogen impacts, probably influences tree growth. Some stratifications for changes in basal area increment as well as in top height increment show larger increases on sites with lower nitrogen supply than on sites with better supply, but the results are in some cases contradictory.

The recorded radial growth trend might be influenced by increasing CO_2 levels in the atmosphere, because the latter would affect all sites for a long time, but the results cannot give any evidence for this hypothesis.

The influence of a large number of effects on increment cannot be distinguished by using monocausal statistic procedures. The data used in this chapter may be used for more sophisticated statistical analyses, but the main problem is the lack of information about exact quantities of the influencing parameters in forest over large areas and the quantifications of the effects of these quantities, including the intercorrelations.

References

Becker M (1989) The role of climate on present and past vitality of Silver fir in the Vosges mountains in Northeastern France. Can J of Forest Research 19, 9: 1110 - 1117

Beers TW (1962) Components of Forest Growth. Journal of Forestry 60 (4): 245-248.

Böhm R (1992) Lufttemperaturschwankungen in Österreich seit 1775. Österreichische Beiträge zu Meteorologie und Geophysik Heft 5, Zentralanstalt für Meteorologie und Geodynamik, Wien

Grosenbaugh LR (1958) Point Sampling and Line Samplimg: Probability Theory, Geometric Implication, Synthesis. Ocas Pap 160, South For. Exp. Stn., USDA Forest Service

Johann K (1977) Eine neue Jahrringmessanlage für Bohrkerne und Stammscheiben. Forstarchiv, 48 Jg, Heft 10: 204-206

Knieling A (1994) Methodische Beiträge zur Auswertung der Österreichischen Forstinventur nach 1980. Dissertation Universität für Bodenkultur, Wien, p 129

Martin GL (1982) A Method for Estimating Ingrowth on Horizontal Sample Points. For Sci, 28 (1): 110-114

Neumann M, Schadauer K (1995) Die Entwicklung des Zuwachses in Österreich an Hand von Bohrkernanalysen. Allgemeine Forst und Jagdzeitung, 166 Jg., H. 12, Frankfurt

Pollanschütz J (1973) Auswertung von Waldinventuren. In: 100 Jahre Universität für Bodenkultur Wien, Bd IV, Teil 1: 355-368

Roesch FA, Green EJ, Scott CT (1989) New Compatible Estimators for Survivor Growth and Ingrowth from Remeasured Horozontal Point Samples. For Sci, 35: 281-293

Volume and Height Growth of Fully Stocked Mature Beech Stands in Slovenia During the Past Three Decades

Marijan Kotar
Biotechnical Faculty
Jamnikarjeva 101
Ljubljana, Slovenia

Abstract

This chapter presents results of a study conducted in beech stands on 18 different site units, with stand age ranging from 94 to 196 years. Over 4000 stem analyses were made and more than 32 000 tree discs were examined. The stands were naturally regenerated in a shelterwood system and may be considered more or less uneven-aged. The structure of the stands is quite uniform since the crowns of most trees are in the upper stand canopy. Trees of the understory stratum have virtually all died away.

A comparison of current annual volume increment (CAI), height and diameter growth with yield tables reveals extremely positive deviations. Furthermore, top height of currently younger stands is, at the same age, higher than top height of currently older stands on the same site unit. However, no clear results concerning site productivity changes can be presented because of methodological problems. Yield tables do not represent sites and silvicultural conditions of the analysed stands in detail and the comparison of growth data at the same age of trees which have regenerated at different periods was hindered on many site units because their age did not differ enough.

1. Introduction

Results of forest surveys, conducted for forest management planning in Slovenia since 1947, show that in commercial forests current annual volume increment (CAI) increased from 3.1 m^3 ha^{-1} y^{-1} in 1947 to 5.3 m^3 ha^{-1} y^{-1} in 1990. This increase may be attributed especially to a higher standing volume per hectare and possibly also to changes in site conditions.

High values of current volume increment determined by increment analyses of some older beech and Norway spruce suggest possible changes in site productivity

European Forest Institute Research Report No. 5
Growth Trends in European Forests
Edited by H. Spiecker, K. Mielikäinen, M. Köhl and J. P. Skovsgaard

(Kotar 1994). Studies conducted in some European countries in recent years also report changes in site productivity. For example, a Swedish study showed that younger stands have higher site indices than older stands under the same site conditions (Elfving et al. 1995). Similarly, several German studies report increasing height growth and volume increment (e.g. Pretzsch 1995). Maximum values greatly exceed values given in yield tables. An analysis of height growth in two beech forests in Denmark revealed an increase in site productivity between two generations of about 3.6 m at a reference age of 100 years (Skovsgaard and Henriksen 1995).

In recent years, an extensive investigation of mature beech forests in Slovenia was started to determine site productivity and to investigate the applicability of site indices as indicators of site productivity. Uneven-aged stands, older than 90 years and fully stocked, in which thinnings and other tending tasks were only exceptionally carried out, were analysed for that purpose. The data obtained may also be used to verify the assumption that forest site productivity has changed over the past decades.

2. Research Obejctives and Subject of Investigation

2.1. Research Objectives

The objectives of the study are:

- to determine whether a trend of current annual volume increment (CAI) has occurred over the past three decades to such an extent that a change in forest site productivity can be concluded,
- to determine whether top height development in younger stands is greater than it was in older stands on the same site unit,
- to determine whether height growth of top height trees differs from data given in yield tables and
- to determine whether radial increment of top height trees indicates any possible changes in forest site productivity.

2.2. Research Subject and Data

The study was conducted on sites with natural plant communities belonging to beech and beech-Norway spruce syntaxon units, covered with beech stands (Table 1). Stand age ranges from 94 to 196 years (Table 2). At these ages stands are at their optimum developmental stage for the initiation of natural regeneration and height growth is usually very small. Most understory trees have already died by this time due to lack of light, resulting in a stand with a homogeneous height structure. In most of the analysed stands the coefficient of variation (CV%) in height was less than 10%. In this developmental stage of a stand self-thinning due to increase in

Table 1. Site units and locations

no.	site unit	parent material	latitude	longitude
1	Querco-Luzulo- Fagetum	eoc. flysch	45°30'	14°23'
2	Adenostylo -Fagetum - I	dolomite	45°34'	14°21'
3	Luzulo-Abieti- Fagetum prealp.	flint keratophyre	46°34'	14°44'
4	Luzulo-Fagetum	sandstone	46°17'	15°42'
5	Abieti-Fagetum din maian	limestone dolomite	45°40'	14°23'
6	Querco-Fagetum	limestone	45°47'	14°56'
7	Hacquetio-Fagetum	dolomite	45°41'	15°10'
8	Festuco drymeiae- Fagetum	sandstone	46°14'	15°44'
9	Abieti-Fagetum din.typ.	dolomite	45°40'	14°39'
10	Blechno-Fagetum	slate and sandstone	46°04'	14°53'
11	Lamio orvalae- Fagetum I.	dolomite and limestone	46°04'	13°58'
12	Luzulo niveae Fagetum	limestone	46°21'	13°15'
13	Anemone-Fagetum	limestone	46°23'	13°53'
14	Seslerio-Fagetum	limestone	45°31'	14°10'
15	Lamio orvalae- Fagetum II.	limestone	46°23'	15°00'
16	Enneaphyllo- Fagetum I.	dolomite and limestone	45°39'	15°05'
17	Enneaphyllo- Fagetum II.	marlaceous limestone	45°45'	15°18'
18	Adenostylo- Fagetum II.	limestone	45°34'	14°29'

crown size has rarely been observed. Though the structure of such forests seems to be homogeneous, these stands are more or less uneven-aged since they developed from natural regeneration over regeneration periods of different length. They were regenerated within a shelterwood system and therefore the difference in tree age in the same stand ranges from a few years to four decades or even more in exceptional cases. In most of the stands, no thinnings or other tending measures were undertaken. If so, they were confined to the understory. Therefore, the stands are fully stocked and have very high standing volumes per hectare (up to 1000 m^3 ha^{-1}, Table 2).

3. Methods

Two approaches for detecting growth trends were applied. The first is a yield table comparison. Yield table data are used as reference growth data. When the applicability of different yield tables available in Central Europe was tested, the tables of Halaj (1987) for Slovenia proved to be the most suitable (Kotar, 1987). Several measures of growth as CAI, height and diameter growth of top height trees compared with yield table data considering site indices, stand age and age of top height

Table 2. General characteristics and data of sample plots.

site unit no.	plot no.	elevation	aspect	mean stand age	standing volume	SI_{100}	CAI (0-10) (yield table)	site productivity (SI_{100})	CAI (0-10)	CAI (11-20)	CAI (21-30)
		m.asl.		year	m^3/ha		$m^3\ ha^{-1}\ year^{-1}$	$m^3\ ha^{-1}\ year^{-1}$	$m^3\ ha^{-1}\ year^{-1}$	$m^3\ ha^{-1}\ year^{-1}$	$m^3\ ha^{-1}\ year^{-1}$
1	1	670	W	131	773	30	7.5	8.8	14.0	14.7	16.7
	2	680	W	134	804	30	7.5	8.8	13.4	14.1	15.0
	3	640	N	134	761	28	7.1	8.0	16.6	16.2	14.0
	4	650	SW	133	593	26	6.7	7.4	12.7	9.6	8.7
	5	650	S	129	645	26	6.9	7.4	13.1	10.9	9.8
2	1	1240	S	143	764	20	5.6	5.4	11.3	10.6	11.2
	2	1230	SE	143	628	20	5.6	5.4	9.0	8.2	9.1
	3	1240	SE	143	537	20	5.6	5.4	9.3	9.1	9.4
	4	1250	SE	154	596	20	5.2	5.4	9.9	8.9	8.6
	5	1265	SW	146	571	18	5.2	4.8	9.1	8.0	8.2
3	1	1040	NW	160	913	28	6.2	8.1	10.2	11.4	13.2
	2	960	NW	146	904	30	7.0	8.9	12.0	15.1	16.9
	3	920	NW	133	901	34	8.3	10.4	11.7	15.2	18.9
	4	880	NW	123	733	32	8.3	9.7	10.0	12.1	13.9
	5	900	NW	145	796	32	7.4	9.7	9.6	12.6	16.1
4	1	530	N	107	1002	36	9.8	11.2	18.8	19.6	15.9
	2	560	NE	110	891	36	9.8	11.2	15.2	18.7	17.2
	3	540	NE	98	754	38	10.8	12.2	18.1	21.3	21.2
	4	550	NE	107	704	32	8.9	9.7	14.3	18.0	19.8
	5	600	NE	99	576	32	9.4	9.7	16.8	20.9	16.6
5	1	980	SE	129	499	26	6.9	7.4	10.7	10.3	10.1
	2	980	SE	129	614	28	7.3	8.1	11.7	11.9	11.0
	3	1020	NE	137	559	26	6.6	7.4	10.9	12.1	13.0
	4	1010	W	135	542	26	6.7	7.4	11.9	10.4	9.3
	5	1020	SW	131	490	26	6.7	7.4	9.4	10.2	9.9
6	1	510	NW	161	804	28	6.2	7.2	9.2	10.0	10.0
	2	520	NW	157	768	28	6.2	7.2	9.3	11.0	11.8
	3	530	NW	146	961	32	7.3	8.7	11.8	12.8	14.0
	4	540	NW	160	895	28	6.2	7.2	9.3	10.4	11.6
	5	540	NW	160	754	28	6.2	7.2	10.4	11.4	11.6
7	1	800	SW	155	1006	26	6.0	7.4	15.8	17.9	16.2
	2	780	SW	142	914	26	6.4	7.4	15.3	16.6	14.3
	3	770	SW	130	637	26	6.9	7.4	13.4	15.4	12.8
	4	760	SW	159	797	26	6.0	7.4	13.6	16.1	13.8
	5	740	SW	137	830	30	7.2	8.9	15.0	16.6	13.9
8	1	500	SE	121	1045	36	9.1	11.2	15.7	17.7	15.1
	2	500	SE	132	870	36	8.7	11.2	15.1	17.3	15.3
	3	510	SE	94	872	32	9.7	9.7	19.8	20.4	18.0
	4	500	SE	123	845	34	8.7	10.4	17.2	19.6	15.9
	5	510	SE	102	1058	34	9.5	10.2	22.1	21.1	19.9
9	1	900	NE	156	759	28	6.2	8.1	11.4	12.4	10.8
	2	890	NE	158	724	30	6.6	8.9	11.7	12.1	11.8
	3	910	NE	174	590	24	5.8	6.7	10.6	11.5	12.6
	4	1010	SW	184	825	24	5.8	6.7	11.7	15.1	16.4
	5	1000	SW	196	778	24	5.8	6.7	11.7	13.4	11.7

SI_{100} = site index
CAI = current annual volume increment ($m^3\ ha^{-1}\ year^{-1}$)
h = top height (100 largest trees per ha)
i_r = 10 year radial increment of top height trees

$h_{t(0)}$	$h_{t(10)}$	$h_{t(20)}$	$h_{t(30)}$	i_h (0-10)	i_h (11-20)	i_h (21-30)	i_r (0-10)	i_r (11-20)	i_r (21-30)	i_r (31-40)
m	m	m	m	m	m	m	cm	cm	cm	cm
33.39	31.34	29.24	27.41	2.05	2.10	1.83	2.156	2.144	3.133	1.944
34.13	32.31	30.63	29.30	1.82	1.68	1.33	1.978	2.333	3.267	1.989
31.79	29.44	27.18	25.15	2.35	2.26	2.03	2.422	2.967	2.844	2.167
29.94	27.77	25.64	24.24	2.17	2.13	1.40	2.400	1.800	1.944	1.467
27.81	25.31	22.97	20.59	2.50	2.34	2.38	2.178	2.078	1.789	1.589
24.14	22.87	21.61	20.63	1.27	1.26	0.98	1.233	1.400	1.633	1.978
23.25	22.02	20.79	19.56	1.23	1.23	1.23	1.122	1.056	1.456	1.744
24.13	22.93	21.73	20.46	1.20	1.20	1.27	1.378	1.478	1.489	1.422
23.92	22.78	21.62	20.31	1.14	1.16	1.31	1.356	1.311	1.300	1.578
21.87	20.84	19.81	18.72	1.03	1.03	1.09	1.100	0.922	1.044	1.456
34.34	33.68	32.52	31.22	1.16	1.16	1.30	1.244	1.311	1.300	1.389
35.64	34.26	32.89	31.17	1.38	1.37	1.72	1.844	2.111	1.656	1.867
37.39	35.77	34.15	32.30	1.62	1.62	1.85	1.456	1.756	2.033	2.144
35.19	33.83	32.44	30.64	1.36	1.39	1.80	1.244	1.344	1.422	1.567
35.93	34.39	32.83	31.12	1.54	1.56	1.71	1.189	1.678	1.833	1.756
37.38	35.01	32.51	30.40	2.37	2.50	2.11	2.256	2.622	2.344	1.889
37.70	35.40	32.91	29.75	2.30	2.49	3.16	1.667	1.967	1.889	2.222
37.75	35.49	32.87	29.00	2.26	2.62	3.87	2.644	2.267	2.144	2.467
32.56	30.40	27.98	25.01	2.16	2.42	2.97	2.589	2.656	3.200	2.778
31.31	29.26	26.71	23.48	2.05	2.55	3.23	3.122	4.033	4.056	3.189
29.82	28.10	26.31	24.07	1.72	1.79	2.24	2.022	1.944	1.922	2.000
31.72	30.14	28.56	26.78	1.58	1.58	1.78	1.733	1.733	1.589	1.800
29.79	28.06	26.26	24.42	1.73	1.80	1.84	1.933	1.867	1.522	1.800
28.86	27.11	25.25	23.36	1.75	1.86	1.89	2.189	1.722	1.533	1.811
28.20	26.39	24.52	22.68	1.81	1.87	1.84	1.822	1.911	1.578	1.856
33.80	32.09	30.16	28.06	1.71	1.93	2.10	1.333	1.600	1.567	1.200
34.59	32.75	30.84	28.78	1.84	1.91	2.06	1.056	1.378	1.467	1.356
36.77	35.08	33.07	30.72	1.69	2.01	2.35	0.956	1.311	1.344	0.956
34.94	33.28	31.58	29.25	1.66	1.70	2.33	0.933	1.122	1.300	1.067
33.64	31.79	29.38	26.92	1.85	2.41	2.46	1.278	1.444	1.667	1.067
31.98	29.97	27.87	26.13	2.01	2.10	1.74	1.956	2.444	2.422	1.911
31.12	29.47	27.41	26.03	1.65	2.06	1.38	1.967	2.389	2.278	1.433
29.76	27.65	25.31	23.16	2.11	2.34	2.15	2.467	3.133	3.056	2.111
31.61	29.44	27.18	25.47	2.17	2.26	1.71	1.622	2.011	1.978	1.444
33.74	31.89	29.29	27.57	1.85	2.60	1.72	2.344	3.067	3.267	2.400
39.47	37.71	35.48	32.62	1.76	2.23	2.86	1.700	1.978	1.756	1.578
39.18	37.33	35.44	32.47	1.85	1.89	2.97	2.344	2.867	2.522	2.111
31.25	29.34	27.08	24.47	1.91	2.26	2.61	2.622	2.889	2.878	2.511
37.47	35.67	33.16	30.42	1.80	2.51	2.74	2.656	3.144	2.444	2.178
33.53	31.59	29.04	26.11	1.94	2.55	2.93	1.956	2.156	2.011	2.144
33.39	32.54	31.65	30.14	0.85	0.89	1.51	1.556	1.700	1.367	0.967
35.02	33.73	32.43	31.27	1.29	1.30	1.16	1.811	1.789	1.756	0.911
29.80	28.02	26.47	24.77	1.78	1.55	1.70	1.489	1.344	1.422	0.978
30.30	28.99	27.45	25.74	1.31	1.54	1.71	1.267	1.522	1.356	1.167
31.08	29.57	27.99	26.07	1.51	1.58	1.92	1.444	1.833	1.622	1.411

Table 2. Continued

site unit no.	plot no.	eleva- tion	aspect	mean stand age	standing volume	SI_{100}	CAI (0-10) (yield table)	site produc- tivity (SI_{100})	CAI (0-10)	CAI (11-20)	CAI (21-30)
		m.asl.		year	m³/ha		m³ ha⁻¹ year⁻¹	m³ ha⁻¹ year⁻¹	m³ ha⁻¹ year⁻¹	m³ ha⁻¹ year⁻¹	m³ ha⁻¹ year⁻¹
10	1	490	NW	187	693	30	6.6	8.0	8.6	8.2	7.8
	2	490	NW	190	694	28	6.2	7.2	12.2	9.6	7.3
	3	500	NE	184	802	30	6.6	8.0	10.9	9.3	8.6
	4	500	NE	188	764	30	6.6	8.0	12.0	8.9	9.2
	5	490	NW	185	633	26	6.0	6.6	12.2	10.8	8.9
11	1	600	SW	139	730	34	8.1	9.4	9.9	10.8	10.1
	2	660	SW	147	847	32	7.3	8.7	11.7	11.6	12.0
	3	680	W	143	684	30	7.0	8.0	8.4	8.3	8.9
	4	880	NW	119	904	32	8.4	8.7	11.9	11.9	12.7
	5	890	NW	136	997	30	7.2	8.0	14.9	14.4	14.2
12	1	1200	SE	137	439	22	6.0	5.3	6.4	10.2	10.5
	2	1200	SE	147	479	20	5.4	4.8	6.4	8.9	9.5
	3	1260	SE	161	410	14	4.7	3.3	5.3	6.6	6.5
	4	1270	SE	155	405	16	4.9	3.7	5.3	9.2	9.5
	5	1270	SE	145	455	18	5.3	4.3	5.4	7.0	7.3
13	1	900	NW	154	653	26	6.2	5.8	7.7	8.5	9.1
	2	900	NW	151	628	30	6.8	7.1	7.0	8.1	8.6
	3	920	NW	152	731	32	7.1	7.7	8.7	10.4	11.6
	4	870	NW	145	669	26	6.4	5.8	9.3	11.3	11.6
	5	890	NW	149	566	28	6.6	6.4	7.6	9.8	11.0
14	1	610	NE	112	364	22	6.7	5.3	8.0	12.9	14.0
	2	650	E	110	345	24	7.3	6.0	8.7	11.5	11.3
	3	650	NW	124	349	22	6.5	5.3	8.5	10.1	9.4
	4	700	SE	120	456	26	7.3	6.6	9.1	12.8	14.2
	5	580	E	123	366	24	6.8	6.0	8.4	10.5	10.5
15	1	605	NE	105	643	32	9.1	9.7	14.7	13.6	15.2
	2	605	NE	105	916	32	9.1	9.7	17.6	17.4	19.7
	3	600	N	102	879	30	8.7	8.9	17.8	16.2	16.6
	4	570	NE	109	753	32	8.9	9.7	16.1	14.9	15.6
	5	540	N	102	558	34	9.5	10.4	13.0	12.8	12.6
16	1	870	NE	146	852	26	6.3	7.4	13.4	13.9	14.1
	2	860	NE	134	824	28	7.1	8.1	12.7	12.9	12.6
	3	850	N	123	548	26	7.1	7.4	9.6	10.0	10.1
	4	840	N	132	677	26	6.7	7.4	12.0	13.1	13.2
	5	900	N	147	724	26	6.3	7.4	11.6	11.6	11.8
17	1	680	N	128	810	36	8.9	11.2	13.2	15.6	20.4
	2	680	N	127	788	34	8.5	10.4	13.4	17.0	21.7
	3	720	N	129	828	34	8.5	10.4	15.1	17.1	20.6
	4	700	N	132	951	34	8.3	10.4	14.9	16.1	18.2
	5	730	N	132	891	34	8.3	10.4	13.7	15.0	16.8
18	1	1380	SW	156	405	16	4.8	4.2	7.0	7.5	7.0
	2	1390	NW	178	405	14	4.6	2.8	5.4	6.9	6.0
	3	1375	SW	129	410	20	5.8	5.4	9.3	9.7	8.5
	4	1420	NW	166	414	16	4.7	4.2	7.1	7.1	7.2
	5	1420	NW	155	366	16	4.8	3.6	7.3	7.4	7.4

SI_{100} = site index
CAI = current annual volume increment (m³ ha⁻¹ year⁻¹)
h = top height (100 largest trees per ha)
i_r = 10 year radial increment of top height trees

$h_{t(0)}$	$h_{t(10)}$	$h_{t(20)}$	$h_{t(30)}$	i_h (0-10)	i_h (11-20)	i_h (21-30)	i_r (0-10)	i_r (11-20)	i_r (21-30)	i_r (31-40)
m	m	m	m	m	m	m	cm	cm	cm	cm
39.12	37.39	35.73	33.96	1.73	1.66	1.77	1.422	1.189	1.211	0.878
37.60	35.51	33.34	31.47	2.09	2.17	1.87	2.200	1.567	1.200	1.011
40.01	38.58	37.16	35.61	1.43	1.42	1.55	1.489	1.478	1.111	0.900
38.71	37.07	35.43	33.42	1.64	1.64	2.01	1.556	1.289	0.956	0.811
34.06	31.69	29.31	27.04	2.37	2.38	2.27	2.622	2.178	1.522	1.256
40.11	37.50	34.86	32.25	2.61	2.64	2.61	1.122	1.711	1.867	1.611
38.37	36.41	34.37	32.38	1.96	2.04	1.99	1.567	1.733	1.878	1.533
35.29	33.81	32.26	30.29	1.48	1.55	1.97	1.133	1.144	1.322	1.189
34.26	32.66	30.96	28.88	1.60	1.70	2.08	1.211	1.289	1.556	1.289
33.53	32.20	30.87	29.05	1.33	1.33	1.82	1.600	1.778	1.711	1.344
25.12	23.99	22.76	21.04	1.13	1.23	1.72	0.978	1.256	1.167	1.011
24.81	23.32	21.91	20.38	1.49	1.41	1.53	0.933	1.367	1.300	1.100
20.00	18.85	17.53	16.07	1.15	1.32	1.46	0.889	1.144	1.156	1.189
20.40	19.23	18.09	16.90	1.17	1.14	1.19	0.878	1.289	1.289	1.233
21.79	20.57	19.25	17.94	1.22	1.32	1.31	1.022	1.200	1.222	1.089
32.50	30.98	29.49	27.89	1.52	1.49	1.60	0.867	1.089	1.333	0.911
35.62	34.04	32.42	30.82	1.58	1.62	1.60	1.056	1.067	1.367	1.100
37.06	35.50	33.98	32.28	1.56	1.52	1.70	1.133	1.300	1.422	1.333
31.05	29.45	27.80	25.99	1.60	1.65	1.81	0.933	1.333	1.333	1.311
32.45	31.15	29.71	27.55	1.30	1.44	2.16	0.889	1.100	1.100	0.933
22.88	20.92	19.02	17.53	1.96	1.90	1.49	1.578	2.200	2.467	2.744
25.88	23.91	21.77	19.35	1.97	2.14	2.42	2.122	2.622	2.911	2.589
24.21	22.28	20.14	18.34	1.93	2.14	1.80	1.933	2.411	2.389	1.722
28.09	26.38	24.64	23.05	1.71	1.74	1.59	1.711	1.833	1.611	1.600
25.99	24.31	22.38	20.21	1.68	1.93	2.17	1.778	1.756	1.911	1.800
38.73	36.38	33.59	30.31	2.35	2.79	3.28	2.511	2.478	3.200	2.722
39.59	37.57	35.16	32.00	2.12	2.41	3.16	2.289	2.111	2.844	2.867
38.55	36.32	33.86	30.23	2.23	2.46	3.63	2.811	2.889	3.500	2.944
39.99	37.40	34.56	31.19	2.59	2.84	3.37	2.300	2.278	2.633	2.611
35.07	32.29	28.44	24.58	2.78	3.85	3.86	2.578	2.567	3.333	3.211
30.62	28.35	26.19	23.41	2.27	2.16	2.78	1.044	1.133	1.111	1.100
31.73	29.63	27.75	24.67	2.10	1.88	3.08	1.067	1.122	1.522	1.356
28.77	26.90	24.87	22.17	1.87	2.03	2.70	1.144	1.456	1.689	1.822
28.37	26.55	24.41	21.92	1.82	2.14	2.49	1.322	1.711	2.089	2.000
30.57	28.78	26.78	23.85	1.79	2.00	2.93	1.611	1.822	2.089	1.978
40.03	38.29	36.03	33.42	1.74	2.26	2.61	1.878	1.522	1.889	1.700
37.56	35.95	33.86	30.84	1.61	2.09	3.02	1.922	2.178	2.400	2.044
37.98	36.33	33.68	30.88	1.65	2.65	2.80	1.522	1.500	1.800	1.733
38.68	36.57	33.94	31.02	2.11	2.63	2.92	1.578	1.589	1.989	1.756
38.93	37.02	34.68	31.22	1.91	2.34	3.46	1.756	1.656	1.811	1.544
20.01	18.90	17.77	16.54	1.11	1.13	1.23	1.067	0.889	0.800	0.989
18.19	17.39	16.62	15.78	0.80	0.77	0.84	1.189	1.322	1.222	1.389
21.84	20.40	19.07	17.52	1.44	1.33	1.55	1.433	1.344	1.133	1.333
20.88	19.80	18.67	17.49	1.08	1.13	1.18	1.511	1.389	1.333	1.467
18.44	17.16	15.93	14.61	1.28	1.23	1.32	1.300	1.200	1.067	1.378

trees respectively. The second approach is a comparison of different growth parameters of plots within site units at the same age. Ages of comparison were reached at different points in time because of different germination periods of the stands. This approach can only be applied to a part of the available data because it requires a large difference in age between several observations, which was not given in all cases (Spiecker et al. 1994).

The investigation was designed as a factorial experiment with five replications, where the factor site unit consisted of 18 levels. From each site unit fully stocked stands were selected in which 5 sample plots were located at random. Sample plots, 30 x 30 m in size, were chosen according to the following criteria (Kotar 1989):

a. In a sample plot beech must comprise at least 90% of the standing volume.
b. The stands in which sample plots are located must be fully stocked; their basal area must be approximately the same as for a stand of the same age on an identical site unit with natural development.
c. In the selected stands no thinnings or only low thinnings should have occurred in the last decades.

Using these criteria, 5 sample plots were selected from each site unit, giving 90 plots in all. In each sample plot all trees were felled and sawed into 6-9 sections depending on tree height. A disc was taken from each section for a dendrochronological analysis. The first disc was taken at stump height and the next at breast height. Further discs were taken at different stem heights. Towards the top of a tree, section length was shorter. Consequently, height growth in the past decades was determined more accurately. For the purpose of this study, more than 4000 trees were cut and analysed, and more than 32 000 discs were examined. For each of the felled trees, 10 year height growth, 10-year diameter growth and volume growth was carefully calculated.

In each sample plot, all traceable stumps of trees removed from the stand in the last 30 years were measured. In a great number of sample plots only stumps of trees which had died due to self-thinning were present, since no tending measures such as thinnings had been undertaken. The stumps were classified into 3 groups according to the time of their origin. The stumps of trees which had been removed or had died in the past 10 years were placed into the first group (t_1 - t_{10}) the stumps from 11 to 20 years ago were placed into the second group (t_{11} - t_{20}), and the stumps from 21 to 30 years ago were placed into the third group (t_{21} - t_{30}). Stumps older than that were not found. The stump classification was used for determining the volume of trees removed in the last three decades. Their volume was determined as follows: In each sample plot diameter at stump height ($d_{0.3}$) and diameter at breast height ($d_{1.3}$=dbh) were measured for all standing trees (i.e., trees which were cut down within the framework of this study). On the basis of these values the regression relation $d_{1.3} = ad_{0.3}^{b}$ was calculated and used to estimate dbh of tree stumps. Their height was estimated using the stand height curve which was determined by calculating the regression relation between dbh and h (tree height) of the analysed trees from the past three decades. For the regression relation the following function was determined: $\ln h=\alpha+\beta \ln d_{1.3}+\gamma \ln^2 d_{1.3}$ (h = trees height, ln = natural logarithm). Then the volume of trees removed in the last three decades was calculated using a two-way volume equation $v=f(d_{1.3}, h)$. The volume of these trees was used for calculating CAI over the past three decades.

CAI of the past decade was calculated by adding the CAI of dead (and thinned) trees to the CAI of trees felled during this investigation (hereafter called: standing trees in a sample plot). We assumed that these trees died mid-decade, that is, 5 years prior to this analysis; CAI was determined using a 10-year periodic increment of volume divided by 10. CAI of trees dying during the past decade (t_1 - t_{10}) was calculated by multiplying the volume of these trees (calculated by the use of stumps as mentioned above) by volume increment percent for the past decade of standing trees in a sample plot. The volume increment for other periods (t_{11} - $t_{20,}$ t_{21} - t_{30}) was calculated in the same way. It should be noted that volume increment of trees which had died in the last three decades amounts to a very small proportion of total CAI in a sample plot. This is evidently from the fact that sample plots were selected from stands in which no thinnings or only low thinnings were undertaken, whereas self-thinning is rather slow at the ages of the stands analysed.

Top height of a sample plot was determined by using the height of 9 trees with the largest tree diameter (100 trees with the largest tree diameter per 1 hectare) and calculating their mean.

When stand age was determined, it was taken into account that, in their young stages, trees of the analysed sample plots grew within a shelterwood system. Consequently, their height and diameter growth was reduced by the effect of shade from overstory trees. We distinguish between chronological or cambial tree age and stand age, which is in fact their actual age, and growth age (effective age), which is the age of a free-grown tree of the same height grown on the same site. Using growth age, we can compare stands with each other which grew in their young ages under a shelterwood system for periods of different length. Thus we can compare stands with each other which developed during regeneration periods of different length. The shading effect was determined so that on each site unit a stand was first identified which developed without shade (except for the first 5 years, when shade is needed for successful regeneration of the stand). Then, stand age at the height of 10 m was calculated. (Height curve was determined: height versus age, and age at the height of 10 m was calculated.) This age was subsequently subtracted from the age shown by the height curve (height versus age) of the analysed stand in a sample plot. The difference in years represents the shading effect; the number of years it took a tree growing under a shelterwood to grow to the height of 10 m in comparison to trees which developed without shade. This calculation was based on the assumption that the cambial age and growth age run parallel when a tree reaches the height of 10 m.

The shading effect varies from 0 to 20 years or more. It is interesting to note that it is positively correlated with the difference between the average age of top height trees and the average age of all trees in a stand (stand age). Age of top height trees is usually a little older than the average age of all trees. This difference is smaller if the average age of top height trees is compared with the average age of trees in the upper three social classes according to Kraft (Kotar 1991).

To determine height growth in the stands in the past three decades, 10-year height increment of top height trees was examined. This was based on the assumption that these trees were top height trees throughout the past three decades; a reasonable assumption since trees with the largest diameter usually retain this social position at ages such as those of the analysed stands. The most vigorous trees in a stand (at the age of 100 years or more) rarely succumb to self-thinning. Additionally, thinnings which can affect the structure of top height trees were either not done in most of the analysed stands or confined to the understory. Hence, height growth of top height-trees at the time of this analysis also shows the development of stand top height in the past 30 years.

Similarly, radial increment of the top height trees for the past three decades was investigated. Since these are trees with the largest diameter, which have usually the largest crowns, their radial increment is less affected by competition of adjacent trees. Trees removed in the past 30 years had little effect on the radial increment of top height trees as the process of self-thinning affects mainly trees of lower social classes and trees with small crowns in the upper classes. Also, the self-thinning process was very slow and thus of little consequence since the analysed stands are rather old. This is substantiated by a small number of stumps from the past 30 years. Therefore, possible changes in diameter growth of top height trees may also be indicative of changes in forest site productivity.

4. Results and Discussion

4.1. Current Annual Volume Increment in the Past Three Decades

Stem or disc analyses and stump analyses of trees removed in the past three decades provide good information on CAI. CAI values of individual sample plots for the past three decades are shown in Table 2 which also shows site productivity values determined using top height, site index (SI_{100}) and yield tables. For comparison, the corresponding CAI given in the yield tables with the highest yield level is presented. This leads to the highest CAI values for a specific site index.

- The variation of CAI values within the three decades is high with an overall tendency to decline (Fig. 1).
- A comparison of CAI_{0-10}, CAI_{11-20}, and CAI_{21-30} of the sample plots with the corresponding CAI given in yield tables shows that all CAI values are higher in all 90 sample plots. CAI exceeds yield table values between 2% and almost 300% (Fig. 1).
- CAI values of all three decades are usually higher than the corresponding site productivity values of the yield tables (3 exceptions out of 270 observations) (Table 2).

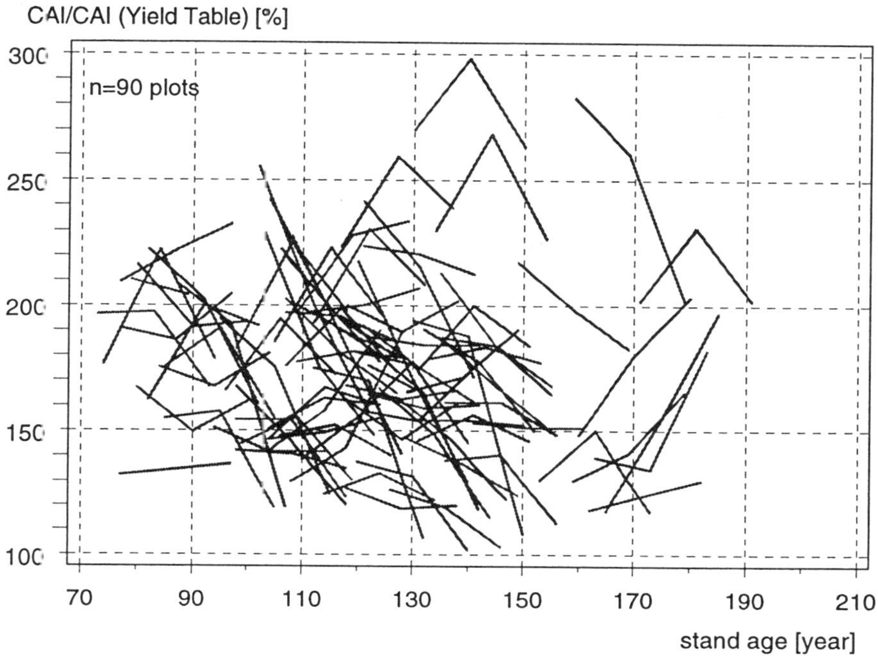

Fig. 1. Comparison of current annual volume increment of the last three decades with CAI from yield tables. High variation of CAI can be observed with an overall tendency to decline. However, CAI values of the sample plots exceed yield table values in all three decades (2-300%).

No trend can be observed by a comparison of CAI of different plots with the same age within a site unit. The plots reached their age of comparison at different points in time (Fig. 2).

The high CAI deviations from CAI values of the yield tables and from site productivity determined by yield table application may be a hint for site productivity changes and cannot be attributed simply to unsuitable yield tables. The high CAI values in stands over 100 years old may only partly be accounted for by the fact that in most of the stands thinnings were not taken. On the other hand, the second trend-detecting method did not show any growth trend.

4.2 Height Growth During the Past Three Decades

Height growth of all trees in sample plots was determined using stem analyses. Possible changes in site productivity affect height growth of top height trees or the top height of a stand. Table 2 shows the values of stand top height in each sample plot, at the time of the analysis ($h_{t(0)}$), 10 years ago ($h_{t(10)}$), 20 years ago ($h_{t(20)}$) and 30 years ago ($h_{t(30)}$).

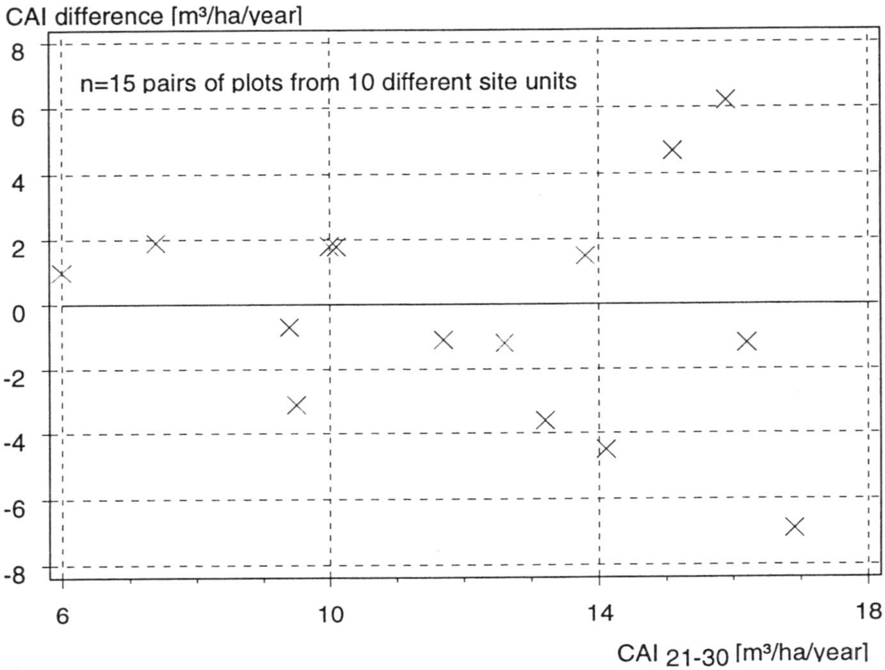

Fig. 2. Difference of current annual volume increment between pairs of sample plots within site units. The plots have approximately the same age at two different points in time (first and the third decade before present, difference of 20 calendar years). The maximum age difference within a pair of two is 7 years (not plotted). The difference between CAI does not deviate significantly from zero; therefore no trend is evident.

Table 2 also shows 10-year increments of top height in the past three decades ($i_{h(0-10)} \ldots i_{h(21-30)}$). The lowest increase in top height in the past decade is 0.80 m in sample plot 2 of site unit 18, whereas the highest increase is 2.78 m in sample plot 5 of site unit 15. In 26% of the sample plots the increase in top height in the past decade is over 2 m, which is considerably more than the increase of top height in the highest site index class in yield tables with the highest yield level at the age of 100 years or more (Halaj 1987). Only in two sample plots (2.2%) top height increment is less than 1 m.

Top height curves (trajectories) cannot reveal a change in site productivity because an increase or a decrease in site productivity is "hidden" in the shape of the curve. These changes can be detected by graphing top height curves of many stands with different ages (at the time of this analysis) on the same site unit. This entails a comparison of the top height of stands regenerated at different points in time on identical site types (Spiecker et al. 1994). The method could not be applied to all site units because ages of the different plots within a site unit are not spread widely enough (Figs. 3 and 4).

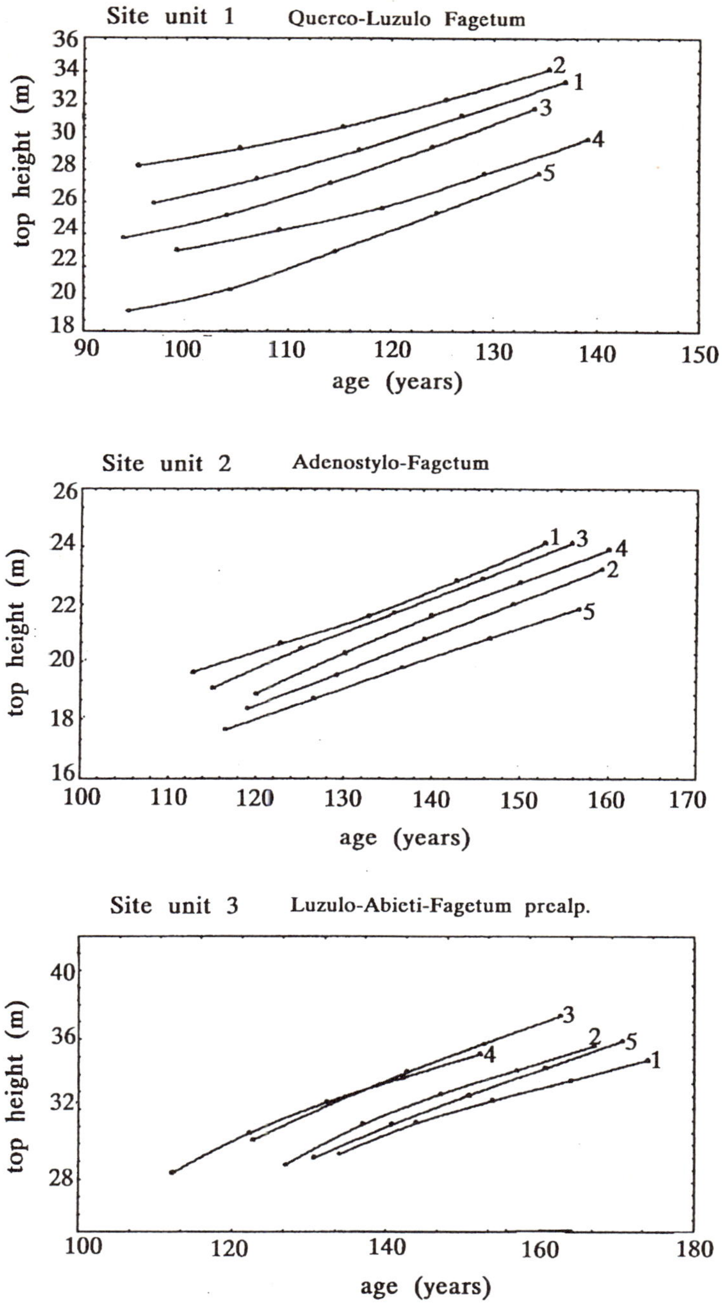

Fig. 3. Top height trajectories of analysed stands for the last 40 years (1., 2., . . . 5 = number of sample plots within site units)

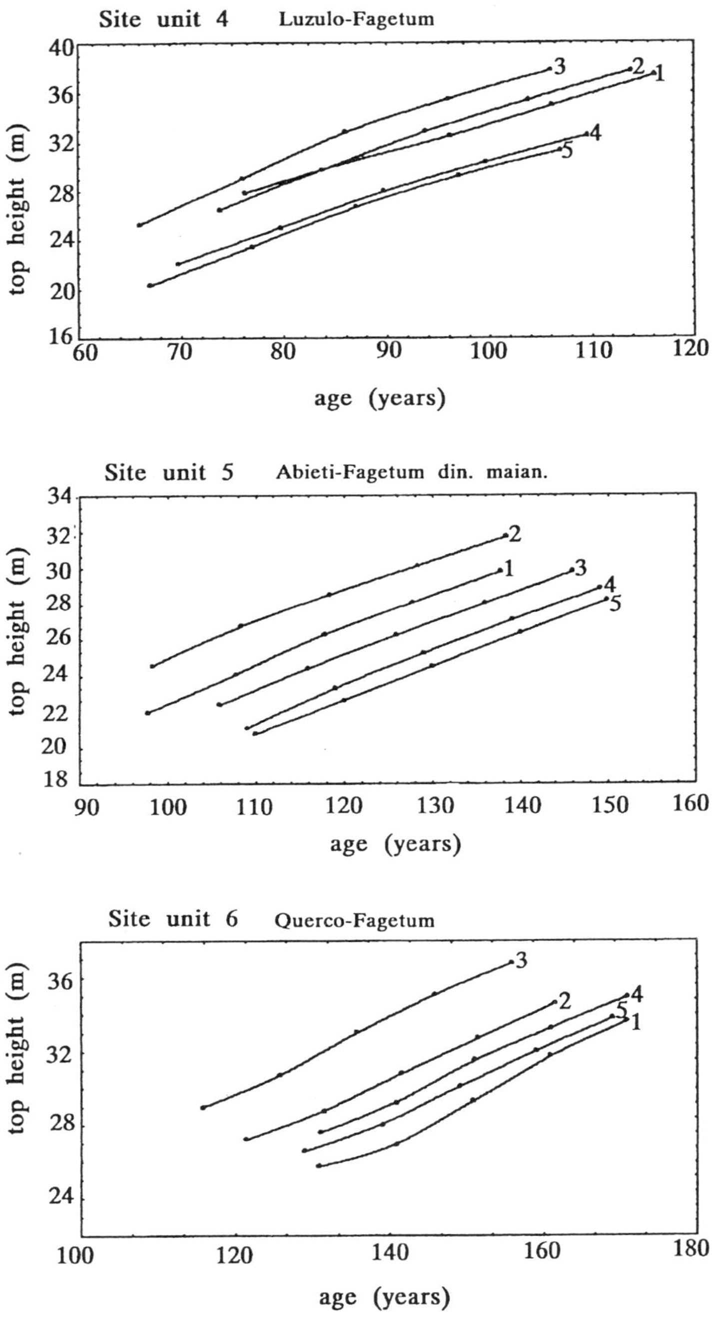

Fig. 3. Continued (Site units 4-6)

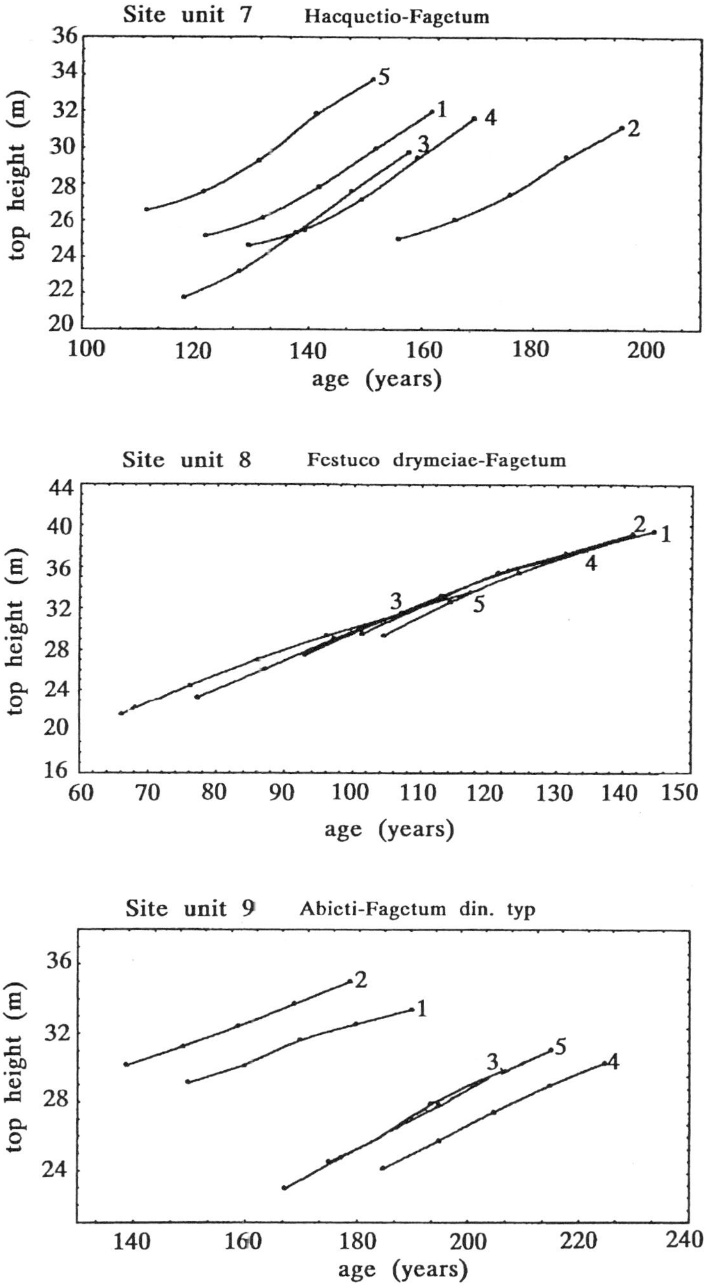

Fig. 3. Continued (Site units 7-9)

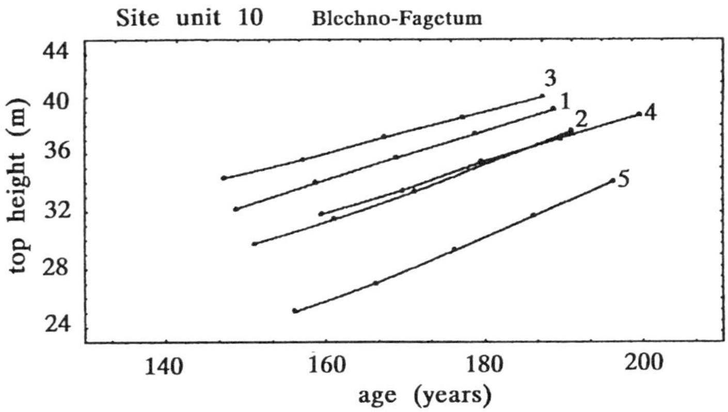

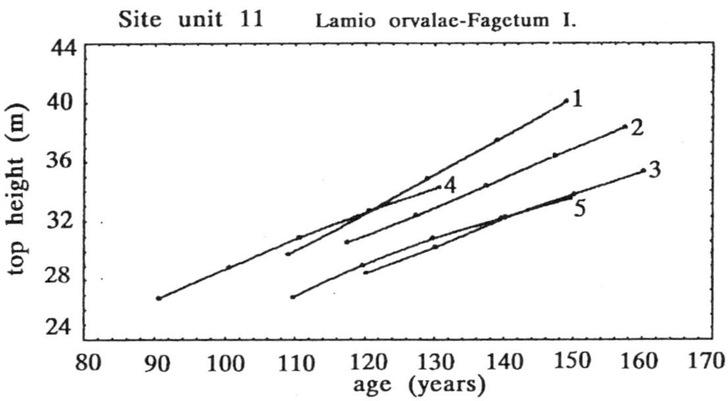

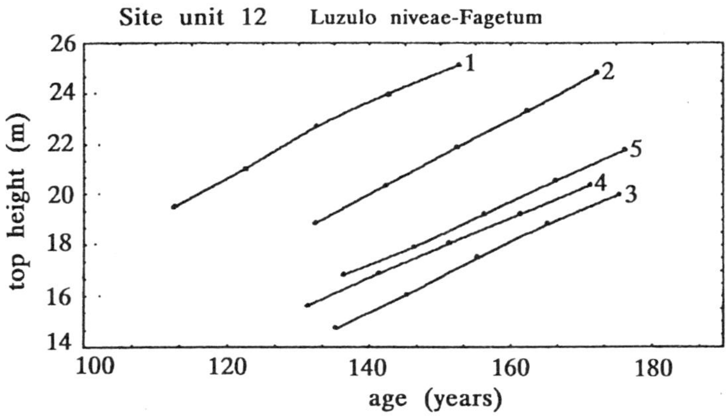

Fig. 3. Continued (Site units 10-12)

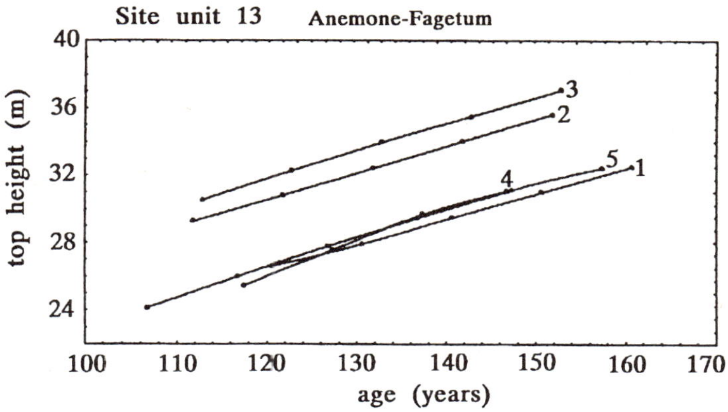

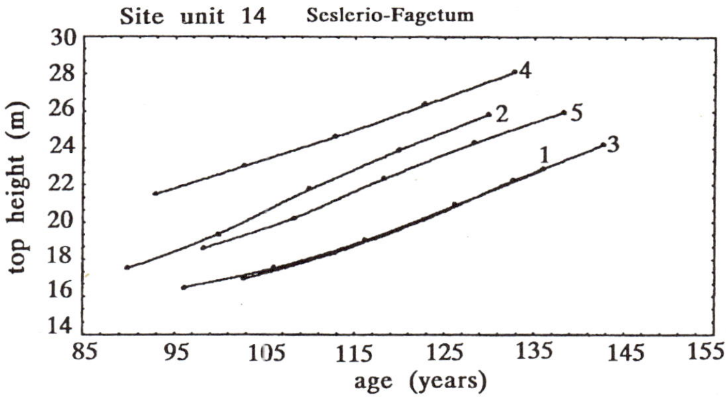

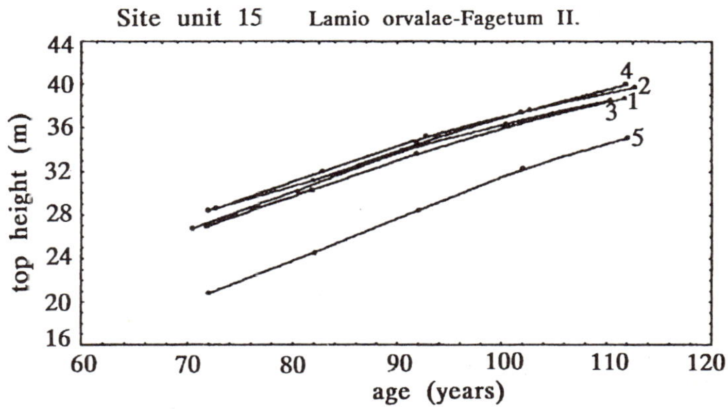

Fig. 3. Continued (Site units 13-15)

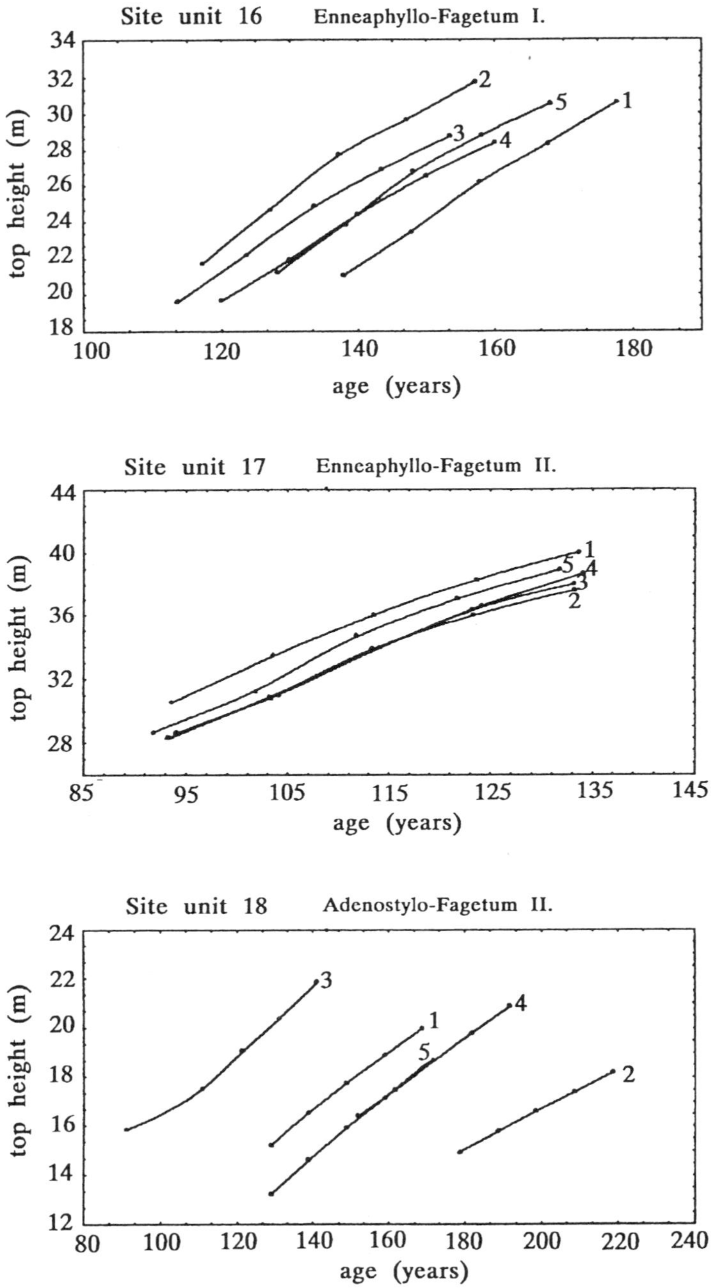

Fig. 3. Continued (Site units 16-18)

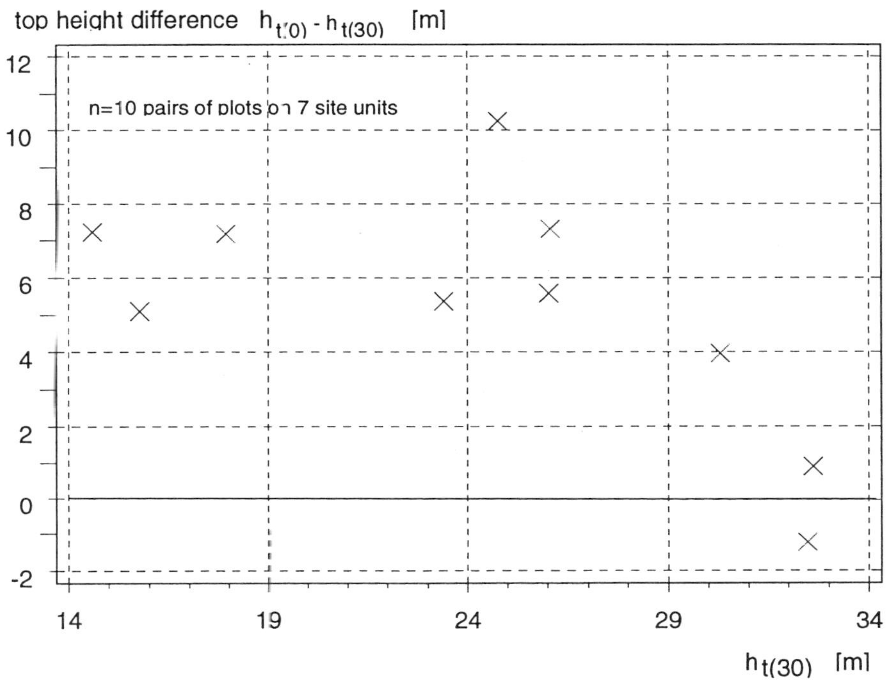

Fig. 4. Difference of top height between pairs of sample plots within a site unit. Tree age is approximately the same at two different points in time (difference of 30 calendar years). The maximum age difference within a pair of two plots (not plotted) is 6 years. The top height difference deviates significantly from zero. A positive trend is evident.

From the data on higher top heights in younger stands on the same site unit at the same age it may be inferred that there has been a change in height growth of stands which regenerated later. On the other hand, height increment data of the past three decades do not show a positive trend (Fig. 5). Therefore, height differences must have developed before that time.

4.3. Radial Growth in the Last Three Decades

Stem analyses offered an opportunity to gain an insight into radial growth at breast height of all trees in sample plots at the time of the analysis. Dominant trees are less influenced by competition than other trees. Therefore, radial growth of the top height trees during the last three decades was examined in more detail. The data are given in Table 2.

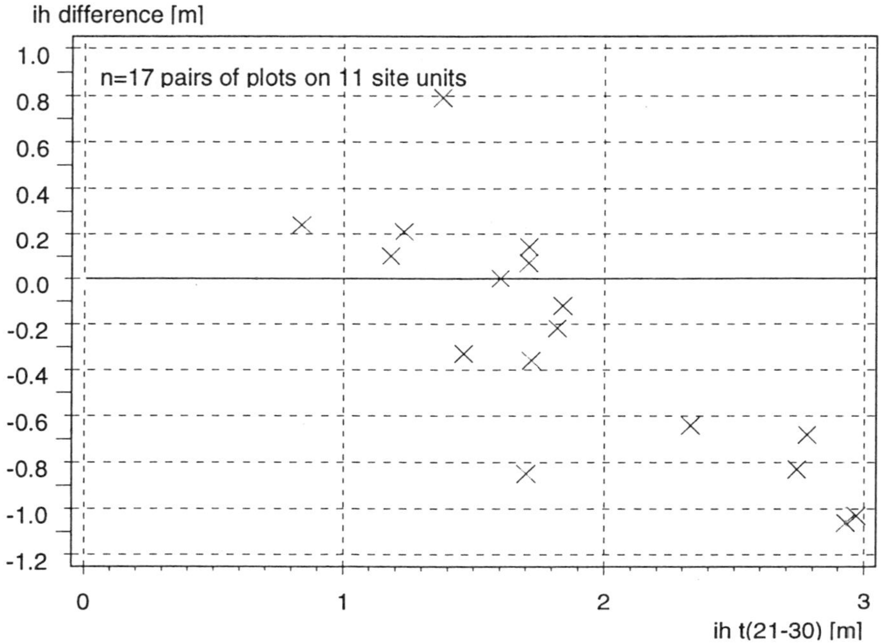

Fig. 5. Difference of 10-year height increment of top height trees between pairs of sample plots within a site unit. Tree age is approximately the same at two different points in time (decade 0-10 and 21-30, difference of 20 calendar years). The maximum age difference within a pair of two plots is 7 years (not plotted). Height increment seems to be smaller in the decade 0-10, smallest increments are linked with high increments in the decade 21-30.

The analysis of data on radial growth shown in Table 2 indicates that radial increment in the last decade (t_{0-10}) is higher than in the decade before (t_{11-20}) in 29 sample plots. This can also be observed in the third decade (t_{21-30}) in comparison to the second decade (t_{11-20}) in 42 sample plots. In the most recent decade (t_0-t_{10}) radial increment of top height trees increased in 33% of sample plots. In an even greater number of sample plots (46%) the increment of top height-trees increased in the second decade (t_{11}-t_{20}) if compared with that in the third decade (t_{21}-t_{30}). Thus, an increment increase already occurred in these plots more than 10 years ago. This increase may be affected by changes in competition, although the analysed stands were generally not thinned, and by climatic influences.

Figure 6 shows the results from applying the second method of trend detection to the radial increment data. No positive trend can be observed. Radial increment seems even to be smaller in the decade 0-10, smallest increments are linked with high increments in the decade 21-30.

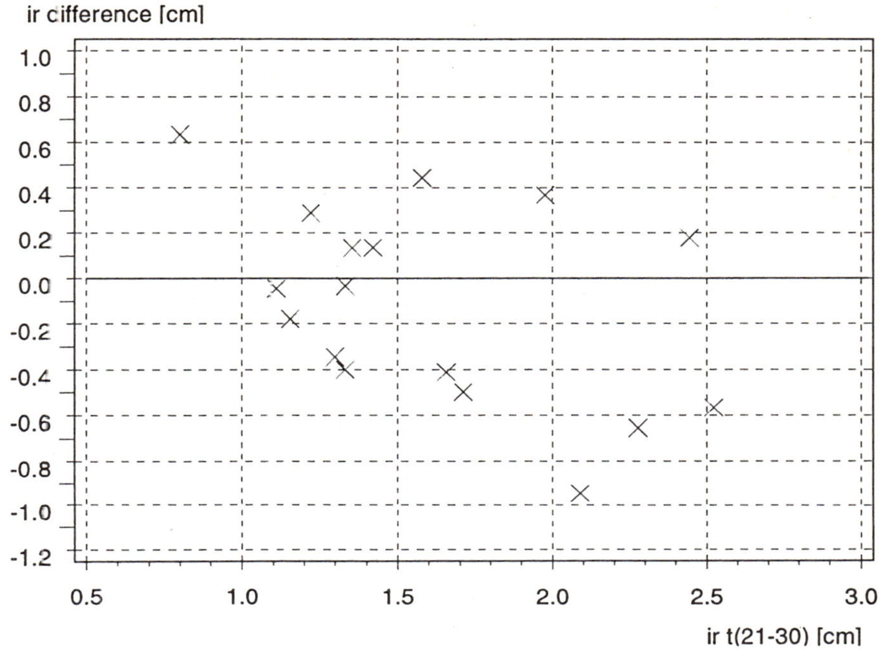

Fig. 6. Difference of 10-year radial increment of top height trees between pairs of sample plots within a site unit. Tree age is approximately the same at two different points in time (difference of 20 calendar years). The maximum age difference within a pair of two plots is 7 years (not plotted). Radial increment seems to be smaller in the decade 0-10, smallest increments are linked with high increments in the decade 21-30.

5. Conclusions

No clear answers to the question of whether site productivity changes occurred in beech forests in Slovenia can be given due to methodological problems. A yield table comparison shows clear increases in CAI and height increment but yield tables do not represent sites and silvicultural conditions of the analysed stands in detail. The second approach for the analysis of growth trends, the comparison of growth parameters at the same age of trees or stands regenerated at different periods, only showed a positive trend for top height. CAI, height and radial increment remain unchanged or even decreased with later regeneration period. The method could not be applied on many site units because the ages of the plots on them were not extensive enough.

Generally, it has to be considered that in many parts of the country, diminishing tree growth has been observed due to a high level of $S0_2$ pollution. This is especially true of areas in the vicinity of coal-fuelled power plants (Kolar 1989).

References

Elfving B et al. (1995) Growth Trends of Forests in Sweden and Norway.This volume.

Halaj J (1987) Rastove tabulky hlavnych drevin CSSR (Yield Tables for Main Tree Species in CSSR) Priroda, Bratislava

Kolar I (1989) Umiranje smreke v gozdovih Saleske doline [Spruce die-back in the forest of the Saleska valley], Zbornik gozd. in les. 34 (1989) pp 121-198, Slovenia

Kotar M (1987) Prüfung der Verwendbarkeit der Ertragstafeln in den Buchen- und Fichtenbeständen in Slowenien [Suitability of Yield Tables in Beech and Spruce Stands in Slovenia]. Conference Proceedings: Functionally integrated management of forests, pp 117-130 Zvolen, Slowakien

Kotar M (1989) Prirastoslovni kazalci rasti in razvoja bukovih gozdov v Sloveniji [Growth and yield indicators of growth and development in beech forests in Slovenia]. Zbornik gozd. in les., Ljubljana, 33. (1989), pp 59-80 + 16 pril. Slovenia

Kotar M (1991)Zgradba bukovih sestojev v njihovi optimalni razvojni fazi [The structure of beech stands in their optimum developmental phase] Zbornik gozd. in les., 38 (1991), pp 15-40, Slovenia

Kotar M (1994) Proizvodna sposobnost gozdnih rastišè, ki jih porašèajo smrekovi in bukovi gozdovi ter njihova proizvodna zmogljivost v optimalni fazi [Determining site productivity on sites which are overgrown with more or less uneven-aged stands in Slovenia], Zbornik gozd. in les. 44 (1994) pp 125-148, Slovenia

Pretzsch H (1995) Growth Trends of Forests in Germany. This volume.

Skovsgaard JP, Henriksne HA (1995) Increasing Site Productivity during Consecutive Generations of Naturally Regenerated and Planted Beech (Fagus sylvatica L.) in Denmark. This volume.

Spiecker H, Mielikäinen K, Köhl M, Untheim H (1994) Growth trends of European forest - Has site productivity changed? European Forest Institute Working Paper 4. Joensuu, Finland

Case Studies of Growing Stock and Height Growth Evolution in Spanish Forests

Gregorio Montero[1], Alberto Rojo[2] and Ramón Elena[3]
[1]Area Selvicultura y Mejora, CIFOR-INIA, Madrid, Spain
[2]E.P.S., Universidad de Santiago, Lugo, Spain
[3]Dpto. Silvopascicultura, EUITF-Universidad Politécnica de Madrid, Spain

Abstract

The objective of this chapter is to analyse the importance of the *increment of forest growth rates* in Spain during the last 150 years by giving a retrospective view of growth trends of Spanish forests. Two studies were carried out: (1) study of growing stock evolution of the Spanish forests during the past century; (2) study of forest growth evolution at site level of a forest in Central Spain. Data from inventories of large forests all around Spain are used for the first purpose and individual tree stem analysis data (*Pinus sylvestris*) from Central Spain are used for the second study. Through both studies it is possible to conclude that growing stocks and forest growth followed increasing trends. Increases in growing stocks were caused by changes in silvicultural practices, but changes in site productivity could be associated with changes in climatic and soil conditions. This chapter discusses the relationships between environmental factors and recorded growth increment, concluding that further research efforts are necessary to characterize the processes that drive growth changes.

1. Introduction

From the northern European foresters' point of view, forest growth is mainly understood in terms of wood production. This prevalence is difficult to be applied to most of the mediterranean forests, where wood is one among other direct produces; in many cases it is actually one of the least important ones. Therefore, forest growth is a more complex concept in mediterranean forests; important direct produces (cork, cone, pinyon, resin, firewood, mushroom, pasture, etc.) as well as indirect ones (soil protection, erosion control, water supply, fire prevention, recreation) must be taken into consideration. Consequently, evaluation of forest growth and site productivity is not a simple task in the mediterranean context, and currently the complex methodology needed for its evaluation has not been fully developed.

European Forest Institute Research Report No. 5
Growth Trends in European Forests
Edited by H. Spiecker, K. Mielikäinen, M. Köhl and J. P. Skovsgaard

In the case of Spain, the mentioned conceptual and methological limitations affect the evaluation of 80% of the total country area, where true mediterranean climatic conditions prevail. On the other hand, wood is the main produce of forests located in both Northern Atlantic and Central Mountain regions of Spain with very important oceanic influences: these forests amount to 36% of the total forested area of Spain (9.3 million ha), and are the main contributors to the total national wood production. The Spanish wood production has increased sixfold from 2.3 million m^3 in 1950 to 14.1 million m^3 in 1993. This important increment was allegedly caused by several factors: (1) extensive application of silvicultural practices that resulted in higher density and growing stock of the stands; (2) increment of forest growth rates; (3) reforestation in forest gaps; (4) progressive intensification of silvicultural practices; (5) afforestation of new lands and gradual harvesting of the areas afforested since 1940, mainly fast-growing species plantations.

Although the forest administration in Spain is more than 150 years old, it is necessary to emphasize that neither surveys nor studies aiming to assess forest growth trends have been carried out. Consequently, there are no specific data available for undertaking the analysis of forest growth at site level. Nevertheless, there is forest information available, obtained periodically from inventories of the most important forests, since their management planning started at the end of the nineteenth century. In order to reach, at the Spanish level, some of the objectives of the EFI Project on *Growth Trends of European Forests*, this chapter analyses the inventory data available together with tree stem analysis data.

The main objective of this chapter is to analyse how important the *increment of forest growth rates* has been in Spain during the past century by giving a retrospective view of the development of forest productivity. In order to fulfill this objective, the paper has been divided in two parts: (1) Analysis of the evolution of the Spanish forests in terms of growing stock during the last 100 years. For this analysis, case studies are presented using data from five important forests, periodically inventoried since the 1890s. (2) Study of forest growth evolution at site level of a *Pinus sylvestris* forest, using stem analysis of trees with different ages growing in stands of the same site quality.

The results of the two components of the study might allow us to establish if an increment of forest growth has occurred and to analyse possible causes for the growth trends that may be found.

2. Evolution of Growing Stocks: Five Case Studies

2.1. Case Study Location and Description

The first step in our approach has been to analyse the growing stock evolution of Spanish forests. Obviously, increment in growing stock does not mean increment in site productivity, but it can give a preliminary view of its evolution. Five case stud-

Table 1. Case study ecological description

Case	Tree species	Latitude / Longitude	Phytoclimate	Bedrock
Case 1	*P. sylvestris*	41° 03' N 3° 49' W	Supramediterranean	Granite & Gneiss
Case 2	*P. nigra*	40° 06' N 1° 57' W	Supramediterranean	Calcareous
Case 3	*P. sylvestris*	40° 54' N 4° 03' W	Supramediterranean	Gneiss
Case 4	*Quercus suber/ Q. canariensis*	36° 35' N 5° 28' W	Mesomediterranean	Sandstone
Case 5	*Fagus sylvativa*	42° 56' N 1° 59' W	Mountain Eurosiberian	Calcareous

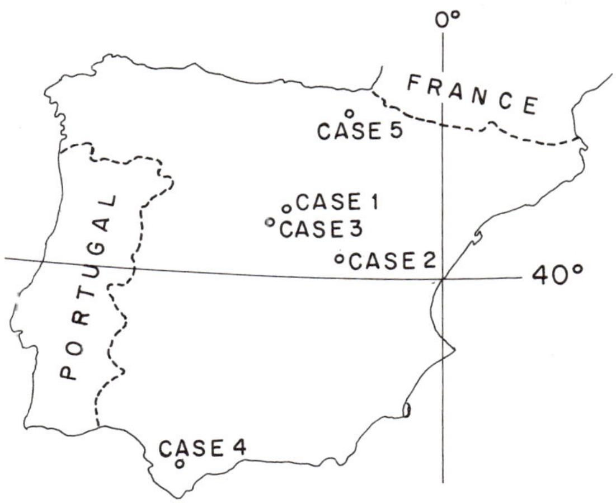

Fig. 1. Map of Spain showing the study case locations.

ies representative of the most important Spanish tree forest species will be presented showing trends in forest growing stocks. Their location and ecological characteristics are shown in Fig.1 and Table 1. In all the cases, the forested area remained stable during the studied period.

As far as the tree species is concerned, three Central European species (*Pinus sylvestris, Fagus sylvatica and Pinus nigra*) and two mediterranean species (*Quercus suber and Quercus canariensis*) are analysed.

2.2. Woodstock Evolution Since the 1890s

The oldest data concerning the cases come from their management plan; most of these were started in the 1890s. Permanent compartment method and regular shelterwood system with artificial regeneration, if needed, were applied. Data of study cases 1, 2, 3 and 5 come from the inventories carried out each 10-year period. Data include number of trees, growing stock, allowable cut and real stock harvested during the period of applicability. Tables 2, 3, 4, and 6 show the above described information.

On the other hand, in case 4 the main forest species is *Quercus suber*, devoted to cork production: consequently, Table 5 shows only data of number of trees, lacking any other woodstock information. As number of debarked trees is highly related to cork production, it could be used for assessing changes in site productivity.

Study cases 1 and 2 are examples of *Pinus sylvestris* and *Pinus nigra* forests where more intensive silviculture with aided regeneration has been applied. In contrast, more extensive silviculture has been applied to the *Pinus sylvestris* forest of the study case 3 and *Fagus sylvatica* forest of the study case 5. In the study case 4, a very typical mediterranean forest with cork as the main direct produce, a specific extensive silviculture not aiming for wood but rather for cork production was applied.

The diagram in Fig. 2 shows the growing stock evolution (m^3/ha) in study cases 1, 2, 3 and 5. In all the cases, with the exception of case 2 (*Pinus nigra* forest), woodstocks have progressively increased, being compatible with their harvesting (Table 2 to 6). On the other hand, case 2 shows a quite stable trend.

2.3. Discussion

After 100 years of forest management planning, the results - analysed through the evolution of stocks and allowable cuts calculated in the decennial revisions - show an increase in growing stock in most of the cases, both in forests where silviculture was more intensive (cases 1 and 2; Tables 2 and 3) and in those where regeneration aid was less and silviculture more extensive (case 3; Table 4). In case 4 (Table 5), with cork oaks (*Quercus suber* L.) and Mirbeck's oaks (*Quercus canariensis* Willd.), intensification is much less evident because the main yield is cork.

Table 2. Evolution of number of trees, stocks, allowable cut and harvesting in case 1: *Pinar de Navafria Forest.* Species: *Pinus sylvestris* (2.760 ha total; 2.505 ha forested).

Project	Period of applicability	Number of trees		Stocks	Allowable cut during the period	Harvested during the period
		Non metric 10-20 cm ϕ	Metric > 20 cm ϕ	$m^3\ ha^{-1}$	m^3	m^3
Management	1896-1905	-	476.825	93.63	45.527	45.596
1st Revision	1907-16	-	579.266	105.15	46.755	46.856
2nd Revision	1918-27	-	559.469	108.21	53.031	53.166
3rd Revision	1928-37	-	619.251	112.92	58.750	58.760
4th Revision	1938-47	-	707.100	128.45	65.749	73.416
5th Revision	1948-57	478.386	664.254	136.62	71.418	70.969
6th Revision	1958-67	476.080	718.412	162.88	90.092	109.008
7th Revision	1968-77	606.513	722.676	179.25	112.928	111.859
8th Revision	1979-88	774.134	736.532	184.08	115.541	110.655
9th Revision	1989-98	925.109	743.752	186.64	135.183	-
Total					659.791[1]	680.285

[1]In order to compare allowable cut and stock harvested total values, the last decade is not included.

Table 3. Evolution of number of trees, stocks, allowable cut and harvesting in case 2: *Los Palancares y Agregados Forest* Species: *Pinus nigra* (4.848 ha total; 4.561 ha forested).

Project	Period of applicability	Number of trees		Stocks	Allowable cut during the period	Harvested during the period
		Non metric 10-20 cm ϕ	Metric > 20 cm ϕ	$m^3\ ha^{-1}$	m^3	m^3
Management	1896-1905	-	429.292	45.62	31.911	32.608
1st Revision	1905-15	-	466.776	46.67	34.286	34.286
2nd Revision	1915-25	-	511.890	55.37	44.376	44.062
Annual plan	1925-28	-	-	-	12.000	12.860
3rd Revision	1928-38	-	462.285	46.39	40.380	42.697
Annual plan	1938-41	-	-	-	5.862	6.867
4th Revision	1941-51	-	456.779	46.49	35.176	37.233
5th Revision	1951-60	605.613	521.625	67.64	44.318	40.630
Prorogation	1960-65	-	-	-	52.350	59.292
6th Revision	1966-75	733.507	506.809	60.16	67.650	70.291
7th Revision	1976-85	748.481	493.555	52.88	64.620	71.184
8th Revision	1986-95	558.532	450.501	47.72	62.128	-
Total					432.929[(1]	452.010

[1]In order to compare allowable cut and stock harvested total values, the last decade is not included.

Table 4. Evolution of number of trees, stocks, allowable cut and harvesting in case 3: *Pinar de Valsain Forest.* Species: *Pinus sylvestris* (7.622 ha total; 7.192 ha forested).

Project	Period of applicability	Number of trees		Stocks	Allowable cut during the period	Harvested during the period
		Non metric 10-20 cm f	Metric > 20 cm f	$m^3\ ha^{-1}$	m^3	m^3
Management	1890-99	-	-	238.01	294.481	192.140
-	1900-09	-	-	-	294.481	142.044
-	1910-19	-	-	-	294.481	140.054
-	1920-29	-	-	-	294.481	147.593
-	1930-39	-	-	-	294.481	119.453
1st Revision	1941-48	852.726	1.854.212	194.40	152.104	156.120
2nd Revision	1948-57	708.619	1.862.408	215.20	218.700	205.962
3rd Revision	1958-64	808.044	1.868.122	219.27	167.692	163.193
4th Revision	1965-77	945.402	1.804.284	270.55	466.338	427.699
5th Revision	1978-87	1.151.749	1.841.368	-	320.190	232.932
6th Revision	1990-99	2.114.976	2.098.618	275.72	270.000	-
Total					2.797.429[1]	1.927.190

[1]In order to compare allowable cut and stock harvested total values, the last decade is not included.

Table 5. Evolution of number of trees in case 4. *El Robledal Forest.* Species: *Quercus suber* and *Quercus canariensis* (5.024 ha)

Project	Date	Number of trees			
		Cork oak			Mirbeck's oak
		Undebarked	Debarked	Total	
Management	1894	0	55.874	55.874	26.278
1st Revision	1904	87.721	117.823	205.544	40.088
2nd Revision	1914	135.239	139.295	274.534	37.298
3rd Revision	-	-	-	-	-
4th Revision	1936	184.963	269.275	454.238	48.642
5th Revision	1944	139.908	276.287	416.195	40.251
6th Revision	1955	161.498	283.222	444.720	35.187
7th Revision	1963	56.357	309.182	365.539	31.981
8th Revision	1971	69.180	324.762	393.942	29.768
9th Revision	1981	109.871	318.537	428.408	27.764
10th Revision	1990	189.767	308.426	498.193	52.002

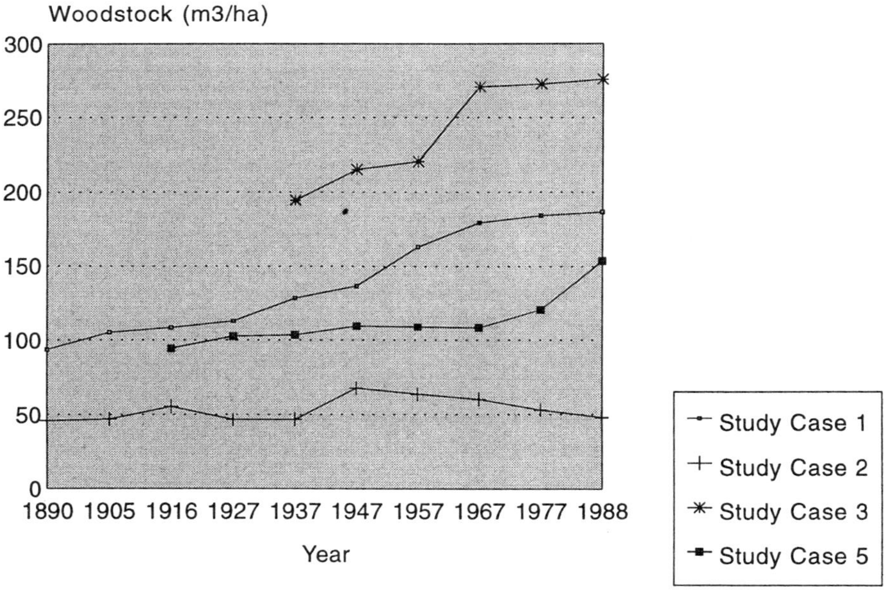

Fig. 2. Wood stock evolution in cases 1, 2, 3 and 5.

Table 6. Evolution of number of trees, stocks, allowable cut and harvesting in case 5: *Aralar forest*. Species: *Fagus sylvativa* (2177 Ha total; 1144 Ha forested)

Project	Period of applicability	N° of trees $\phi > 20$ cm	Stocks $m^3\ ha^{-1}$	Allowable cut during the period of applicability m^3	Harvested during the period m^3
Management	1900-1922	230.790	50.66	43.357	45.787
1st Revision	1923-1936	249.726	53.94	49.215	54.157
2nd Revision	1937-1944	250.387	54.46	32.320	29.947
3rd Revision	1945-1957	251.854	57.48	37.500	41.485
4th Revision	1958-1972	258.428	56.97	32.325	35.666
5th Revision	1973-1993	255.145	60.33	59.240	50.703
6th Revision	1994-2008	255.879	80.75	51.735	-
Total				253.957[(1]	257.745

[1]In order to compare allowable cut and stock harvested total values, the last decade is not included.

The following results can be drawn from Tables 2 to 6:

- *Pinar de Navafría* Forest (Table 2): After 100 years of forest management, wood stocks (in m^3) have increased twofold and the number of metric trees (diameter > 20 cm) increased by 1.6. Non metric trees (10 cm < diameter < 20 cm) doubled between 1948 and 1989. These increases occurred after extracting 680 285 m^3 between 1896 and 1989, which means that the wood harvested was 2.6 times the initial stock. Another indication of the grade of planning fulfillment and the level of the applied silviculture is that actual fellings are almost exactly as proposed in the decennial plans.
- *Los Palancares y Agregados* Forest (Table 3): Current stocks (1985), both in number of metric trees and in m^3, are slightly higher than the initial stocks (1885). The amount of wood produced by the forest and harvested in 100 years is twice as much as the initial stocks. Data show that the silviculture applied is almost as intensive as in Pinar de Navafría with regard to harvesting, but regeneration achieved has not been so intense, shown by the fact that the number of metric trees and stocks remain nearly the same as the initial ones, whereas in Pinar de Navafría they have been multiplied by the factors 1.6 and 2, respectively. As in the previous case, the stock harvested coincides almost exactly with what was planned in the management.
- *Pinar de Valsaín* Forest (Table 4): Current stocks (1990) remain almost the same as the initial ones (1890). Extractions have been of 1 927 190 m^3, slightly higher than the initial volume (1 814 150 m^3). This fact, and the fact that only 70% of what had been planned has been extracted, shows that an extensive or less intensive silviculture has been and is being applied in this forest, compared to the two discussed above.
- *El Robledal* Forest (Table 5): Debarked trees («segunderos», producers of first quality cork) were multiplied by a factor of 5.5 between 1894 and 1990; undebarked trees («bornizos») by factor 2, and the total number of trees by factor 9. Cork yield - according to the current managers - was tripled (enough information to elaborate a complete series of yields is not available).
- *Aralar* Forest (Table 6): In 100 years, the growing woodstock (m^3) was multiplied by a factor of 1.6 and the number of metric trees (diameter > 20 cm) by 1.11. These increments were reached after harvesting 257 745 (m^3). This growth trend shows that the wood harvested during these 100 years is 2.34 times the initial woodstock (110 290 m^3). Although the length of revision periods was rather irregular, the planning objectives were reached surprisingly. There are no data on non-metric trees, but regional foresters are sure that the current number is twice the initial figure. The major portion (2177-1144=1033 ha) of area in *Aralar* Forest not covered by trees is a consequence of grazing practices: sheep and cattle are responsible for the fact that the silvicultural plans are not completable at their highest efficiency. The already mentioned irregularity of revision periods is a result of the grazing pressures.

Results are outstanding in all the study cases and similar to those obtained in other works on the evolution of managed Spanish forests (Montero et al. 1994; Rojo and

Manuel 1992; Jimenez 1992; Manuel et al. 1993). Although these data are aggregated by forest, they can be considered significant enough and representative of the woodstock evolution in Spanish forests during the past 100 years.

Forest primary production (wood or cork) has increased and other products - environmental or even landscape protection - have not decreased since management started. In no case, not even in the forests with the most intensive silviculture, was soil degradation observable; on the contrary, yields per hectare are either the same or higher compared to the time when forest management plans started, since the forests have more trees and these are more uniformly distributed. Certain European authors (Susmel 1980) in Germany, Switzerland, Austria, Italy etc., have mentioned and in some cases confirmed a possible loss of fertility due to clearcutting followed by artificial regeneration and intense soil preparation. This fact has not been evident in Spanish forests where the silvicultural treatments applied were much less intense than in the cases referred to by this author, although it often included small compartment clearcutting. There is not enough information to confirm - not even to suggest - a certain degradation which should be evident by a decrease of growth and yield of the forests treated with this type of cutting in rotations of nearly 100 years.

3. Evolution of Forest Site Growth

3.1. Theoretical Bases and Methodological Background

From an ecosystemic conceptual view of wood stands, it is generally assumed that any forest site type, defined by its geomorphological, climatic, soil and plant components, has a determined efficiency for biomass production. Such production does not fully coincide with wood production, traditionally considered in economical terms as the only true forest produce.

For traditional foresters, the more accurate methodological approach to evaluate the forest site efficiency for wood production is the repetitive measurement of growing, dead and harvested stocks, along the stand rotation period. This approach is a time- and cost-intensive method, especially when it is applied to long and medium rotation species: consequently, foresters had to develop indirect indicators of forest production, based on measurements of stand attributes such as top height, maximum mean growth, final woodstock or interception. They allow assessment of stand production efficiency in a fast and simple way using data well known to forest managers.

Among these indicators, top height is the most accurate attribute used for wood stand productivity. This accuracy has been empirically demonstrated for determined species and age stand where top height reaches the highest correlation to total volume and consequently wood production. Following Eichhorn's Law established in 1904, the relationship between total production and top height does not depend on the age, site quality or stand density. Later researchers have analysed several aspects related to the applicability of the Eichhorn's Law: (1) the law applies better to conifer species than to hardwood species stands; (2) the relationship is more accurate when using stand total volume (growing plus harvested woodstock) than using only stand existing volume; (3) the law is applicable to stand density ranging among the basal area limits established by Wiedemann (1939) and Assmann (1961).

The former considerations give the theoretical bases for using stand top height as indicator of site productivity. If stand top height has varied, and no significant variation in its basal area has occurred during a certain period of time, it is because site productivity has changed in similar quantity.

If site quality curves (top height/age) exist for the studied species, the simplest method for analysing forest growth variation should be to apply these to dominant trees of stands whose site quality has periodically been assessed long beforehand, through inventory measurements. Nevertheless, the lack of mathematical accuracy of most of the site quality curves does not allow the use of this method. Comparison between data of stem analysis of dominant trees growing in the same site quality stands but of different age is as the alternate appropriate method for evaluating changes in forest growth at site level.

3.2. Results from *Pinar de Valsaín* Forest

Following the previous considerations in the Spanish case, we obtained dominant tree stem analysis data from *Pinar de Valsaín* Forest (case study 3) stands of different ages but located on contiguous compartments with the same site quality.

According to the distribution shown in Table 7, two separate samples of dominant trees were selected and measured in 1992: (1) eight old dominant trees, located in different quality sites, aged more than 170 years, from natural regeneration during the period between 1820 and 1840. (2) nine young dominant trees, belonging to stands contiguous to the oldest ones but with their same site quality, aged from 40 to 70 years, from natural regeneration during the period between 1920 and 40.

Both samples were highly representative of the site diversity existing in this 7622 ha forest of *Pinus sylvestris*.

The selected trees were felled and stem analysed, counting tree rings at every 2 m height. From these data, curves of tree height evolution were drawn as shown in Figures 3, 4, 5 and 6.

Table 7. Number of sample trees distributed among site qualities. site index 100 is the stand top height at 100 year age

Site quality	Site index 100	> 170 years	> 70 years
I	> 24.5 m	2	3
II	18.5 to 24.5	3	3
III	< 18.5 m	3	3

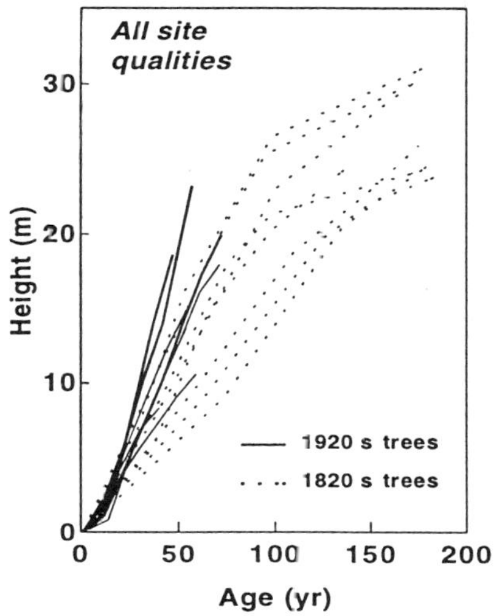

Fig. 3. Stem analysis curves of all sample trees in *Pinar de Valsaín* Forest.

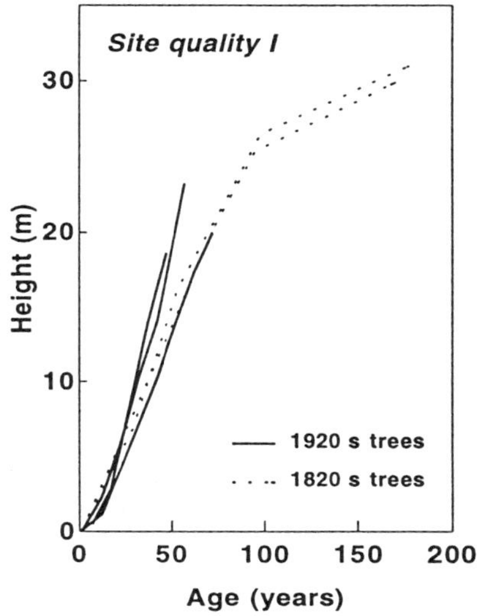

Fig. 4. Stem analysis curves of site quality I trees in *Pinar de Valsaín* Forest.

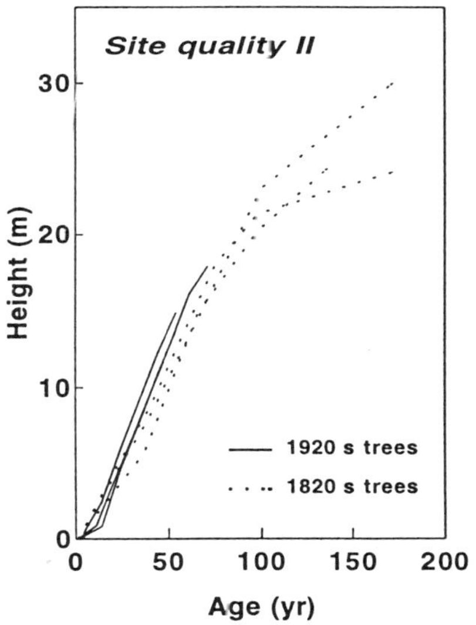

Fig. 5. Stem analysis curves of site quality II dominant trees in *Pinar de Valsaín* Forest.

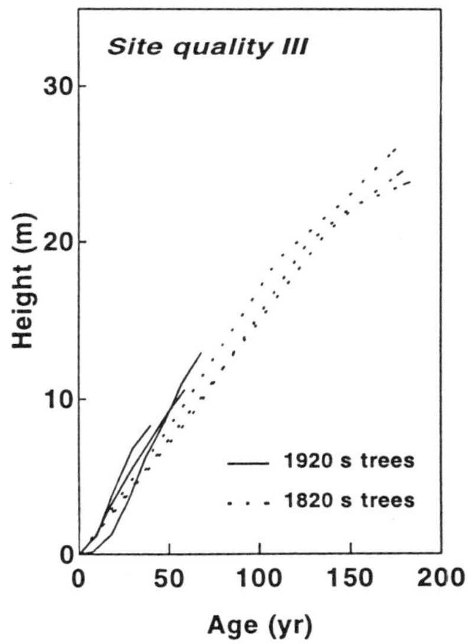

Fig. 6. Stem analysis curves of site quality III dominant trees in *Pinar de Valsaín* Forest.

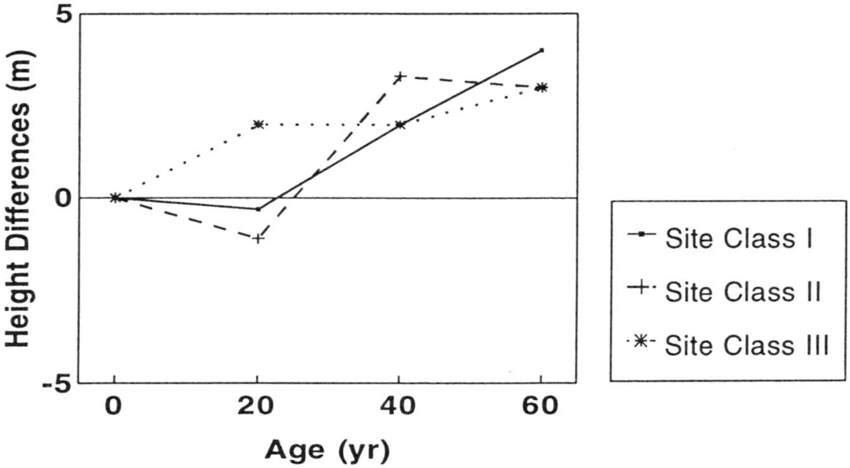

Fig. 7. Comparative analysis of height growth between sample 1820s and 1920s dominant trees. Each curve shows the height differences for the three site qualities.

Figure 3 presents altogether the height variation curves of the 17 trees studied, without distinguishing site qualities. Thus, differences between old and young trees become evident. Old tree curves are clearly grouped into three categories, coinciding with the site quality where they grew. Young tree curves present the same pattern, but considerably higher than the old ones. This trend can be seen in greater detail by drawing separately the curves of trees grown on sites of the same quality. Figures 4, 5 and 6 show the tree height evolution curves corresponding to I, II and III site qualities, respectively.

From these, it appears that, for all site qualities, dominant trees seeded during the 1820-40 period present a slower height evolution than dominant trees seeded 100 years later, during the 1920-1940 period.

When in-depth analysis of these figures is carried out (Fig.7), two well-defined phases appear: (1) At lower ages (< 20 years), trees from the 1920s grew more slowly than those from the 1820s. (2) At medium ages (> 20 years), trees from the 1920s regained height growth and clearly surpassed the height of the trees from the 1820s. In the exception made in the case of one tree of site quality I, the described pattern is evident for all analysed trees.

3.3. Discussion

Accepting the existence of a relationship between top height and total wood production (Assmann, 1961), the data from *Pinar de Valsaín* suggest that stands have followed a positive growth trend, probably due to a site productivity increment. This trend became evident when the young sample trees were 20 years old, but not earlier. Taking this into consideration, the probable increment on forest productivity can be dated after the 1940s.

There are no data available about soil or climatic conditions in the site where the trees were felled. Consequently, no evidence about the real causes of the changes in site productivity can be obtained. Nevertheless, it is possible to explore some preliminary hypotheses about them: *Pinus sylvestris* species has its southern limit of natural distribution in Spain, where it grows only in mountainous regions on sites higher than 1500 m.a.s.l. In *Pinar de Valsaín* Forest sites neither water supply nor nutrient deficiencies have been reported during the past 100 years by their forest managers. Recent studies on climate change at Spanish level (Allue, 1995) have reported a slight increase in temperature levels in Central Spain over the last 30 years. This increase can be the factor responsible for an extension of the *Pinus sylvestris* growth period length. Further research especially designed with the objective of studying growth trends and their causes has to be undertaken.

4. Conclusions

1. Growing stock and allowable cut increments have been recorded in all Spanish forests since the 1890s when management planning was initially implemented.
2. The application of sound silvicultural treatments seems to be an important factor responsible for the recorded increment of forest growing stocks.
3. At a site level, *Pinus sylvestris* stands in a forest in Central Spain have shown important increments in top height which might have been caused by changes in site productivity.
4. Site productivity increments are not fully understood, but changes in temperature regime, and consequent changes in growth period length could be responsible for this.
5. Further research efforts have to be made, covering other Spanish forest types and thoroughly analysing the real causes of forest growth variations at site level.

References

Allue JL (1995) El Cambio Climático y los Montes Españoles, Cuadernos de la Sociedad Española de Ciencias Forestales, N. 2. pp 35-64

Assmann E (1970) The Principles of Forest Yield Study. Pergamon Press ltd. Oxford. 506 p

Elena-Rossello R, Sanchez O (1991) Los pinares españoles de *Pinus nigra Arn.* Síntesis ecol6gica. Monografía nº 81. INIA, Madrid

Garcia Diaz E (1963) España y la selvicultura. II Asamblea Técnica Forestal. Ministerio de Agricultura, Madrid. pp 81-84

Jimenez MJ (1992) Evolución del monte "Pinares Llanos" a través de su ordenación. E.T.S.I. de Montes, ETSIM. Madrid. (not published)

Manuel C, Rojo A and Montero G (1993) Intervención dasocrática en los pinares de Cercedilla y Navacerrada. Congreso Forestal Español. Tomo II, pp. 693-698- Lourizan

Monteiro A (1982) Técnicas de produçao florestal. INIC, Lisboa

Montero G (1992) Aspectos ecológicos y productivos de la selvicultura. Ecología, 6: 111-121

Montero G, Rojo A, y Hernandez A (1994) Theory and practice of silviculture. IUFRO Meeting on mountain silviculture 1993. Spain. Investigación Agraria, Fueca de Serie 3: >3-18

Rojo A. and Manuel C (1992) La intervención dasocrática en los montes públicos españoles. El caso del "Pinar y agregados de Cercedilla", Madrid. Agricultura y Sociedad n. 65: 415-452

Susmel L (1980) Normalizzacione delle foreste alpine, Basi economoche-equilibrio modelli culturali-producttivitá. Liviana Ed. S.P.A.: 427. Italia

Velez R (1982) Forest fires in mediterranean countries. In *Actas del seminario sobre métodos y equipos para la prevención de los incendios forestales*. ICONA/FAO, Madrid

Wiedemann E (1939) Untersuchungen der Preuß. Versuchsanstalt uber Estragsta felfragen. Mitt.a.Fw.u.Fw. 10.3

Growth Trends in Portuguese Forests: An Exploratory Analysis

Margarida Tomé[1], Fernanda Ribeiro, Fernando Páscoa, Rui Silva, Mário Tavares, Amélia Palma and M. João C. Paulo
[1]Instituto Superior de Agronomia, Departamento de Engenharia Florestal, Tapada da Ajuda, 1399 Lisboa Codex, Portugal

Abstract

The objective of this chapter is to give a retrospective view of growth trends of Portuguese forests to identify possible causes for the observed trends, and, when possible, to quantify their effects. Growth trends were studied for two of the most important forest tree species in Portugal, maritime pine and eucalyptus, using growth data available from other research projects. Poplar (*Populus* sp.) is also included in the study. Different sources of data have been used: (1) individual tree stem analysis data (maritime pine and poplar); (2) permanent plots or long-term experiments (eucalypt); (3) forest inventories of small domains (maritime pine). Meteorological information from four climatic stations was also obtained. The permanent plots and continuous forest inventory data were restricted in order to minimise the heterogeneity in the silvicultural treatments present that could induce changes in growth and therefore lead to incorrect assumptions concerning changes in site productivity. Different methodologies were used for the plot growth data and for the stem analysis data. The first were analysed using multivariate linear regression analysis with the objective of testing the significance of adding the year at the start of the growing period and/or precipitation during the growing season to a base model expressing stand growth as a function of stand variables such as age, site index or stems per ha. The analysis of the residuals from existing growth models was also used to confirm the trends that could be found in the previous analysis. The stem analysis data from maritime pine, representing a relatively long time series, were analysed by fitting a mean height growth curve and plotting the mean annual differences (residuals from the mean curve) over year. Climate-dominant height relationships were studied by modelling the mean annual differences in growth with the metereological data available, using a stepwise regression procedure. The growth of two generations of poplar trees available was compared by fitting the Richard´s function to the whole data set and then separately to the data corresponding to different generations. These analyses indicate that volume growth has been decreasing during the last two decades both for maritime pine and eucalyptus, although the trends in dominant height growth are positive for maritime pine and negative for eucalyptus. The negative trend seems to be related to the decreasing amounts of precipitation during the

European Forest Institute Research Report No. 5
Growth Trends in European Forests
Edited by H. Spiecker, K. Mielikäinen, M. Köhl and J. P. Skovsgaard

growing season (spring and early summer) within the last decades, although the variables used to account for growing season precipitation were not fully able to counter-balance the observed negative trends. The negative trend in dominant height growth of eucalyptus plantations may be related to the occurrence of free growth in eucalyptus. No reason was found for the positive trend in dominant height growth of maritime pine. No significant differences between the growth in the two generations of poplar trees was found, although the effect of spring precipitation in tree dbh growth could be seen graphically.

1. Introduction

The aim of this chapter is to report the Portuguese contribution to the project Growth trends of European forests - has site productivity changed? The project is meant to give a retrospective view of the development of forest site productivity in Europe.

Although the project was begun in 1993, Portuguese researchers first learned about it during the workshop held in August 1994 in Joensuu. The importance of the project was recognised by the Portuguese participants at the workshop, who then began to look for data sources that would enable them to answer the question as to whether or not there were any growth trends in Portuguese forests. If the answer is that there are, then a second question arises: "Is it possible to identify the causes of the observed trends?" The results reported here are by no means exhaustive, as the time available both to find and analyse appropriate data sources was too short. Therefore we think that they may only be considered as indicative of the growth trends of Portuguese forests. However, this preliminary analysis indicates the importance that must be given to this subject in future research programs to be carried on with a wider range of data, namely with data specifically gathered for the purpose.

As an introduction to the paper, brief characterisations both of the environmental conditions and of the present forest cover in Portugal are given. It is important to be aware of the differences between forest types and environmental conditions in Southern Europe and those observed in Central and Northern Europe. Growth trends for each of the more important tree species for which growth data were available are then presented.

2. Environmental Conditions in Portugal

Portugal is situated in the south-west of Europe, approximately between 37° and 42° N latitude and between 6° 20′ and 9° 30′W Gr. longitude. Accordingly, the environmental conditions in the country are different from those in most of Europe. The average altitude is 310 m, but there is a sharp contrast between the regions to the north and to the south of the Tagus River, the altitude rising to more than 400 m in 48% of the area to the north of the Tagus River, while the average altitude in the south is 169 m, not rising to 200 m in 61% of the region. The climate varies strongly according to latitude, distance from the sea and relief features. The average annual precipitation is higher in the mountains of the northern region (higher than 1000 mm, but usually less than 1500 mm), gradually decreasing towards the east and the south in parallel with the orography. South of the Tagus River it is less than 700 mm in ca. 80% of the areas. The variation in precipitation throughout the year is very high, with heavy rainfall in the winter and a dry summer. Precipitation values of the humid half-year (from October to March) are ca. 75% of the annual average. The April-September half-year is rather dry and warm, causing summer droughts, particularly in the eastern and southern regions. Precipitation distribution throughout the year, as well as total annual rainfall, varies a lot from year to year. Annual average temperature is higher to the south of the Tagus River (mostly over 16 °C) and lower to the north, mainly in the higher lands and in the inland regions, where it may drop to less than 12 °C. The amount of annual evapotranspiration is also variable, the highest values occurring in regions with the lowest precipitation originating sub-humid and semiarid climates, namely in the southern inland. In contrast, in the northern highlands there is an evident surplus of precipitation when compared to the evapotranspiration associated with humid and wet climates.

Such environmental characteristics lead to a strong differentiation in the spatial distribution of the native vegetation types, dominated mainly by species of the genus *Quercus*. The more humid and cooler northern regions are characterised by the dominance of deciduous oaks, *Q. robur* in the coastal regions and *Q. pyrenaica* in the interior. *Q. faginea* prevails in the coastal natural vegetation of the centre. The warmer and drier inland regions to the south of the Tagus River are characterised by forests of evergreen oaks dominated by *Q. rotundifolia*. In the coastal southern regions and in the Tagus Valley, the cork oak (*Q. suber*) prevails as a result of the careful selection made by successive generations of farmers. Another important native species is *Pinus pinaster* (maritime pine) which occurs naturally to the north of the Tagus River, from the coastline to altitudes of 700-900 m, and in the upper half of the coastal strip of the south.

3. Present Forest Cover

Present land use, in particular forest cover areas, is the consequence of human activities. There had been a gradual decrease in the forest area until the last century, when large afforestation programs led to an inversion of the decreasing trend of forest areas. The natural vegetation, dominated by the genus *Quercus*, has been greatly reduced (mainly to the north of the Tagus River) as a consequence of overgrazing, wildfire (grazing is traditionally managed by fire) and felling for agricultural use (wheat cultivation), followed by soil erosion. Consequently, most soils are usually thin, often associated with rock cropping-up, low in organic matter and normally low in fertility. Soils in the northeast are usually better for they have a higher organic matter content and are not as susceptible to erosion. According to the official maps for actual and potential land use, only 27% of the Portuguese soils are appropriate for agricultural use. Assuming 3% for social areas, at least 70% of the country should be forest or forest areas.

The first Portuguese foresters, graduated late in the nineteenth century, prepared afforestation programs in order to recuperate the eroded soils in the northern and central mountains and also to cover the sand dunes from the coastal strip. The maritime pine (*Pinus pinaster* Ait.) was the species able to grow best under such adverse environmental conditions. It was regenerated by broadcast sowing in most of the afforestations made to the north of the Tagus River, up to an altitude of 800 m. One of the biggest afforestation efforts occurred in the 1940s (Plano de Repovoamento Florestal), in which extensive areas of pine were regenerated with a great emphasis on state areas (3% of the country forested areas) or on areas belonging to the local communities (baldios, 12% of the country forested areas). In the better soils, local and exotic broadleaves were planted. From the 1960s onwards, the eucalyptus became one of the species most frequently used in the afforestation programs, increasing its area from 100 000 ha in 1960 to 550 000 ha at present. Of these, pulp and paper companies own ca. 50% (7% of the country forested area). Afforestation programs in private areas (representing 78% of the country forested area) were begun in the 1970s, taking advantage of the subsidies for afforestation considered in new legislation. These considered a broader range of species, some of them exotic (e.g. *Pseudotsuga menziesii*, *Pinus nigra*, *Cupressus lusitanica*, *Quercus rubra*), but still most of the areas were planted as monocultures of conifers, with patches of broadleaves in the better sites. Some mixed-species stands were regenerated at that time, line mixtures being the preferred composition. Most of these plantations are now in bad shape, as they were not subject to any management after planting. Big afforestation programs are still going on. As an example, between 1987 and 1991, an area of 74 000 ha was regenerated, 46% of which was with maritime pine. To sum up, forest area increased from 1.9 million ha in 1920 to more than 3 million in 1990 (last NFI 1880/90).

These facts lead us to the current state of forest land in Portugal. From the 8.9 million ha of continental Portugal, approximately 1/3 is covered by forests, 16% is forestable land and some of the area used for agriculture is being converted to forest as a consequence of the set-aside community policies. The main species are maritime pine, cork oak, evergreen oak and eucalyptus, which, according to the last NFI (DGF 1992), occupy 39, 22, 15 and 15% of the forested area, respectively. These species appear mainly in pure stands, but in 23 % of the areas mixed-species stands can be found, dominating one of the above species in half of the mixed-species stands.

4. Analysis of Growth Trends

Available growth data in Portugal were analysed in order to select those which could be useful in detecting changes in site productivity. Initially, the more important tree species, such as maritime pine, cork oak and eucalyptus, were considered. However, as it was impossible to obtain cork oak growth data in due time, this species was not included in the study. Some interesting growth data were also found for poplar. Although it is not an important species in terms of area, it may become more interesting in the near future as an alternative to the afforestation of some of the former agricultural lands, so it was also considered.

4.1. Data Available

Maritime Pine. Two different sources of growth data were used to study growth trends of maritime pine: continuous forest inventory for management purposes in the National Forest of Leiria and stem analysis of individual dominant trees growing in the sand dunes along the coast.

The National Forest of Leiria is a state property with an area of 11029 ha located in the coastal dunes of Marinha Grande in central Portugal (Fig. 1). Most of the forest area (8702 ha) is made up of pure maritime pine stands managed for high quality timber production (production area). A small percentage along the coast is protection forest with the maritime pine as dominant species (1976 ha). Some areas with different land uses can also be found other protection forests (150 ha), agricultural (2 ha) and social areas (16 ha). The National Forest of Leiria is divided into 343 approximately rectangular management units that are some times subdivided into two or more stands. For management purposes, the National Forest Service designed a continuous forest inventory that has systematically covered the forest since 1970 with an intensity of one 500-1000 m^2 plot per ha (two plots per ha until 1988) every 5 years. The inventories are carried out before thinning, for thinning control, beginning around the age of 10 years, when the first (non-commercial) thinning occurs. The inventories provide the diameter distribution (frequencies by 5 cm dbh classes), the dominant height and a sampling of heights for each plot. From these, the density, the mean height and diameter, the basal area and the stand volume per ha can be easily obtained. Thinning information is available only on a stand basis. Consequently, the inventory data were aggregated on a stand basis for growth and yield modelling purposes. Small stands with an area less than 10 ha were discarded to avoid stand parameters estimated on a small number of plots. Other uncharacteristic stands such as those that had been affected by forest fire were also excluded from the analysis. The growth periods with occurrence of thinnings were also eliminated in order to study growth trends. Table 1a and b briefly characterises the data set available after the elimination of the unwanted stands. As can be seen, the number of growth periods during the decade 1970-1980 is much higher than in the

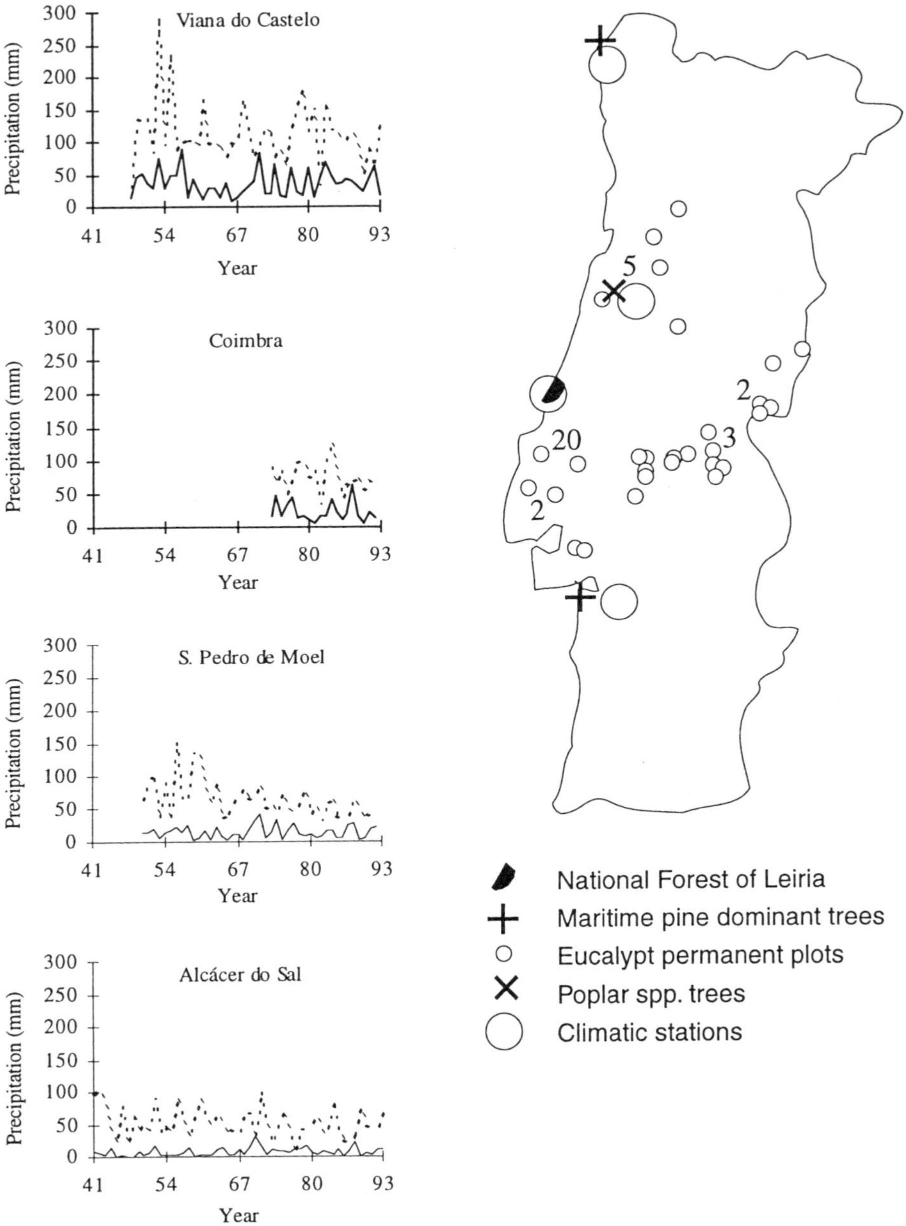

Fig. 1. Location of the eucalyptus permanent plots, of the National Forest of Leiria (maritime pine CFI data), of the sites where individual trees (maritime pine and poplar) were harvested and of the climatic stations used in the study. The numbers indicate that more than one plot was used from the same site. The insets show the spring (dotted line) and summer (full line) precipitation (means of 3 months) for each one of the climatic stations.

Table 1a. Characterisation of the maritime pine continuous forest inventory data.

Variable	1970-1980 (n=174)			1980-1990 (n=71)		
	Min	Mean	Max	Min	Mean	Max
SI	14.00	18.94	24.00	14.00	18.25	23.00
T	9.00	45.38	105.00	12.00	45.89	81.00
N	156.00	366.70	1332.00	156.00	508.70	2073.00
H_{dom}	4.80	17.10	24.90	6.00	16.53	25.10
V	6.90	128.00	323.90	16.80	140.60	352.80
G	2.31	17.30	32.52	5.30	18.83	34.20
i_G	0.06	0.87	2.50	0.00	0.74	2.00
i_{Hdom}	0.00	0.26	0.90	0.04	0.30	0.80
i_V	0.10	7.76	12.78	0.00	6.72	13.15
$year_{sgp}$	70.00	74.89	79.00	80.00	82.37	85.00

where SI is site index; T is stand age; N is number of trees per ha; H_{dom} is dominant stand height; V is volume per ha ; G is basal area per ha; i_G, i_{Hdom}, i_V are current annual growth in basal area , dominant height and volume and $year_{sgp}$ is the year at the start of the growth period.

Table 1b. Number of growth periods by age class, site index and decade for the maritime pine continuous forest inventory data set.

Decade	Site Index Classes (base age 50)	Age Classes					
		0-15	15-30	30-45	45-60	> 60	Total
70-80	<18	4	8	12	27	18	
	18-20	5	11	11	22	17	
	>20	0	11	21	4	3	
Total							174
	>18	6	3	2	12	13	
80-90	18-20	4	6	1	7	6	
	>20	-	3	1	6	1	
Total							71
Total							245

decade 1980-1990 as a consequence of the fact that the data from the second half of the last decade (year at the start of the growth period greater or equal than 1985) are not yet included in the data set.

Data from the stem analysis of 32 individual dominant trees harvested from pure stands of maritime pine located in three different regions - north (National Forest of Camarido), centre (National Forest of Leiria) and south (Forest of Comporta) - from the coastal dunes in Portugal (Fig. 1) were available from an ongoing project on wood quality modelling. All the selected stands are located in the coastal dunes, with similar soils (sandy soils), exposition and slope (flat land) and altitude (ranging from 10 to 50 m). Stand age ranges from 10 to 73 years of age and stand density from 220 to 693 trees per ha. Plot area was defined according to stand age and ranges from 750 m^2 to 3000 m^2, to include ca. 50-60 trees. Height growth in the sampled trees was assessed through the location of the whorls and respective measurement of the annual shoots; dbh growth was obtained by measuring the rings in wood samples at dbh level for trees older than 50 years of age or from disks at the same level for younger trees. As dbh growth was not yet available for every tree, only dominant height growth analysis is included in this study. Table 2 characterises the 32 dominant trees that were used in the analysis.

Table 2. Characterisation of the maritime pine dominant trees harvested for stem analysis.

Site	Age Classe	Number of trees	Site index (base age 50)	dbh	height
	10	2	24/25	17.0-20.0	8.9-9.1
	15	2	19/21	14.0-18.6	6.9-8.6
Camarido	20	2	21/23	20.0-21.2	11.4-13.1
(north)	30	2	21/22	30.0-35.0	15.3-18.4
	50	1	27	49.0	24.3
	70	1	19	49.5	22.3
	10	2	19/25	13.0-13.7	5.8-8.9
	15	2	21/23	16.6-19.0	10.1-10.8
Leiria	20	2	17/21	19.0-23.4	8.9-12.9
(center)	30	2	23/23	32.5-33.6	17.4-18.0
	50	2	14/24	29.0-41.0	12.4-24.0
	70	2	14/20	38.7-50.1	18.5-24.0
	10	2	20/21	12.5-14.0	6.0-6.1
	15	2	20/22	14.1-20.0	8.0-9.2
Comporta	20	2	20/21	22.0-23.0	11.0-12.5
(south)	30	2	17/17	27.0-30.0	14.4-14.5
	50	1	17	38.4	17.8
	70	1	21	47.0	23.6

Eucalyptus Plantations. Eucalyptus data came from permanent plots established since 1970 in first rotation eucalyptus plantations by CELBI´s pulpwood enterprise. These plots were located mainly in the central coastal region of Portugal. In addition to the data from these plots, data from a set of more recent permanent plots, established by ACEL (an association of Portuguese pulp companies) in 1989 were also used. These last plots were established to complement the data already available, covering different ecological zones, different site preparation methods, different rotations and a significant sampling of young ages (not always present in the CELBI's data). CELBI´s plots have been remeasured annually several times; in a few instances, more than a year went by between measurements. In each permanent plot, diameter at breast height (dbh) has been measured for all trees, while the heights have been measured in sample trees. Heights have also been measured at the ratio of one per 100 m^2 among the largest dbh trees in order to determine dominant heigh. The ACEL´s plots were measured two or, in some cases, three times using a similar methodology.

In order to obtain a more homogeneous data set, the following kinds of plots were eliminated: plots with too many (≥1250) or too few (<900) trees per ha; plots established using site preparation techniques different from those used at present; plots measured at unusually high ages (>15 years); plots in the 2nd or later rotations; and, finally, plots located anywhere except in the central (coastal and inland) region of Portugal. Data available for the study included 59 permanent plots, originating 344 growth periods (Table 3a and b).

Poplar Plantations. Poplar data came from an ongoing project whose main purpose is to develop growth and yield models for this species in a farm forestry system. In connection with this project, some plantations, located in former agricultural lands of the Mondego River, are being studied. Accordingly, late in 1993 and early in 1994, 44 permanent plots were established in these stands with ages ranging from 8 to 22 years. As the stands are small in size, plots were installed in order to quantify also the edge effect on the growth. At the time of writing, the plots have been measured twice, but some trees were cut for stem analysis purposes. Of these, two stands, both located in the lower valley of the Mondego River, were selected. The stands have a similar site index (25.6 and 25.3 m at 12 years, respectively) and were established with the same spacing (6x6 m - 278 trees ha^{-1}) and clone. Both sites were formerly used for agriculture (corn and rice crops); they present similar soil types (alluvial structured sandy soils), exposition and slope (flat land) and altitude (15 and 40 m, respectively). The two stands, of 12 and 20 years of age, represent two different generations of very similar stands and were therefore considered appropriate for the growth trend analysis. Two 500 m^2 plots, corresponding to 16 trees each, were available from each stand. As the plots were established at wide spacing and with clones, no clear competitive hierarchy was present at the time the plots were established; consequently, all the trees were considered to be competition free (dominant); however, the selection of the two trees to be harvested for stem analysis was based on the definition of dominant trees, namely those trees with the largest dbh. Tree dbh, total height, four crown radii and crown depth were measured

Table 3a. Characterisation of the eucalypt permanent plot data set.

Variable	1970-1980 (n=194)			1980-1990 (n=150)		
	Min	Mean	Max	Min	Mean	Max
SI	13.10	20.32	27.60	13.10	20.73	30.30
T	1.20	7.68	13.80	1.00	10.17	15.00
N	529.00	970.90	1125.00	533.00	1016.00	1235.00
H_{dom}	2.10	16.95	32.70	3.90	20.77	34.30
V	0.10	117.00	437.70	0.60	164.10	490.00
G	0.05	14.26	37.09	0.39	18.01	39.84
i_G	0.12	1.93	5.44	0.00	1.30	3.38
i_{Hdom}	0.00	1.75	5.00	0.00	1.07	3.80
i_V	1.50	23.60	87.14	0.00	17.40	69.40
$year_{sgp}$	70.00	74.96	79.00	80.00	85.11	91.00

where symbols are as in Table 1a.

Table 3b. Number of growth periods by age class, site index and decade for the eucalypt permanent plot data set.

Decade	Site Index Classes (base age 10)	Age Classes					
		0-3	3-6	6-9	9-12	12-15	
	<15	3	5	12	1	-	
70-80	15-20	7	18	22	14	7	
	20-25	5	19	26	21	12	
	>25	2	5	8	4	3	
Total							194
	<15	1	1	1	7	4	
80-91	15-20	-	5	11	9	16	
	20-25	3	17	15	16	31	
	>25	1	2	3	2	5	
Total							150
Total							344

before harvesting in one site and only tree dbh and height were measured in the other. Dbh growth was assessed after harvest by measuring radial increment in two radii (north and west) with an Eklund positionometer. Table 4 summarises the data available from the eight poplar trees used in the study.

Climatic Data. As mentioned previously, the available data sets were restricted, as much as possible, in order to obtain a set of increments from stands with a fairly homogeneous silviculture. For instance, the permanent plots and continuous forest inventory data sets selected for the study excluded all the plots with characteristics thought to be able to modify stand increment such as different silvicultural treatment (e.g. spacing, site preparation). Although there are still some variations impossible to eliminate, they are minimised and will probably not be important in the growth trends that can be found. Climate is another factor that may be responsible for observed growth trends (e.g. Becker 1989). Fig. 1 shows the location of the four climatic stations - Viana do Castelo, Coimbra, S. Pedro de Muel e Alcácer - that were used in this study. Monthly precipitation data were available, with a few exceptions, for every one of the climatic stations. As far as temperature is concerned, mean monthly temperature was available for one of the stations (Coimbra) whereas only the monthly maximum and minimum temperatures were available for the other three stations. Some data are missing in each meteorological time series. Spring and summer rainfall are usually considered as important driving variables of stand basal area growth. Accordingly, spring and summer rainfall, as well as their sum, were computed for each climatic station. Spring and summer temperatures were also calculated in a similar manner.

Table 4. Characterisation of the poplar trees harvested for stem analysis.

Site	dbh (cm)	height (m)	Crown radii (m)				Crown depth (m)
			N	S	E	W	
	33.15	26.0	3.6	2.9	3.1	3.3	6.0
Quinta	30.39	19.5	3.0	3.1	3.2	3.7	9.5
das	28.97	26.5	3.0	2.8	3.0	3.1	8.0
Areias	31.51	25.0	3.5	3.3	3.4	3.7	8.5
	39.78	30.0	[1]	[1]	[1]	[1]	[1]
Vale	40.76	31.0	[1]	[1]	[1]	[1]	[1]
Travesso	39.12	28.5	[1]	[1]	[1]	[1]	[1]
	38.42	29.0	[1]	[1]	[1]	[1]	[1]

[1]not available

In the plot growth data, the use of the rainfall data was problematic, particularly in the eucalyptus permanent plot data set, as a consequence of the irregularity of the measurements. The dependent variable used in all the statistical analyses was the current annual growth, computed as the difference between two successive measurements divided by the difference between the respective ages. While this way of computing growth was acceptable because every growth period that did not include at least one complete growing season was discarded from the analysis, it was more difficult to establish correlations with climatic data as the three spring and/or summer months were not always included in the period and because some periods included more than one growing season. In fact, mean spring (or summer) rainfall had to be computed as the mean of the respective values for the springs (or summers) included in the growing period. This problem is minimised in the maritime pine data as the continuous forest inventory in the National Forest of Leiria is routinely performed in September, and is irrelevant for the stem analysis data. The monthly values of precipitation and temperature available for each climatic station as well as the computed values for the summer and spring were used in this case.

4.2. Methods

Detection of Trends or Changes. The data available were not gathered with the objective of estimating growth trends or changes in site productivity; consequently, the statistical methods specifically designed for time series analysis cannot be used. As stressed in the EFI report, at least 20 to 30 consecutive observations in time should be available in order to allow for time series and trend analysis; this is not the case for most of the available data sets. Even for the individual tree stem analysis data, tree age ranges between 10 and 73 years of age, so that the required 20 to 30 observations at dbh level could not be made for all the trees. Different methodologies were designed for the analysis of permanent plots and continuous forest inventory data sets (plot growth data) and for the stem analysis data.

In the case of plot growth data, several statistical analyses were selected: (i) multivariate linear regression analysis between increment and several parameters presumably affecting it; (ii) analysis of the residuals from existing growth models representative of all data; (iii) dominant height growth modelling using growth functions or their difference forms. Multiple linear regression analysis was performed to select multiple linear models of the volume, basal area and dominant height increments over several stand variables. An all possible subset regressions algorithm was used to select the subset of regressors (all significant at least at the 0.05 level) that best predicted each one of the increments referred to above. The selected models were used as base models for the subsequent analysis. The significance of adding the year at the start of the growth period ($year_{sgp}$) to each one of the selected base models was then tested. The analysis of the residuals (observed-predicted) from existing growth models was used to confirm any trend that could be found in the previous analysis. The dominant height and basal area growth models developed by

Falcão (1992) for unthinned stands in the National Forest of Leiria with the data set used in this chapter were selected for the analysis of the residuals. Residuals from the model were computed for each individual growth period and annual means of the residuals were computed by age class and analysed by means of graphs. The same analysis was carried out for the eucalyptus permanent plot data in relation to the basal area growth model developed by Carvalho (1992) and the site index curves presented by Tomé (1990). Finally, the dominant height growth models already selected for the residual analysis were fitted separately to the different decades considered and then to all data. Simulations obtained with the three curves were used to assess the importance of the differences found, from a practical standpoint.

Stem analysis data from maritime pine dominant trees were analysed using a procedure similar to the one suggested by Becker (1989). The mean height growth curve was obtained by fitting the Richard´s function to the data from all the trees available for the study. As can be seen in Table 4, there is some variation in site index among the stands where the pine trees for stem analysis were harvested. To avoid this problem, the parameters of the Richard´s function were expressed as linear functions of site index. After analysing some alternative formulations, corresponding to every parameter or a subset of the parameters expressed as a linear function of site index, it was decided to express only the k parameter as a linear function of site index. The significance of the differences between alternative formulations was assessed using the Gallant´s test (Gallant 1975). After obtaining the mean height curve, the annual differences in growth (residuals from the mean curve) were computed for each individual height growth series. The new data were then studied without reference to the cambial age. The plot of the mean residuals obtained for each year against the year was used to detect the growth trends present in the data. This analysis was carried out for all data and also separately for each location.

The same methodology was not appropriate for analysing the stem analysis data from poplar, as there are no long-term data available. Therefore, these last data were first analysed by means of graphs and then by fitting the Richard´s function to all data and then separately to the data corresponding to different generations. The significance of the differences between the two generations present was assessed using the Gallant´s test (Gallant 1975).

Reasons for Change. As mentioned in the previous section, the permanent plots and continuous forest inventory data sets selected for the study excluded all the plots with characteristics that could conceivably have modified stand increment, such as different silvicultural treatment. Although there are still some variations impossible to eliminate, their significance in explaining any growth trends that might be found will be minimal. Spring and summer precipitation as well as their sum (growing season precipitation) were considered as one of the factors that can be correlated with the increment. Accordingly, spring, summer and growing season precipitation ($prec_{spring}$, $prec_{summer}$ and $prec_{combined}$) of the S. Pedro de Muel climatic station were also considered as regressors in the multiple linear regression analysis described in the previous section. The significance of adding each one of them to the model selected was tested both with and without the year at the start of the growing period present in the model.

Climate-dominant height growth relationships for stem analysis data from maritime pine were established using a linear regression model with monthly, spring and summer meteorological data of the current and the 2 preceding years, regressed on the mean annual differences in growth as the dependent variable (Becker, 1989). The stepwise procedure was used to select a subset out of the 139 meteorological variables that were tested. In addition to the meteorological variables, growth in the preceding year and site index were also tested as independent variables. The stepwise procedure was used for all data (1952-1993) and for the period 1952-1975 as a calibration period.

Precipitation-diameter growth relationships for stem analysis data from poplar were assessed only graphically.

5. Results

5.1. Maritime Pine

Continuous Forest Inventory Data. Table 5 shows the results of the multiple linear regression analysis performed for the increments in the three variables considered (dominant height, stand basal area and stand volume), both including or not including precipitation data. Basal area and volume increments showed significant negative regression coefficients with the year at the start of the growing period $year_{sgp}$ (evidenced by a significant student t-value), indicating that a negative growth trend may be present in the data. On the other hand, there is a significant positive trend in dominant height increment, not large enough, however, to counter-balance the negative trend observed in basal area increment, so that the trend in volume increment must ultimately be described as negative. The variable $year_{sgp}$ itself is, of course, not responsible for this trend. As previously mentioned, one of the factors that probably affects growth is the amount of rainfall during the growing season. Although the correlation between precipitation and the variable $year_{sgp}$ is fair (e.g. -0.211 for $prec_{spring}$, 0.041 for $prec_{summer}$ and -0.147 for $prec_{combined}$ in the S. Pedro de Muel climatic station), the contribution of these variables to the multiple linear regression model seems to be related. Precipitation is not significant in explaining variation in dominant height growth, but is significant in the multiple linear models of basal area and volume growth without the variable $year_{sgp}$. In the model with both variables, however, precipitation is significant only for dominant height growth.

Fig. 2 shows the graphics of the mean annual residuals for stand basal area and dominant height for the age class 45-60, one of the age classes with a broader range of $year_{sgp}$. As can be seen, the trends in basal area growth (negative) and in domi-

Table 5. Multilinear regression analysis performed with the maritime pine forest inventory data set.

Increment	Variables in the model	t-student values				
		Base model	with yearsgp	with precipitation	with both	
i_G	SI	10.27	8.93	9.05	8.84	
	N_1	4.20	6.63	5.88	6.55	
	$Hdom_1$	-8.24	-5.88	-6.75	-5.91	
	yearsgp	-	-6.08	-	-2.84	
	$prec_{combined}$	-	-	5.37	0.96	n.s.
	r^2	0.700	0.740	0.732	0.7407	
i_{Hdom}	SI	5.58	6.01	5.47	6.51	
	N_1	2.87	1.53	3.19	1.99	
	$Hdom_1$	-4.74	-5.22	-4.50	-5.40	
	T_1	2.09	2.19	2.11	2.51	
	yearsgp	-	2.66	-	5.21	
	$prec_{spring}$	-	-	1.40	4.65	
	r^2	0.599	0.611	0.602	0.643	
i_V	SI	12.72	10.95	11.26	10.87	
	N_1	3.04	5.37	4.44	5.33	
	G_1	4.91	5.97	5.49	5.94	
	V_1	-4.62	-4.31	-4.36	-4.29	
	yearsgp	-	-5.56	-	-3.20	
	$prec_{combined}$	-	-	4.47	0.39	n.s.
	r^2	0.501	0.558	0.539	0.558	

where i_G, i_{Hdom}, i_V are current annual growth in basal area , dominant height and volume; $year_{sgp}$ is the year at the start of the growth period; $prec_{spring}$, $prec_{summer}$ and $prec_{combined}$ are respectively the spring and summer precipitations and their sum; and n.s. indicates a non-significantt-value at the 0.05 significance level

nant height growth (positive) established by means of the multiple linear regression analysis may also be established through analysis of the residuals from existing models.

Fig. 3 shows that the existing growth trends are important from a practical standpoint. In fact, the use of a common model that does not take the differences between decades (median frequency level growth trend) into account can lead to over or underestimating dominant height. For a site of medium productivity, deviation from the real values may be as much as 2 m (11%) at 50 years of age.

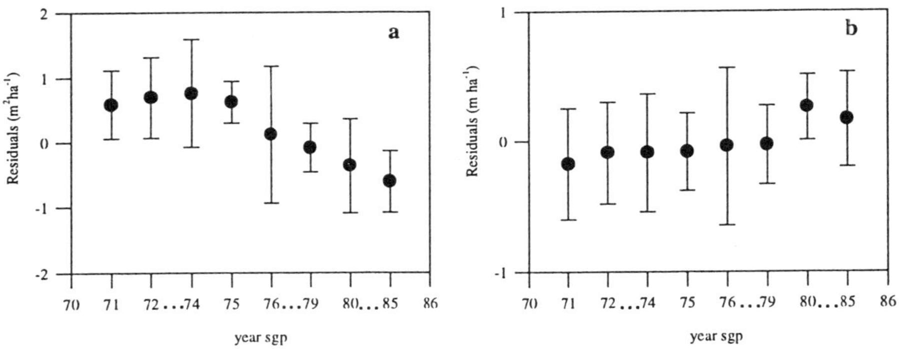

Fig. 2. Plot of mean annual residuals from existing growth models for maritime pine (*a*: basal area growth; *b*: dominant height growth) over the year at the start of the growing period for the age class 45-60.

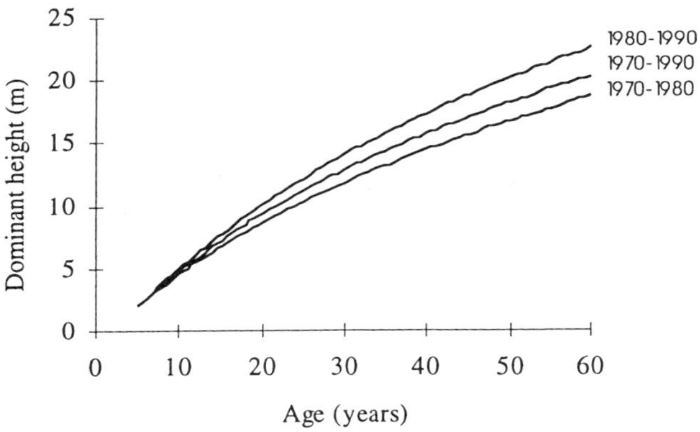

Fig. 3. Simulated dominant height growth of a maritime pine stand of medium site index with 2 m at 5 years of age, using growth models fitted to the whole data set and separately to the data from each decade.

Stem Analysis Data. Fig. 4 shows the mean height growth of dominant trees for the 32 maritime pine trees. Three different site indices are represented and, as expected, the maximum occurs earlier in the best sites. When plotting (decade scale) the mean annual differences for each site individually and for all data combined, similar medium frequency level patterns were obtained, indicating a positive trend during the period 1955-1961, followed by a more or less stable period, a slight decrease from 1971-1977 and, finally, a positive trend during the two last decades (1977-1992). Fig. 5 shows these trends for all data and there, too, the positive trend

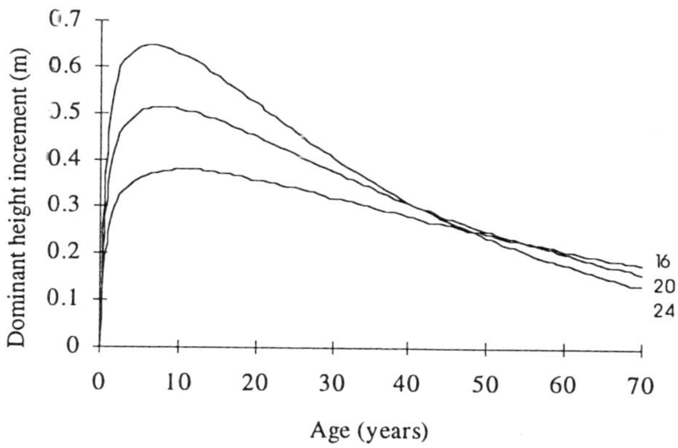

Fig. 4. Mean height growth of maritime pine dominant trees for three site index classes, simulated with a Richard's function fitted to the stem analysis data from the three sites.

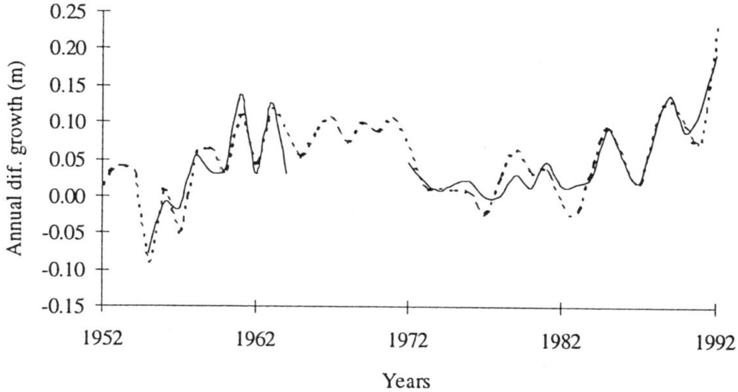

Fig. 5. Observed mean annual differences in height growth (dotted line) and the respective estimated values with the model fitted to the whole set of data.

observed in the continuous forest inventory data for the last two decades also becomes apparent. Table 6 summarises the results of the two stepwise regressions that were performed, with the data from the 32 trees, in order to model the mean annual differences with the meteorological information. This analysis was not carried on by site because of the small number of trees available for each site. Figures 5 and 6 show the observed mean annual differences in height growth in comparison to the values estimated by the two models. As can be seen in Table 6, the model for all data includes a great number of variables and some of the regression coefficients

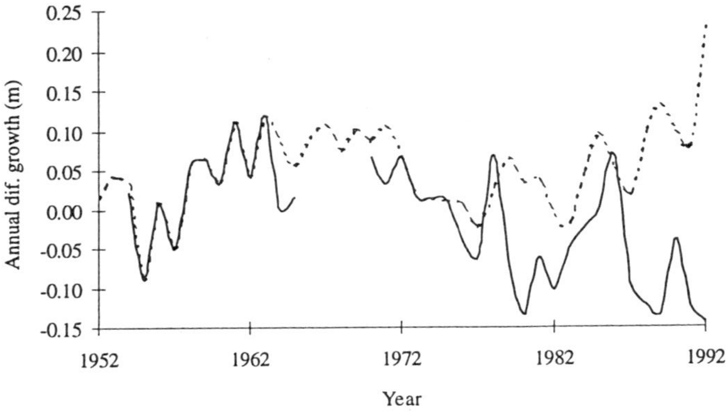

Fig. 6. Observed mean annual differences in height growth (dotted line) and the respective estimated values with the model fitted to the period 1952-1975.

Table 6. Model explaining mean differences in dominant height growth of maritime pine for all data and for the calibration period.

	Period 1952-1992			Period 1952-1975	
	Variable	t-student		Variable	t-student
Prec4	(y-2)	108.75	Prec5	(y-2)	113.74
Tmin7	(y-2)	97.38	Tmin5	(y-2)	-69.72
Tmin1	(y-1)	87.66	Tmin8	(y)	-45.42
Tmin10	(y)	-85.77	Tmax2	(y-2)	42.92
Tmin7	(y-1)	-77.64	Tmin7	(y-2)	42.71
Tmax10	(y)	73.38	Prec4	(y)	24.67
Prec9	(y)	-54.64	Precma	(y-2)	14.46
Tmin9	(y-1)	49.16	Precs	(y-2)	9.22
Tmin4	(y-1)	32.11	Tmin11	(y-1)	4.81
Tmin4	(y-1)	28.84			
Tmin5	(y)	27.95			
Tmax2	(y-2)	23.25			
Tmax2	(y-2)	16.7			
Tmin9	(y)	15.38			
Tmaxsp	(y)	-9.14			
Tmin7	(y)	7.85			
Tmax10	(y-2)	5.33			
r^2	0.8672			0.5060	

where Tmax*i* is the maximum temperature in the *i* month; Tmin*i* is the minimum temperature in the *i* month; Prec*i* is the mean precipitation in the *i* month; Tmaxsp is the mean Tmax*i* for spring months ; Precma is the mean annual precipitation; Precs is the mean summer precipitation; y is the year$_{sgp}$ and (y-1) and (y-2) are the preceding years

exhibit unexpected signs. This may be due to the presence of multicolinearity, to be expected in a model with 17 variables. Generally, both models show the importance of precipitation mainly in spring (positive correlation). Maximum and minimum temperatures seem to play an important role, although the models obtained cannot be easily interpreted from a biological standpoint The mean annual differences in growth estimated with the model fitted to the data set as a whole almost coincide with the observed values, both the medium-frequency and the high-frequency level (annual variations) patterns could be predicted. The model fitted to the calibration period (1952-1975) estimates annual differences in growth nearly coincident with the observed ones; in the verification period, although the high-frequency level pattern is present in the estimated values, the model was unable to predict the medium-frequency positive growth trend present in the data. This fact may be due to the small number of years included in the calibration, to the fact that the meteorological variables available are not the most appropriate to explain the medium-frequency trends, or to the fact that climate is not the explanation for the positive trend in dominant height growth that seems to have occurred over the last two decades.

5.2. Eucalypt

Table 7 shows the results of the multivariate linear regression analysis performed with the permanent plot data of eucalypt. Growth in all the variables considered shows a significant negative trend with the year at the start of the growth period, even including dominant height, which had shown a positive trend in the maritime pine. Spring and summer precipitation are positively correlated with growth in all the variables. However, precipitation is not able to counterbalance the negative trend observed over time, as is shown by the significant t-values for the variable $year_{sgp}$ and $prec_{combined}$ in the model with both variables. It is important to stress that the method used to compute the variables that account for growing season precipitation has some drawbacks, as was previously explained. The negative trend in the growth of all the variables can also be observed in the analysis of the residuals from existing models (Fig. 7) and from the fitting of the dominant height growth model to the data set as a whole and then separately to each decade (Fig. 8). This last analysis indicates that the use of a common model which does not take the medium-frequency level growth trend effects into account can lead to over or underestimating dominant height. For a site of medium productivity, deviation from the real values may be as much as 2 m (10%) at 10 years of age.

5.3. Poplar

As can be seen in Fig. 9, trees planted in 1980 exhibit a mean dbh slightly higher than those planted in 1973 during the first 10 years. The dbh increment decreases after the age of 10 years, being the dbh growth curves for both generations almost coincident from 11 years of age until the harvest. However, there is no statistically

Table 7. Multilinear regression analysis performed with the eucalypt permanent plot data.

Increment	Variables in the model	t-student values			
		Base model	with yearsgp	with precipitation	with both
i_G	SI	10.62	10.85	10.63	10.96
	T_I^2	-12.75	-8.07	-12.71	-8.22
	N_I	2.06	4.26	2.49	4.39
	$year_{sgp}$	-	-7.25	-	-6.78
	$prec_{combined}$	-	-	4.00	3.48
	r^2	0.476	0.547	0.500	0.562
i_{Hdom}	SI	7.59	8.42	7.31	7.94
	T_I	2.85	4.09	2.60	3.57
	$Hdom_I$	-5.84	-6.66	-5.52	-6.17
	$year_{sgp}$	-	-4.20	-	-3.16
	$prec_{combined}$	-	-	3.75	2.55
	r^2	0.292	0.327	0.320	0.340
i_V	SI	4.10	4.79	3.96	4.47
	T_I^2	-5.23	-3.23	-5.44	-3.72
	G_I	6.88	6.20	7.30	6.62
	$year_{sgp}$	-	-4.13	-	-2.92
	$prec_{combined}$	-	-	4.30	3.16
	r^2	0.382	0.412	0.414	0.429

where symbols are as in Table 5

Table 8. Results of the Gallant´s test performed to evaluate the advantages of fitting separate models to the data from each generation present in the poplar stem analysis data

Data set	i	SSR_i	d.f.	Gallant test $SSR_3/(SSR_2+SSR_1)$	
1973-1993	1	55055.1	73	0.999	n.s.
1980-1993	2	23304.5	21		
Combined	3	78250.0	121		

n.s.- non significant at the 0.05 level

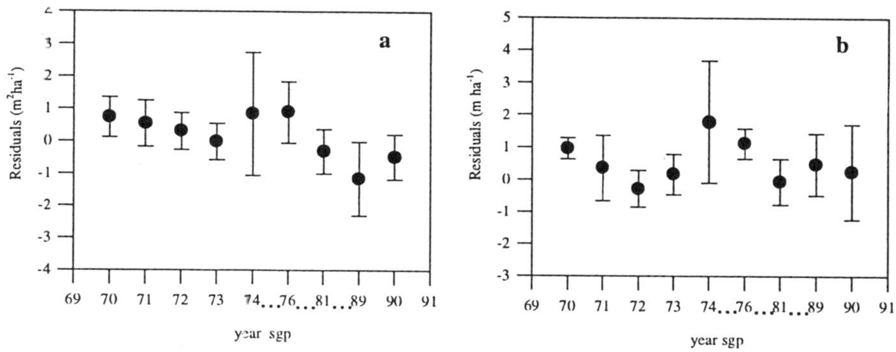

Fig. 7. Plot of mean annual residuals from existing growth models for eucalyptus (*a*: basal area growth; *b*: dominant height growth) over the year at the start of the growing period for the age class 0-5.

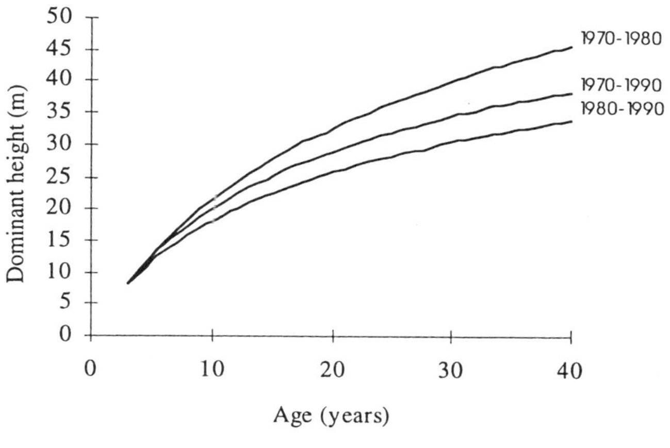

Fig. 8. Simulated dominant height growth of an eucalyptus stand of medium site index with 8 m at 3 years of age using growth models fitted to the whole data set and separately to the data from each decade.

significant advantage in fitting separate growth curves to the data from the two generations considered (Table 8). Although the differences between the two generations are not statistically significant, it can be seen in Fig. 9 that the initial higher dbh increment observed for the trees planted in 1980 is related to the higher spring precipitation registered in 1984, which also influences the growth of the trees planted in 1972 at the age of 12 years.

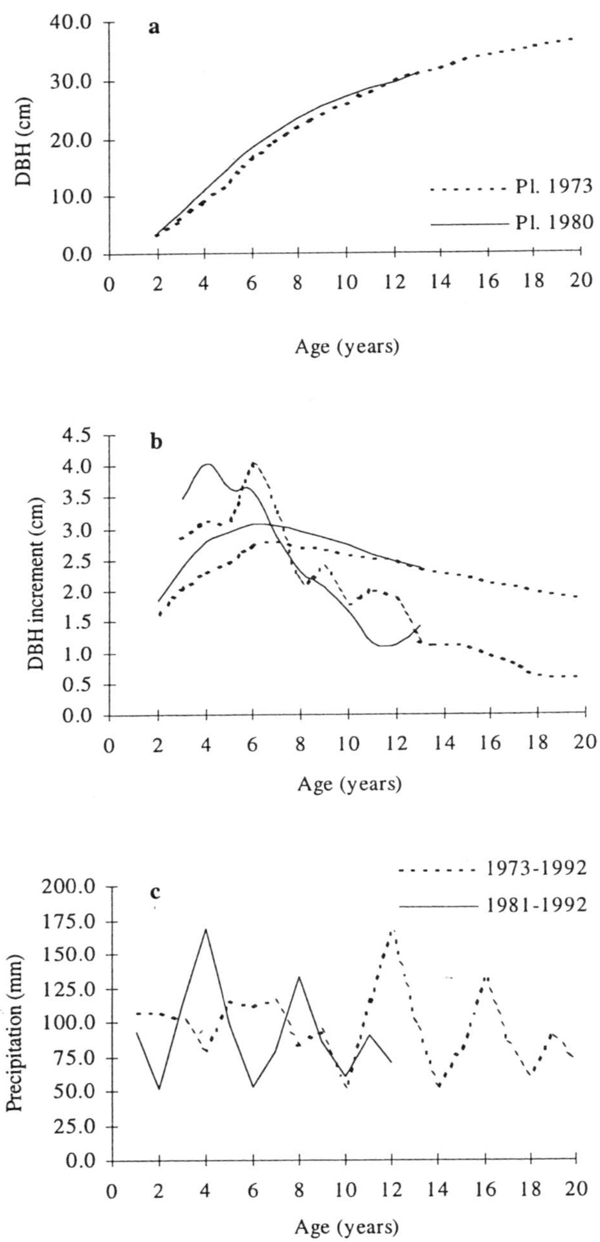

Fig. 9. Dbh growth (*a*), mean annual and current dbh increments (*b*), and spring precipitation over age (*c*) for the poplar stem analysis data

6. Discussion

Different data sets from three species were analysed in order to gain insight into the problem of changes in forest site productivity and their causes. Our discussion of forest site productivity is based upon the definition given in Spiecker et al. (1994) for this project, namely, the wood production potential of a site for a particular tree species or forest type. Growth in volume, basal area or height are given as examples of indicators of site productivity.

Several methodologies were defined according to the characteristics of each data set. In most cases, the analysis was restricted to the two past decades (from 1970).

6.1. Maritime Pine

If dominant height is considered to be an indicator of site productivity, the results of our study suggest that there has been a positive growth trend during the two last decades. This conclusion is supported by the analysis of the two data sets available - the continuous forest inventory of the National Forest of Leiria and stem analyses of 32 dominant trees harvested in three stands from the coastal dunes - and by different methodologies. Several meteorological data were tested as possible causes of the observed growth trend, but, although it has been proved that they play an important role in explaining annual differences in growth, it is by no means clear that they are the only explanation for the increasing growth rates in dominant height over the last few decades. A sound conclusion must be based on more complete climate information and on a larger data set.

In spite of the positive trend observed in dominant height growth, stand volume growth has decreased during the last decade because of a negative trend in basal area growth. One explanation for this negative trend may be that as there has been relatively little precipitation during the growing season (mainly spring and early summer) over the last few decades, the length of the growing season has been effectively decreased. The multivariate linear regression analysis that was performed shows that basal area growth, as well as volume growth, is positively affected by higher levels of precipitation in the growing season (both spring and summer), although precipitation cannot fully counterbalance the decreasing growth trend of basal area. Perhaps the variables that were used to account for precipitation during the growing season were not the most appropriate. Becker (1989), for example, used the monthly hydric balance successfully in order to explain differences in growth. As in the case of dominant height, a sound conclusion can only be obtained in the future when more data are available for a more complete analysis.

When analysing the results of the multivariate linear regression analysis, it is interesting to point out that the effect of precipitation on dominant height growth is not as important as its effect on basal area and volume growth, precipitation not being significant when added to the base model (without $year_{sgp}$). This may be explained by the different physiological mechanisms that control the growth of individual trees in dbh and height. Height growth in maritime pine is predetermined, which means that it is not very dependent on the annual variations in climate, whereas dbh growth, dependent on the carbohydrates available for the cambium, is strongly controlled by the environmental conditions present during the growing season.

One conclusion that can be drawn from the study of growth trends in maritime pine is that it is important to study more than one indicator (stand variable) of site productivity. In fact, the trends observed in dominant height and basal area growth are opposite, in this case, leading to a negative trend in volume growth. Had the study taken into account only the dominant height growth (the stand variable more commonly used as a productivity indicator), the results would have been quite different.

6.2. Eucalyptus Plantations

The results obtained with the permanent plot data of eucalyptus are similar to those obtained from the continuous forest inventory of maritime pine, although the trend in dominant height is negative, as it is for stand basal area and volume growth. This may be related to the fact that height growth in eucalyptus is characterised by free growth and is thus much more influenced by the climatic conditions of the current growing season than height of maritime pine, whose growth is predetermined. Again, precipitation during the growing season (spring and summer) is significantly correlated with growth; decreasing amounts of rainfall during the last decades partly explain the observed negative trends. More complete climatic data must be used in order to discover whether there are other causes for the observed negative trends. It is important to stress again that the irregularity of the measurements made it very difficult to define variables that took into account precipitation during the growing season. Had the measurements been regularly spaced over time, precipitation levels during the growing season might have been able to fully explain the observed negative trends in growth.

6.3. Poplar Plantations

The data from poplar plantations, although small in number, were ideal for studying growth trends, as we were able to eliminate the effects of silviculture and genetics by comparing trees from two generations of plantations on agricultural land. The trees, of the same clone, were planted at the same spacing in both plots. The statistical analysis indicates that there are no significant differences in the growth of the two generations, although the effect of spring precipitation in tree dbh growth could be seen graphically. It was to be anticipated that the effect of precipitation would be less evident in poplar plantations than in eucalyptus plantations or maritime pine stands, as poplar is planted only in sites where water is readily available, while most eucalyptus and pine stands are planted in coastal dunes or in poor forest soils where less water is available.

Acknowledgements

Although the research described in this chapter has not been given any specific funding, the data used were collected under the scope of several ongoing research projects, namely Improvement of Eucalyptus Management. An Integrated Approach: Breeding, Silviculture and Economics (CE AIR 92.1678), Relations between Intensification of Silviculture, Stability and Health of Forests (CE RECITE - Compostela Forest Network, action 4: research), Upgrading of Southern European Pines (CE Forest MA2-CT91-0040) and Alternative Agricultural Land Use with Fast Growing Trees (CE AIR3-CT92-0134).

References

Becker 1989 The role of climate on present and past vitality of silver fir forests in the Vosges mountains of northeastern France. Canadian Journal of Forest Research 19: 1110-1117

Carvalho 1992 Modelação dinâmica do crescimento em área basal de povoamentos de *Eucalyptus globulus* Labill. na região centro do país. Selecção e validação de modelos. Relatório do estágio de Engenharia Florestal, Instituto Superior de Agronomia, Lisboa

DGF 1992 Distribuição da floresta em Portugal continental. Áreas florestais por distrito. Direcção-Geral das Florestas, Estudos e Informação 299, Lisboa, Portugal

Falcão 1992 Contribuição para um modelo de produção para a Mata Nacional de Leiria. Modelos para a altura dominante e área basal. Relatório do estágio de Engenharia Florestal, Instituto Superior de Agronomia, Lisboa

Gallant AR 1975 Testing a subset of the parameters of a nonlinear regression model. J. Am. Stat. Assoc., 70: 927-932

Spiecker H, Mielikäinen K, Köhl M and Untheim H 1994 Growth trends of European forests. Has site productivity changed? EFI Working paper 4, Joensuu, Finland

Tomé M 1990 Modelação do crescimento em altura dominante de eucaliptais com recurso à análise multivariada e à análise de regressão. In: M. A. Turkman and M. L. Carvalho (eds), Actas da 1ª Conferência em Estatística e Optimização. Centro de Estatística e Aplicações e Departamento de Estatística e Investigação Operacional, Universidade de Lisboa, Lisboa, Portugal, pp. 443-462

Discussion

H. Spiecker, K. Mielikäinen, M. Köhl and J.P. Skovsgaard

1. Detecting Growth Trends

It is difficult to separate growth trends caused by long-term changes in site productivity from episodic growth changes caused by extreme events such as frost, drought, snow, storm, fire, insect or fungal diseases or a combination of several events. Even a single event can modify tree growth for up to several decades. Such an event may change competition as trees with greater vitality gain growing space and resources from their weakened neighbors. These changes in stand dynamics initiate long lasting changes in tree growth.

An example of long lasting cyclic growth variation is shown in Fig. 1 which illustrates how increasing and decreasing trends may last for several decades. These trends can be due to stand dynamics as well as climatic fluctuations. One possible cause for these effects may be due to the cyclic variation of sunspots, which are widely discussed (Sinkevich and Lindholm, in this volume), but not as widely accepted.

There are several methodological problems in detecting trends. The separation of long-term trends from seasonal variations is one of the major topics in the analysis of growth changes. As a rule of thumb at least 20 to 30 observations over a given period should be available for time series and trend analysis. Statistical analyses for testing the significance of trends can, however, be misleading. Depending on the beginning and end of the observation period, and the geographic region, one may find contradictory trends in a long chronology of growth.

Growth trends in this book were identified by comparing actual growth with expected growth. Expected growth was generally derived from past experience. Preferably growth data from earlier growth periods were used. In some cases yield tables and other growth models were used as a reference (Wenk and Vogel; Pretzsch; Zingg, in this volume). However, a deviation from expected growth can be interpreted as a change of site productivity only when the reference was based on the same constraints (stand characteristics, management regime and comparable locations) as the growth pattern under concern.

The use of old yield tables as references can be misleading as they may lack representativity and seldom meet the required constraints. Although it is interesting

European Forest Institute Research Report No. 5
Growth Trends in European Forests
Edited by H. Spiecker, K. Mielikäinen, M. Köhl and J. P. Skovsgaard

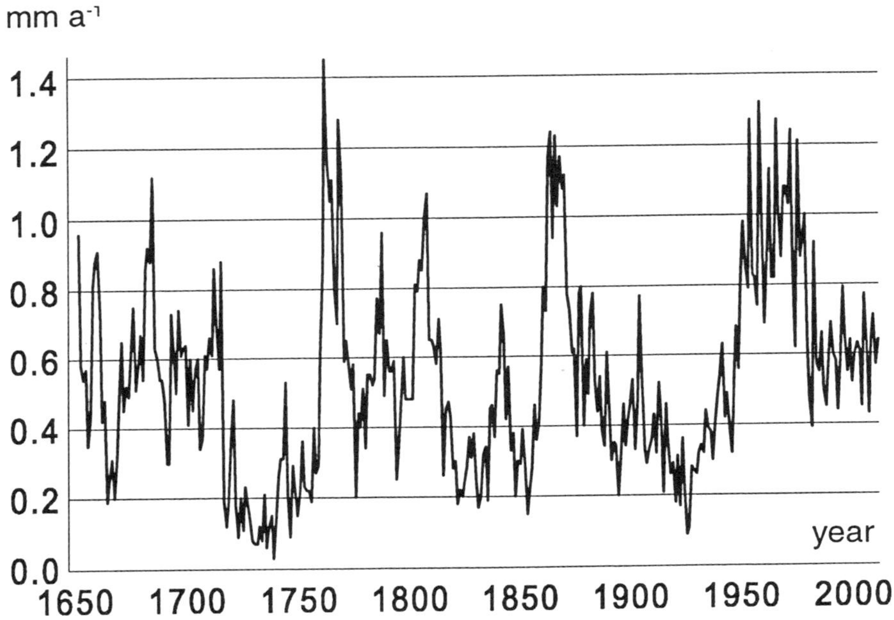

Fig. 1. Variation of ring width of a Scots pine tree in Finnish Lapland close to the timber line.

to know whether old yield tables are still suitable to be used in practical forestry, one must be cautious when drawing general conclusions.

When growth data from earlier growth periods were used for reference, growth data of consecutive generations at the same location were a reliable base for the analysis of changes in site productivity (Skovsgaard and Henriksen; Sterba, in this volume; Eriksson and Johansson 1993 cf, in this volume; Kenk et al. 1991), provided that the genetic composition and management regime of these generations have not significantly changed.

2. Growth Trends in Various Regions of Europe

2.1. Fennoscandia and Northwestern Russia

Seven papers describe forest growth in Norway, Sweden, Finland and Northwestern Russia. The studies analysed growth in generally even-aged, single species stands of Scots pine (*Pinus sylvestris* L.) and, in three studies, in stands of Norway spruce (*Picea abies* [L.] Karst.). These studies are based on tree analysis data, permanent research plot data and inventory data. The observation periods were longer than 40 years, some even several hundred years. Stem diameter as well as height, stand basal area and stand volume were used as growth parameters (see Table 1 in Conclusions and Summary, page 370).

The findings show regional differences. A clearly negative growth trend on the Russian Kola Peninsula north of the Arctic Circle was detected by Nöjd and Reams (1995, cf. Mielikäinen and Sennov, in this volume). In this area there are two large nickel smelters with annual sulfur emissions of more than 200 kT each. The impact of air pollution can be measured in the width of annual rings in a restricted area of approximately 30 to 40 kilometers around the smelters. The radius of the circle of dead forests was about 5 km.

In Finnish Lapland, close to the timber line, Zetterberg et al. (in this volume) constructed a continuous tree-ring curve covering a period of more than 2000 years. This chronology was based on living sample trees supplemented with subfossil trees found in lakes. There was both short- and long-term variation in pine growth in this area. The mean ring width, about 0.6 mm, was relatively constant for the whole 2000-year period without any specific trend in this century. The same kind of periodic growth variation for Scots pine, without any clear trends during this century, was also found in the study of Sinkevich and Lindholm (in this volume) in Russian Karelia, approximately 1000 kilometers southeast of the study area of Zetterberg et al.. In the same latitude in Southern and Central Finland Mielikäinen and Timonen (in this volume) found no trend in the ring width or in the stand volume increment of Scots pine in both unmanaged and regularly managed stands. Some of the studied stands, however, showed some signs of accelerating height growth in the southernmost part of the study area close to the city of Helsinki during the last 40 years.

An increasing trend was more visible in the data of Mielikäinen and Sennov, (in this volume) from their study of permanent research plots in the neighborhood of the city of St Petersburg, Russia, situated approximately 200 km south of the study area of Sinkevich and Lindholm. Most of the studied stands showed a clear increase in the current stand volume increment during the last three decades, in spite of the high age of the stands.

The results of the study by Elfving and Nyström (in this volume) of Scots pine plantations in Sweden in stands established after 1940 indicated a slightly higher site index than in stands that were established earlier. At the same latitude in Norway, Elfving et al. (in this volume) did not find any trend in the height development of dominant trees of Scots pine and Norway spruce during the last thirty years. The stand volume growth showed a slightly decreasing trend, possibly due to the growth models used for this study. A very clear positive trend in the diameter and height increment of Scots pine and Norway spruce beginning in the 1950s was found by Elfving et al. (in this volume) in Southern and Central Sweden. This result was based on increment measurements of about 100,000 sample trees in the national forest inventories. In the same region, Eriksson and Karlsson (in this volume) found an increase of 0,6-9,3 % in the site indices of Scots pine and Norway spruce based on dominant height development on permanent research plots during the period 1955 to 1995. The increase was highest in the southernmost part of the country. Previously, an increasing trend in the height and volume development of two successive Norway spruce generations has been found on permanent research plots (Eriksson and Johansson 1993 cf., in this volume).

Summarising the results from Northern Europe, it seems evident that in Northern and Central Fennoscandia the growth of Scots pine has not changed in a discernable manner from the natural long-term growth variation in this area. In the southernmost part of Finland and in the region of St Petersburg there are signs of a slight increase in the dominant height and stand volume growth of Scots pine and Norway spruce. A distinct positive growth trend of Scots pine and Norway spruce was found in Southern Sweden.

2.2. Central and Western Europe

Twelve papers refer to forest growth in Central and Western Europe. In these studies even-aged, single species stands as well as uneven-aged, mixed stands were included. The studies concentrated on four species: Norway spruce (*Picea abies* [L.] Karst.), beech (*Fagus sylvatica* L.), silver fir (*Abies alba* Mill.) and Scots pine (*Pinus sylvestris* L.). Whereas most studies were based on data from tree analysis, permanent research plot data and inventory data were used only to a minor extent. The observation periods ranged from 25 to 150 years.

The investigated variables were height, diameter, basal area and volume on tree and stand levels. On research plots and in inventories generally, the total height of stands and the height of cut trees were measured (Skovsgaard and Henriksen; Wenk and Vogel; Pretzsch, in this volume). For analysis of annual height increments shoot growth of cut trees was measured (Untheim; Kotar, in this volume). In most papers diameter increments of stems were included. Annual diameter increments were derived from tree ring measurements of discs or cores at a height of 1.3 meters, except in the analyses of Schneider and Hartmann and of Bräker (in this volume), who also collected discs from the top end of the boles. Periodic increments in permanent plots were derived from consecutive measurements of trees (Köhl; Zingg; Schadauer, in this volume). In some cases periodic increment data were compared with annual data (Schadauer; Zingg, in this volume). It is well known that diameter growth is affected by competition. Therefore, Badeau et al. (in this volume) analysed long-term changes of diameter growth in forest types with quite different conditions of competition.

For the analysis of possible trends the growth of trees with different germination dates were compared, in most cases. In some cases it was possible to compare the growth of two consecutive generations (Skovsgaard and Henriksen, in this volume) whereas in other cases growth of trees with different ages were compared (Wenk and Vogel; Untheim, Badeau et al.; Schneider and Hartmann; Schadauer; Kotar, in this volume). It was assumed that the sites of the different age classes were similar. However, site characteristics for individual trees (Untheim, in this volume) or stands (Kotar, in this volume) were investigated and compared in a few cases. Only in some studies were yield tables or other growth models used as a reference (Wenk and Vogel; Pretzsch; Zingg, in this volume).

Although data sources, growth parameters and evaluation methods varied considerably, the general findings of the papers are consistent: in Central and Western Europe forest growth has increased in recent decades. The only exceptions, quoted by Pretzsch (in this volume) are in some forests at high elevations of the German Middle Mountains, and the Alps where a significant reduction in volume growth of Norway spruce has been reported. Other findings on some forests in the Middle Mountains of Germany and France, however, showed an increase in growth (Badeau, in this volume; Kenk et al. 1991). Besides differences in local growth conditions methodological aspects have to be considered (Sterba, in this volume) for explanation of these differences. Although growth fluctuations vary with species, location and site, there is clear evidence that in general height growth, as well as diameter growth, have increased in recent decades. The results derived from long-term observations on permanent plots and from tree analysis are supported by inventory results (Wenk and Vogel; Pretzsch; Köhl; Schadauer, in this volume) which are representative for large areas, but cover only shorter observation periods.

In recent years, similar time specific growth variations were observed in Switzerland, France and Western Germany: a growth decrease in the 1970s and a growth increase in the 1980s. The decrease and the increase were more pronounced in the growth pattern of silver fir than in the growth pattern of Norway spruce. There are indications that these growth variations were caused, at least to some extent, by climatic factors (Schneider and Hartmann; Bräker, in this volume)

A consistent increasing growth trend was apparent in Western and Central Europe. Thus, it can be assumed that either the factors causing the growth increase were represented in all areas or different factors or combinations of factors in various areas had similar effects on growth. Possible causes are discussed in section 3.

2.3. Southern Europe

Two papers refer to forest growth in Southern Europe. Even-aged single species forests with Scots pine (*Pinus sylvestris* L.), black pine (*Pinus nigra*), maritime pine (*Pinus pinaster*), beech (*Fagus sylvatica* L.), cork oak (*Quercus suber* L.) and Mirbecks's oak (*Quercus canariensis* Willd.), eucalypt (*Eucalyptus globulus* Labill.) and poplar (*Populus* sp.) were investigated, using inventory data, permanent research plot data and data from tree analysis. The observation period varied between 20 and 100 years. In defense of the limited representation of Southern European studies (Montero et al., in this volume) explain that forest management and silviculture follow practices different from those in Central and Northern Europe, where the regular high forest system dominates, and that inventory and experimental data reflect rather different aspects from those needed for analysing possible growth trends for volume production. Preliminary Spanish studies (Montero et al., in this volume), however, stress the necessity to establish a base for such studies. Portuguese studies (Tomé et al., in this volume) on the medium-term indicate a negative basal area and volume growth trend for maritime pine and eucalypt over the past two decades. This is attributed mainly to decreasing precipitation. Growth of dominant height decreased in eucalypt stands but increased in maritime pine. For diameter increment of poplar no clear trend was evident.

3. Potential Causes

3.1. Limitations of the analysis of causal relationships

The analysis of possible causes and the investigation of their contribution to observed growth trends was undertaken with the objective of distinguishing between causal and non-causal associations. It is widely accepted that causal inference should not be exclusively based on a statistical approach. Schreuder and Thomas (1992) investigated cause-effect relationships concerning growth declines in Georgia, USA. They used three criteria to evaluate cause-effect relationships: (1) consistency, (2) responsiveness and (3) known mechanisms (Mosteller and Tukey 1977).

1. Consistency: The criterion of consistency refers to the fact that there is a consistent, repeatedly observed causal association in time and space between the effect and the hypothetical causal agent. As some causes can be observed only in specific situations or require the presence of other causal agents, the lack of consistency does not exclude causal associations.
2. Responsiveness: When exposed to the hypothetical causal agent under controlled conditions, the exposed object responds in a reproducible and quantitative way.
3. Known mechanism: There is a known mechanism or process by which the observed effect can be related to the hypothetical causal agent.

Besides the criteria used to verify a causal relationship, measurement of the effect is also crucial for causal inference. In quantitative terms, an effect is the difference of growth between two groups or periods in time which can be given as an absolute or relative value. The significance of such a difference depends on the number of observations, which is normally small in research plots and large in forest surveys. Therefore, in research plots generally only large differences can be detected, whereas in forest surveys even extremely small differences may be statistically significant.

Hypotheses cannot be tested with any acceptable degree of precision without an adequate experimental design. Many of the research plots presented in this book were planned before the statistical principles of experimental design were applied, and in most cases the plots were not installed for analysis of growth trends. The principles of randomisation and replication were occasionally violated. Another restriction was the geographic coverage of the studies. The plots did not cover all important forest areas of Europe in a representative way. In addition, the analysed growth parameters may not always adequately represent site productivity. And finally, growth was driven by many natural and human influences which were not always known in detail. In many cases no, or insufficient, long-term data were available for possible factors affecting growth at locations under consideration. These restrictions limit the extrapolation and generalisation of the findings.

Causal inference in a natural environment and large regions bears specific problems. The identification of possible causes was adversely affected by insufficient information on specific causes and their interactions. Causal inference renders more difficult because there are situations in which specific causes are active or a specific constellation of causal agents is necessary to cause reactions. Thus, a direct relationship between a single causal agent and an effect depends on a series of apparently independent variables. Minor changes in doses of single agents can affect the entire system of cause-effects. Antagonistic or synergistic effects can be found for confounding factors and may be difficult to detect. Therefore, a variable of minor importance might be selected as a causal agent, with the result that the real cause may not be detected. Because of these restrictions it was not possible to identify all causes and quantify their effects on forest growth.

3.2. Hypotheses on causal relationships

The spatial and temporal patterns of growth trends were expected to provide some hints regarding possible causes. However, interpretations have to be made rather cautiously because several different factors may simultaneously influence forest growth and their total effects may be rather different from the sum of individual effects. Growth rates of forests were not only modified by various natural influences, but also by human influences, for example by land use history, forest management, natural disturbances, and climate, including atmospheric deposition. The main interest in growth trends in this project referred to changes in site productivity (cf. Chapter 1 Introduction). Possible causes for these changes could be any growth influencing factors that have changed with time. In this sub-section some main factors which may have caused changes in site productivity are briefly described.

Land use history and forest management. In many regions of Europe, the former use of forests, not only for wood production, but for forage, litter raking and pasturing, have a long-term effect on soil conditions. The impoverishment of nutrients by the extraction of woody and herbal biomass had a distinct influence on site quality and affected the growth conditions of trees for decades. In addition, former agricultural land was often abandoned because of poor site productivity and marginal return rates, which, in some cases, may have been caused by excessive utilisation. Successive recovery of the soil may have increased site productivity in recent decades.

Management practices such as regeneration methods, tending, thinning and harvesting regimes have an effect on site productivity and may explain, to some extent, the observed growth trends. Growth of young stands may be higher due to soil preparation, selection of species and provenances, quality of plant material or weed control. Intensified thinning may have an effect on site productivity by altering nutrient cycling and by reducing competition for light, nutrients and water. Species composition has an effect on the root systems, litter quality and nutrient storage.

Fertilisers and lime have been applied to parts of European forests for many decades in order to increase site productivity and to overcome some effects of site degradation caused by former land use. Lime was used since the early 1980s to compensate against acid rain effects. Drainage of peatlands plays an important role for site productivity improvements in Northern Europe. In various parts of Central Europe, flood control and river diversion have had an effect on the ground water levels.

The results of national assessments reflect not only changes in site productivity but also changes in species and age distribution. A large scale shift in age class distribution of forests has occurred. Younger age classes of forests in Central Europe are often comprised of a large portion of conifer stands with high volume increments. On the other hand, especially in remote and inaccessible areas, forests became overmature, resulting in growth recession and instability. In Germany, for example, large forested areas were clear cut for reparation payments soon after World War II and were often afforestated with highly productive even-aged forests. In recent decades the negative impact of these activities on biodiversity and stability of stands encouraged the conversion of stands to so-called close to nature forest types.

Natural disturbances and climate, including atmospheric deposition. Besides human activities, natural disturbances play a decisive role in forest growth. Events like fires, pests, droughts, storms, snow or avalanches cause an incision in the natural succession and change driving factors for forest growth, such as competition for light, nutrients and water and tree species composition. They may change tree competition and may have an impact on selection processes. In addition, soil conditions and nutrient supply may be altered (Hüttl and Schaaf 1995). Damage caused by game, for instance through grazing, mainly affects growth and the survival of young trees and may result in a shift of species and age distribution and may delay stand establishment and tree growth.

Temperature and precipitation are often analysed as influencing factors because data referring to them are widely available. Growth response to climatic influences varies with species, provenance, competition status and site conditions. Temperature affects the length of the growing season. However, not only changes in mean conditions may influence site productivity. Most evident are the effects of extreme and unusual events such as late spring frosts, summer droughts, unusual cold and wet or hot and dry summers as well as extreme and abrupt changes in temperature. Sometimes such events occur locally and are not reflected in variables on climate and weather conditions generally available. These events not only have a direct effect on trees, but also on other organisms such as insects or microbial pathogens, on disturbances such as fire or windthrow, and on biological as well as chemical and physical processes in the soil. Correlations of these factors with growth have been reported in many parts of Europe. Droughts generally caused growth decrease. In Central Europe such a growth recession occurred for example in the late 1940s. The drought that attracted most attention occurred in the mid 1970s at the beginning of the discussion on defoliation and forest dieback. The long-lasting after-effects of climate complicated the detection of possible causes of changes in site productivity.

Climatic variables were tested as predictors of the growth variation in some studies in this book (Bräker; Schneider and Hartmann; Zingg, in this volume). The results show that these variables correlate with short- and medium-term growth variation. Long-term trends, however, could not be predicted using climate variables. The effect of predicted global warming on tree growth is not yet clear. In Portugal (Tomé et al., in this volume) a long drought period has caused growth decrease during the last two decades.

Those trees which grow at higher elevations than the Pyrenees, outside the range of forest management and exposed to low nitrogen deposition showed a clear positive growth trend (Badeau et al, in this volume). In these cases the increased content of CO_2 in the atmosphere (Nicolussi et al. 1995) or other climatic conditions could be major influencing factors. CO_2 may stimulate photosynthesis, reduce respiration and relieve water and low-light stress. However, as the growth response is modified by species and site it is difficult to draw general conclusions regarding their effects on forest growth. In Finland no clear growth trend was found (Zetterberg et al.; Mielikäinen and Timonen, in this volume). Possibly the effect of CO_2 on tree growth varies in different parts of Europe. It is equally possible that several different factors have influenced forest growth simultaneously and that their total effects have changed individual effects.

Nitrogen is known to be an important nutrient for plant growth. In most studies published in this book, the annual nitrogen deposition is regarded as a probable growth promoting factor during the last decades. The positive growth trends observed in many parts of Europe may be at least partially caused by increased nitrogen depositions. The fact that no trend was detected in some parts of Northern Europe (Nöjd and Reams 1995 cf. Mielikäinen and Sennov, in this volume) may be explained by low nitrogen deposition. However, in the Pyrenees growth increased (Badeau et al., in this volume) although nitrogen deposition was low.

The ozone content of the lower layer of the atmosphere has supposedly increased during this century. On sunny days the highest peaks of ozone concentration can be many times higher than average conditions, especially at higher elevations. The problem of the impact of ozone on tree vitality and growth, especially in high elevation forests of Central Europe, is widely discussed. However, no clear evidence of an influence on tree growth under natural conditions exists.

One direct effect around major industrial centers (e.g. nickel smelters in Kola Peninsula and power plants in Ore Mountains) of SO_2 on the foliage of trees, combined with indirect effects of acidification and high content of heavy metals in the soil, is forest dieback. In extreme cases, the area of total destruction reaches a distance of 5-10 km from the pollution source (Mielikäinen and Sennov, in this volume). Effects of SO_2 in lower concentrations on tree growth under natural conditions have not yet been proven. This also holds for other depositions and air pollutants.

4. Implications of growth trends on forest management

In relation to forest regeneration, an increase in site productivity affects forest regeneration in two different ways. Firstly, an increased density of stands reduces competition of weeds during regeneration, due to a decrease in light intensity in stands. This will facilitate regeneration but the competitive pressure of light demanding tree species will be weakened. Secondly, regeneration may be more difficult when site productivity is higher because of stronger competition from ground vegetation, as long as there is sufficient light on the ground.

Growth trends of different tree species and of different size classes of the same species (Spiecker 1994) vary. Changes in the ability to compete have an impact on stand management. The information available about species specific growth trends does not allow general predictions of changing growth relations between species. Competition control may require adjusted tending and thinning methods because different species may react differently to changes in site productivity. An adjusted thinning regime is also required to counterbalance increasing growth rates, otherwise an increase in growing stock will continue and the stability of forests will be reduced.

An increasing growth trend may have an effect on the age of harvesting. Fast growing stands reach the desired dimensions earlier than before; this will lower the economically optimal rotation age. There is some evidence that an increasing growth trend also results in a shift in the growth rhythm of trees (Eriksson and Johansson 1993). Trees may grow faster in their youth but may reach the culmination of current annual increment earlier than in former generations. On the other hand, an increase in site productivity also accelerates the growth of old trees.

Changes in site productivity may alter the width of tree rings, if this effect is not compensated for by changes in competition. Effects of altered ring width on the technical quality of timber varies with tree species and depends on the purpose of its usage. The quality of Scots pine, for example, may decrease if the ring width becomes wider, whereas wider rings of beech or ash often increase wood quality. Changes in site productivity and tree competition may also affect natural pruning processes, and may have consequences for branch size and number.

During the last decades, the wood production of European forests has been 40-50 % higher than the harvest cuttings. This was, in part, due to heavy clear cuts in several countries after World War II. National forest statistics in several countries show that the annual drain has remained at the same level during the last decades (e.g. Kuusela 1994). This is the case both for thinnings and final cuttings. In this situation, the density of young and mid-rotation stands is increasing. In addition, the mean age of forests is becoming higher. The mean standing volume of forest areas exceeds 300 $m^3 ha^{-1}$ in several regions of Central Europe. Such a high level has never been reported earlier in this century. In areas of increasing growth, the density of forests is rising, if removals do not increase at the same time. Dense conifer stands offer less space for light-demanding tree species and ground vegetation to live and therefore biodiversity is probably reduced.

A high standing volume of mature trees provides a high wood supply. On the other hand, it also causes a remarkable risk for the health of trees and forest ecosystems. As stands are higher they are generally more susceptible to storm damages and the economic risk for timber production is higher. Most forests in Europe are man-made; natural forests hardly exist. A conversion to stable more natural forests is only possible through active forest management. This includes the necessity to cut timber. The public should be aware that timber is a renewable material which absorbs CO_2 as a green house gas during its production. Thus, the promotion of timber products as an industrial material and new ways to expand the lifetime of timber products need to be supported.

Since forest planning can not only rely on past experience, efficient inventory systems, and forest management planning tools are needed. In addition, the assessment and mapping of sites has to be continuously updated.

5. Research implications

Public interest in forests has increased since environmental problems in our modern industrial society have become more serious. The principle of sustainability has been a leading management principle in forestry for more than two centuries in Europe. Sustainable forest management requires management tools for economical, ecological and other purposes. Complex information including ecological perspectives, productive techniques, and social needs are required. Unbiased, precise and accurate growth estimation is an essential basis for evaluation of sustainability (Moir and Mowrer 1995). Studies in this book clearly demonstrate that forest growth and growth conditions are changing. Observed growth trends revealed a certain lack of knowledge about the growth of European forests. Not all important European forest areas were covered in this volume. Further empirical studies on forest growth are needed. Existing yield tables and growth models seem to be no longer valid for a prognosis and the control of the sustainability of wood production. Flexible management-oriented growth models that are adjustable to observed environmental changes are needed. In addition, the increasing importance of a continued effort to monitor forest growth, growth conditions and stand dynamics in European forests is evident. Efficient methods for monitoring changes in growth conditions and in forest growth have to be developed. This includes the elaboration of reliable site productivity indicators and efficient sampling techniques.

A general assessment of the implications of environmental changes on ecological aspects such as biodiversity is not yet possible. As tree species and provenances may react differently to changing growth conditions, competition and selection processes may change. Trends inducing growth factors can also have direct and indirect effects on ground vegetation and on soils. Due to increased nitrogen input some plant species adapted to nitrogen-poor soils can vanish and so reduce the biodiversity of forest stands. Long-lasting increased nitrogen deposition may stimulate the biomass production of the ground vegetation and the biomass accumulation in the humus layer. Some results from fertilizer experiments in Central Europe suggest that accelerated tree growth may react more sensitively to drought (Spiecker 1991). Whether this phenomenon is induced by imbalances between the amount of roots and the size of transpirating foliage still needs to be further investigated. When CO_2 is the growth stimulating factor the sensitivity to drought may be reduced.

For the conversion of high productive man-made forests into more stable close to nature forests with a higher degree of diversity and adaptability adequate management practices are essential. Silvicultural strategies have to be developed which are able to maintain or increase stability of forest ecosystems in this transition phase without major ecological damage and economic loss. In addition, research is needed for the development of management strategies to mitigate effects of environmental changes on forest ecosystem functioning. The risks involved in different management options have to be assessed.

Forest growth can be seen as an indicator for environmental changes. The detection of growth trends should, therefore, be followed by an intensive effort to describe and qualify the causes of the observed trends. A better understanding of causal relationships between environmental factors and forest growth would enable scientists to develop scenarios of future forest growth independent of possible environmental changes described by other research disciplines. Thus, long-term experimental studies should be implemented. Also, long-term inventory data with replicated observations, and growth studies on long-term research plots and quasi-experimental retrospective growth studies based on stem analysis can lead to valuable information about causal relationships. Research projects should be designed that enable the separation of confounding factors such as climate, soil type, deposition, air chemistry, silvicultural treatment, historical land use, tree species or genetics. Ideally, a large set of factors and an even larger set of factor combinations have to be investigated. The relationship between site factors and growth needs further investigation. Indicators for the description of growth potential and of yield level are essential and site index systems need to be adapted to changing growth conditions.

Further development of process-based growth models will be helpful for a better understanding of causal relationships between environmental factors and growth. Physiological models should, in the long run, be linked to empirical models. Observations and experiments are required, both in the field and in the laboratory, and interdisciplinary research is needed. The amount and complexity of the scientific problems evolving from the observed forest growth trends show that solutions can only be developed by the cooperation of scientists on a European or world wide level. Cooperation will lead to a more comprehensive understanding and will provide a more realistic and reliable basis. Furthermore, it will reduce overall costs in relation to the expected benefits of improved knowledge.

References

Cannell M (1995) Forests and the global carbon cycle in the past, present and future. European Forest Institute Research Report No. 2: 66 p.

Dixon RK, Brown S, Houghton RA, Solomon AM, Trexler MC, Wisniewski J (1994) Carbon pools and flux of global forest ecosystems. Science 263:185-190.

Eriksson H, Johansson U (1993) Yield of Norway spruce (*Picea abies* [L.] Karst.) in two consecutive rotations in southwestern Sweden. Plant and Soil 154: 239-247.

Hüttl RF, Schaaf W (1995) Nutrient supply of forest soils in relation to management and site history. In: Nutrient uptake and cycling in forest ecosystems; Nilsson LO Hüttl RF Johansson UT editors; Kluwer Academic Press 685 p.

Kauppi PE, Mielikäinen K, Kuusela K (1992) Biomass and carbon budget of European forests, 1971 to 1990. Science 256: 70-74.

Kenk G, Spiecker H, Diener G (1991) Referenzdaten zum Waldwachstum. Kernforschungszentrum Karlsruhe, Projekt Europäisches Forschungszentrum für Maßnahmen der Luftreinhaltung, Forschungsbericht 82: 59 p

Kuusela K (1994) Forest Resources in Europe 1950-1990. European Forest Institute Research Report 1. Cambridge University Press. 154 p.

Nicolussi K, Bortenschlager S, Körner C (1995) Increase in tree-ring width in subalpine Pinus cembra from the central Alps that may be CO_2 -related. Trees 9: 181-189.

Mälkönen E, Kukkola M (1991) Effect of long-term fertilization on the biomass production and nutrient status of Scots pine stands. Fertilizer Research 27: 113-127.

Moir WH, Mowrer HT (1995) Unsustainability. Forest Ecology and Management 73: 239-248.

Mosteller F, Tukey JW (1977) Data analysis and regression. Reading, MA, Addison-Wesley 588p

Nöjd P, Reams GA (1996) Growth variation of Scots pine across a pollution gradient on the Kola Peninsula, Russia. Environmental Pollution (in press).

Schreuder HT , Thomas CE (1991) Establishing cause-effect relationships using forest survey data. Forest Science 37: 1497-1525.

Spiecker H (1991) Liming, nitrogen and phosphorus fertilization and the annual volume increment of Norway spruce stands on long-term permanent plots in Southwestern Germany. Fertilizer Research 27: 87-93

Spiecker H (1994) Effects of environmental changes on growth dynamics in mixed stands. Proceedings from the Symposium of IUFRO Working Groups: S4.01-03 and S4.01 04, April 25-29,1994Lousã/Coimbra:239-246

Conclusions and Summary

H. Spiecker, K. Mielikäinen, M. Köhl and J.P. Skovsgaard

Growth of European forests has changed considerably in recent decades. Although the methods applied in the 22 studies varied according to the data available, most studies showed the same general trend: site productivity has increased on many sites (Table 1). An increasing growth trend has been observed in the southern regions of Northern Europe, in most regions of Central Europe and in some parts of Southern Europe (Fig. 1). No clear trend was found in the most northern part of Europe, in rare cases in Central Europe and in some observations in Southern Europe. A decreasing trend was found in exceptional cases where extreme growth conditions such as intense exposure to pollutants or exceptional climatic conditions occurred.

The general increasing growth trend may have been caused by one factor, a factor combination or by regionally changing factors which finally had similar effects on growth. As possible causes land use history, forest management, natural disturbances, climate, including nitrogen deposition and increased CO_2 content of the atmosphere, and other factors, including a combination of several factors have been discussed. The significance of each factor possibly varies in space and time. Growth responses to the influencing factor are modified by site and stand conditions.

Increased site productivity affects the genetic composition of forest stands and the nutrient cycling. Changes in site productivity have an impact on forest management, for example on regeneration, tending and cutting strategies. Past growth observations may not reflect actual growth adequately. Therefore, we need to improve analyses and prediction tools to ensure continuous sustainable management.

European Forest Institute Research Report No. 5
Growth Trends in European Forests
Edited by H. Spiecker, K. Mielikäinen, M. Köhl and J. P. Skovsgaard

Table 1. Summary. Growth trends in European Forests

	Country/ region	Species	Forest type / Silviculture	Period (years)	No. of locations
Northern Europe	N Fennosc.	Scots pine	unmanaged, natural forest	2150	42
	Finland and NW Russia	Scots pine, Norway spruce, birch	even-aged, single species	100	
	NW Russia	Scots pine	even-aged, uneven-aged, single species	100-300	5
	Finland	Scots pine	unmanaged and managed, even-aged, single species	100/70	
	Sweden and Norway	Scots pine, Norway spruce	even-aged single species	>70	
	Sweden	Scots pine	even-aged, single species	90	
	Sweden	Scots pine, Norway spruce	even-aged, single species	40	
Central Europe	Denmark	beech	even-aged, single species	70	2
	Germany	Norway spruce	even-aged, single species	25-110	3
	Germany	Norway spruce, silver fir, Scots pine, beech, sessile oak	several forest types and silvicultural systems	20/150	125/175
	Germany	Norway spruce, beech	even-aged, single species	70/110	1
	Central Europe	Norway spruce, silver fir, Scots pine, beech	several forest types and silvicultural systems	>30	
	France	Norway spruce, silver fir, beech, sessile & pedunculate oak, Corsican pine, Mountain pine	several forest types and silvicultural systems	150	6
	Switzerland	Norway spruce	uneven-aged, mixed species	90	1
	Switzerland	silver fir, Norway spruce	uneven-aged, mixed species	90	1/44
	Switzerland	Norway spruce, silver fir, beech and other species	several forest types and silvicultural systems	15-25	39
	Switzerland	Norway spruce, silver fir, beech	even-aged, single species and uneven-aged, mixed species	70-100	11
	Austria	all Austrian species together, Norway spruce	several forest types and silvicultural systems	30/100	
	Slovenia	beech	unmanaged and managed, (un)even-aged, single species	30	
Southern Europe	Spain	Scots pine, black pine, beech, cork & Mirbeck's oak	even-aged, single species	50-100	
	Portugal	maritime pine, eucalypt, poplar	even-aged, single species	20/40	64

ps: Preliminary study *[1]: trends in this table appear as recommended by the authors (S): Summary of studies

No. of plots	No. of trees	Data source	Variable(s)	Trend(s) since mid -1900's *1	Author(s)
	1023	single tree data	diameter	0	Zetterberg et al.
		single tree data, research plots	diameter, stand volume	- / 0 / +	Mielikäinen & Sennov (S)
18	237	single tree data	diameter	0	Sinkevich & Lindholm
140		single tree data, research plots	height, diameter, stand volume	0	Mielikäinen & Timonen
		inventory data, research plots	height, stand basal area, stand volume	+	Elfving et al. (S)
816		inventory data, research plots	height	+ / 0	Elfving & Nyström
1792		research plots	height	+	Eriksson & Karlsson
18		research plots	height	+	Skovsgaard & Henriksen
448	2277	inventory data, single tree data	height	+	Wenk & Vogel
269/425	>227000	inventory data, research plots, single tree data	height, diameter, stand basal area, stand volume	+ / -	Pretzsch (S)
76	76	single tree data	height, diameter	+	Untheim
		single tree data, inventory data, research plots	stand volume, height, diameter	+ / 0	Sterba (S)
1421	9366	single tree data	diameter	+	Badeau et al. (S)
200	1113	Single tree data	diameter	+	Schneider & Hartmann
1/153	64/200	single tree data (res. plots)/(inv. plots)	diameter	+	Bräker
	82580	inventory data	diameter	+	Köhl
25		research plots	diameter, stand basal area	+	Zingg
		inventory data, incl. single tree data	diameter, basal area, height	+	Schadauer
90		single tree data	height, volume, diameter	+	Kotar
		inventory data	height, stand volume	ps (+)	Montero et al.
169	40	single tree data, research plots, inventory data	dominant height, stand volume, stand basal area, indiv. tree dbh	- / 0 / +	Tomé et al.

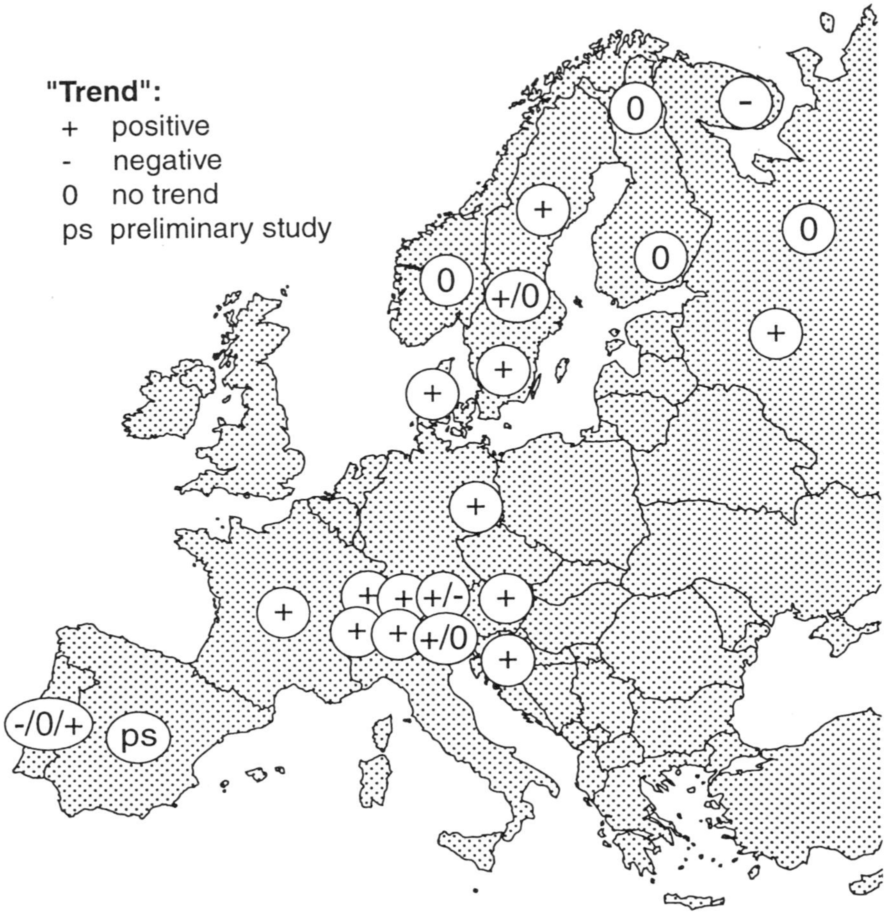

Fig. 1. Map showing the locations and summarized results of the studies